COLLECTED STUDIES SERIES

Cycles of Time and
Scientific Learning
in Medieval Europe

Professor Wesley M. Stevens

Wesley M. Stevens

Cycles of Time and Scientific Learning in Medieval Europe

Routledge
Taylor & Francis Group

LONDON AND NEW YORK

First published 1995 by Ashgate Publishing

Published 2016 by Routledge
2 Park Square, Milton Park, Abingdon, Oxon OX14 4RN
711 Third Avenue, New York, NY 10017, USA

Routledge is an imprint of the Taylor & Francis Group, an informa business

British Library CIP Data
  Stevens, Wesley M.
  Cycles of Time and Scientific Learning in Medieval Europe.
  (Variorum Collected Studies Series; CS 482)
  I. Title II. Series
  509.4

US Library of Congress CIP Data
  Stevens, Wesley M.
  Cycles of Time and Scientific Learning in Medieval Europe /
  Wesley M. Stevens.
  p. cm. — (Collected Studies Series: CS482)
  Includes index.   ISBN 0-86078-471-1
  1. Astronomy, Medieval—Europe. 2. Science, Medieval.
  3. Time—Measurement—History. 4. Calendar—Europe— History.
  5. Church calendar—Europe—History. I. Title. II. Series:
  Collected Studies: CS482.
  QB23.S74 1995   94–37035
  520'.94'0902—dc20   CIP

Transferred to Digital Printing in 2011

ISBN 9780860784715 (hbk)

COLLECTED STUDIES SERIES CS482

CONTENTS

This volume contains x + 320 pages

FOREWORD

Alternations of daylight and dark are readily recognisable, and we adjust to them without much attention, as we do to longer and shorter seasons. Those recurrent phenomena however imply periodicity, and the mind tends to reach for terms by which they may be better understood. The western calendar is a product of thought about light, dark, and seasons in terms of cycles of time. The concentrated analysis of such phenomena during many cycles is *computus* or time-reckoning: its goals are intelligent comprehension of experience, enlightened observation, and accurate prediction of recurrent phenomena.

One result of *computus* is that astronomy is improved by more dependable time-series; another is a common calendar with expected holy days and days of rest. To begin with however, the calendar was not so common or so easy to use as it is now assumed to be. Studies of *computus* have always been difficult and somewhat tedious. For historians however they may be quite interesting in revealing unexpected aspects of life, both expert and common.

The search for surviving Latin computistical manuscripts has led me and my family to interesting cities, towns, and even deserted places. The search also led us to experiences which were often marvelous but sometimes rather difficult. There are at least 9,000 surviving Latin manuscripts which contain computistical tracts and tables transcribed before A.D.1600. I have searched for and held in my hands for study perhaps 1,200 of them, including most of those written from the eighth to the eleventh centuries. Those manuscripts have revealed unexpected evidence that early Christians drew more heavily and more specifically upon Hellenistic mathematics and natural sciences than any lectures, articles, or books had ever recognised, much less explained.

The *computus* is not one thing, of course, and it was never isolated from other studies. It was and is an attempt to make sense of the lives and the communities of Christians in western Europe, an uncompleted endeavour in which an astonishing number of intelligent persons participated. The most important event in the Christian life is the Passion and the Resurrection of Jesus, whom we call the Christ. At first therefore mathematics and science, so far as they were available, were put to the service of faith and piety in search for a descriptive system of lunar and solar cycles which would be sufficient to predict accurately the coming dates of the Passion. Rather soon in those first centuries, the Passion was subsumed to the Resurrection in thought and practice of Christians, so that they came to speak and think of Easter as the eve and day of Easter Sunday. None of the several attempts to predict a series of dates for future Easter Sundays were entirely successful, and it is often difficult to see just why. Yet, those exercises also provided occasions for stubborn scholars, laymen and priests but especially curious monks, to improve their arithmetic, astronomy, and even geometry. Schools were organised for these tasks to a surprising extent, though they were not obviously required by Christian piety. Such early studies occasionally led to new contributions by Christian scholars to science which were significant in their times and contexts. Some of their discoveries are still useful, though their origins may generally not be acknowledged today.

What do we know about the people who sang the liturgies, who wrote the verses, who led the worshiping communities? If we follow manuscript evidence, we must now expect a cantor to be adept not only with modes and neums but also with *annus mundi*, *annus passionis*, and *annus domini* in a variety of systems. We may sympathise with a poet who complains about carrying over the sixes, twelves, and twenties, the tens and hundreds in multiplication, as well as about the restrictive Latin meters. An abbot was just doing his job when he speculated about planetary positions while assigning night watches. Some of the resulting tables of data with columns of *Kalends* and *luna* XIV are discouraging to look at now; counting *epactae* must always have been a bore; and what can one do with surviving diagrams of planetary cycles which surely do not fit Ptolemaic astronomy? We may evaluate more broadly and enjoy the richness of lives revealed by evidence in manuscript.

Wesley M. Stevens
Winnipeg

ACKNOWLEDGEMENTS

The author and the publisher, Variorum of Ashgate Publishing Ltd, acknowledge copyright and express their appreciation for permission to reprint and revise these essays to the editors of the volumes in which they appeared and to the following publishers: The International Society for the Study of Time, Westport, and International Universities Press, Madison, Connecticut (I); The Parish of St Paul's Church, Jarrow, Tyne-and-Wear (II); The History of Science Society, Philadelphia, Pennsylvania, and the University of Chicago Press (III); The Pontifical Institute of Mediaeval Studies, Toronto (IV); The Center for Medieval Studies, University of Minnesota, and the University of Minnesota Press, Minneapolis (V); Harrassowitz Verlag, Wiesbaden (VI); The Bodleian Library, Oxford (VII); Sint-Pietersabdij von Steenbrugge and Brepols Publishers, Turnhout, Belgium (VIII); Hill Monastic Manuscript Library, Saint John's Abbey and University, Collegeville, Minnesota (IX); The Canadian Historical Association, Ottawa (X); Rheinische-Westfälische Technische Hochschule, Aachen, and Birkhäuser Verlag, Basel (XI).

PUBLISHER'S NOTE

The articles in this volume, as in all others in the Collected Studies Series, have not been given a new, continuous pagination. In order to avoid confusion, and to facilitate their use where these same studies have been referred to elsewhere, the original pagination has been maintained wherever possible.

Each article has been given a Roman number in order of appearance, as listed in the Contents. This number is repeated on each page and quoted in the index entries.

Corrections noted in the Addenda et Corrigenda have been marked by an asterix in the margin corresponding to the relevant text to be amended.

I

Cycles of Time: Calendrical and Astronomical Reckonings in Early Science

Abstract Calendar reckonings in early medieval schools were a result of the fruitful meeting of Hellenistic science and Christian piety. The Christian worship of God demanded a common festival for celebrating the life and death of Jesus. But the historical information for setting a date for celebrating either Passion or Resurrection was fragmentary and assumed either Jewish or Roman calendars. Predicting the dates of future celebrations thus required calculations of lunar and solar cycles and of their coordination in terms of calendar usages.

This paper reviews Hellenistic resources for calendar cycles, in so far as they were available to Hippolytus of Ostia, Christian scholars of Alexandria and Milano, Victurius calculator of Aquitaine, the Schythian Dionysius Exiguus, the venerable Bede and others. It emphasizes that their production of various sequences and cycles of time were never satisfactory but always stimulated further calculations of the regularities of the sun and moon (whose cycles are not commensurable), further observations of the irregularities of the wandering stars (planets), the use of Euclidean geometry, and diverse conceptions of the globe of the earth and its continents.

The discipline enveloping such studies was known as computus. Evidence from numerous manuscripts from the second to the eleventh centuries is cited to show that calendrical reckoning or computus was the context for continuing and intensive studies in the mathematical sciences of early Christian schools during the so-called Dark Ages.

Abbreviations Used in the Text

ed	edidit	rpr	reprinted
ep	epistola	s	saeculum, siècle
fl	floruit	seq	et sequentia
f	folio recto	tr	translatus
ms	codex manuscriptus	v	verso

I

Memorable events of our lives may be recalled in past sequences or anticipated in terms of future sequences. This recall may be in terms of stories recounted by experts in each community, but another way to account for past events of significance has been with written annals and chronicles. All of this organizing of social memory by sequences implies some sort of calendar, and all known peoples seem to have created calendars. Once a form of calendar has been found helpful within a culture, it seems to be accepted and applied without very much further thought about its structure and procedures. Before that form becomes common, however, there has usually been a great deal of serious thinking about how to make the calendar and how to use it.

I

The European calendar, which now seems so certain and accurate, has a long history of uncertainty and inaccuracy before it reached the common form in the hands of a Northumbrian monk of the eighth century, the venerable Bede, who lived from A.D. 672 until 735. In the several stories of earlier attempts to develop the European calendar, one may identify many Greek, Latin, African, and Scythian scholars at work during times when studies of this kind were much more difficult to pursue than now. The written product of such scientific and technical work is called *computus,* and it is in computistical tracts and tables that one finds how early mathematicians and astronomers were understanding cyles of time and were doing something useful with them.

The ancient word, computus, is first of all a Latin term denoting reckoning, the practice of arithmetic for any purpose; but it soon became applied especially for reckoning a series of dates or the cycles of time. Thus computus[1] often refers to the use of tables of dates, or to the methods for calculating the dates, or to explanations of the significance of such data as they refer to the circular courses of sun and moon. The earliest Christian literature of computus includes a fifth century *Calculus* by Victurius of Aquitaine (fl.A.D. 450–460)[2] and his *Cursus paschalis* to

[1]Computistical literature from the second to the eighth centuries was surveyed by Jones (1943, pp. 6–122), with bibliographical notes to the chapters of Bedae *De temporibus ratione liber.* A useful guide to the diverse works which bear the name of Bede mistakenly in the volumes of PL XC and XCIV has also been provided by Jones (1939). Computistical and other scientific labors by Scots, Britons, Saxons, and Angles have been discussed by Stevens (1981). See also an excellent survey of computi from the ninth to fourteenth centuries by Borst (1988). All Latin computistical works written prior to A.D. 1200 which have been printed, and those which have thus far only been identified in manuscript but not yet printed, will be included in Stevens, *A Catalogue of Computistical Tracts,* A.D *200–1200* (in preparation).

An interesting survey of the social motivations for making calendars, teminologies, and achievements is Fraser (1987), chapter 2, "For Ever and a Day. The Reckoning of Time."

[2]Various selections from the basic arithmetic of Victurius have been published by Christ (1983, pp. 132–136); Friedlein (1871); and *idem* in *Boncompagni* (1871, p. 443). A partial edition from a pqor manuscript was printed by Herwagen (1563, p. 147 seq.), rpr *PL* XC (1850), coll. 667–680. The entire prologue and text of this *Calculus* should be published in a critical edition.

be discussed below, the sixth century *Cyclus* and nine *Argumenta* by Dionysius Exiguus (A.D. 525–531) working in Italy, the ten items of computistical literature named in the early seventh century by Cummian (A.D. 632–633) in southern Ireland, and the many documents assembled at Wearmouth and Jarrow in Northumbria during the early eighth century, which were transcribed in ms Oxford Bodleian Library Bodley 309 (s.XI 2, from Vendôme) and several other similar collections of texts identified by Jones (1937, 1943, pp. 105–113). In the search for understanding during the medieval centuries, however, computus became something more; for it tended also to attract other mathematical and scientific materials far beyond the requirements of a calendar. Time-reckoning required recognition of the Zodiac and its twelve divisions and 48 or 60 or 360 parts as a basic scale for quantified observations of solar and lunar motions. Computus is also a term which could denote a whole essay about how the calendar works, or an entire codex in which such materials formed a large portion of the contents, though mixed with other subjects such as a catalogue of stars or studies of poetic meter or excerpts of liturgy with musical annotation, all of which assumed or were offered in terms of sequences or cycles of time. Computus may even include brief discussions of the whole order of the cosmos (Stevens, 1979).

Turning the leaves of computistical codices prepared for study in the eighth, ninth, or tenth centuries of our era, one finds even more diverse works of scientific interest; for example, there will be *mappaemundi* displaying the globe of earth with four zones or with three continents or with five parallels of klimata which we might call terrestial latitudes.[3] There will be *planisphaeria* with stereographic projections of stars and constellations for both the northern and the southern hemispheres; some will be projected from an infinite distance above the earth, looking down as Hipparchos imagined in the second century B.C.; and some will be projected from the south celestial pole, looking up as Ptolemy explained in the second century A.D.[4] Some of these books of computus describe and display alternate systems of planetary motion in epicyclic or in eccentric systems. Of course this means that sometimes the earth was shown as the exact center of all planetary motions, which were perfectly circular, consistent with Aristotle. But other diagrams in the early ninth century show planetary motions and their extremes or apses which were irregular and not concentric with the earth.[5] By the end of the ninth and during the

[3]*Monumenta cartographica* (1964), especially chapter 2 for which earlier manuscript examples could be cited: Aujac (1974), Stevens (1980).

[4]Six planispheres used in ninth and tenth century monastic schools have been listed by McGurk (1973, esp. pp. 200–201); other manuscript examples are known in each successive century. The earliest example is found in the Regensburg manuscript, München Bayerische Staatsbibliothek CLM 210 (A.D. 810–816) f.113v, briefly described by North (1975, pp. 386–388); a color reproduction may be seen in Bullough (1965, p. 189, plate 50). The two perspectives have been explained by Stevens (1980).

[5]Ms Paris Bibliothèque National Nouvelles acquisitions lat.1615 (s.IX 1) f.128–193 was part of a large computistical and astronomical collection dating from about A.D. 809 before it was copied at Auxerre about 830. It shows three diagrams: f.159v planets in circles whose distance is set by tonal proportions; f.160v planets on circles with different centers (earth is also shown), and the circles broken with zigzags representing retrograde motion; f.161 planets on other circles with diverse centers and blind areas when a planet is too near the sun to be seen. The drawings were used during ninth and tenth centuries in the school of St. Benoît-sur-Loire at Fleury.

tenth century, such planetary motions were also plotted on linear scales of 30 × 12 parts.[6] These computistical collections often contain multiple *argumenta* on the same topic, that is, not only alternate explanations of how to solve a problem but also irreconcilable data to be analyzed[7]

All of these documents have their origins in centuries which have been called "The Dark Ages," a period during which all learning was supposed to have been lost to the barbarian hordes in the West! For several generations of historians therefore, such materials were not supposed to exist in the pious schools of monks who only prayed and read Holy Scriptures. But these works were indeed written in those times, they were taught in schools for piety, and they do still exist in abundance. Between the decline of Roman control in the West and the flowering of Carolingian schools, there was scientific activity of importance, a small portion of which will be presented here. It is found in the meeting of Hellenistic science and Christian piety. In order to understand the development of computistical science, it is necessary to recall an essential element of all religion and particularly of the Christian religion: the cult or worship of God.

II

Christians should worship together. That is an obvious implication of the apostolic teachings about the historic Jesus and of the Hebrew Scriptures which they inherited and shared with the Jews. But it would be almost impossible to find a time when all or even most Christians have actually worshipped together for their chief festival of the year, *Pesach, Pascha, Easter*. Celebration of the resurrection of Jesus Christ occurred on different dates according to various calendars and according to divergent systems of reckoning within the same calendars (Richardson, 1973). This variety existed already in the Hellenistic society within which Jews and Christians tried to worship, and it is a product of intelligent reflections and decisions, not only

---

Other types of diagrams for studies of planetary motions have been published by Eastwood (1983a, 1983b, 1987).

[6]Ms Bern Burgerliche Bibliothek 347 (s.IX ex) f.22v–24 display the planets in figures similar to the Auxerre/Fleury ms Paris BN N.a.lat.1615, cited in the previous note. But ms Bern f.24v shows a rectilinear coordinate schema for displaying the patterns of latitudes for apogees and perigees of planetary orbits. Illustrations from the Bern, Paris, and other manuscripts were published by Eastwood. Late examples of the rectilinear schema were printed by Pedersen and Pihl (1974, p. 246).

*   [7]Good examples of this practice may be seen in the *Compilatio computistica et astronomica* A.D *DCCCVIIII* which was probably compiled at the palace school of Charlemagne in the form of three books; earliest copies are ms München Staatsbibliothek CLM 210 (A.D. 810–818) f.4–163v and Wien Nationalbibliothek Lat.387 (A.D. 812–816) f.4–165, each of which gave the date 810 in argumenta from their exemplar, rather than the expected year 809. An expanded seven-book version of this *Compilatio* in ms Monza Biblioteca Capitolare f-9/176 (A D. 869) f.7–92v, and other manuscripts. Outline of contents and a survey of the two groups of manuscripts were made by King (1969). It was probably from such a collection of texts for study that Walahfrid Strabo took his series of dates for the Incarnation in ms Sankt Gallen Stiftsbibliothek 878 (s.IX second quarter), pp. 284–292; see Stevens (1972).

about historical events but also about the observed and calculated cycles of lunar and solar motions.

In the Hellenistic tradition the earliest attempt at systematic coordination of lunar and solar cycles in tabular form appears to be an intercalation cycle of eight years, reported by Pliny (Plinius, *NH* II 97, 215). This 8-year cycle or octaëteris may have been received from the ancient Babylonians and used by Chaldeans and Greeks. It was based upon a solar year of 365 days and upon lunar months of either 29 or 30 days. If you observe the moon over a season or two, you will find that it reappears above the horizon sometimes after 29 days and sometimes after 30 days, and you may recognize a mean lunar month of 29½ days. Eight years will include 99 cycles of the moon; but if you multiply 29½ days by 99 cycles, the result will be half a day more than if you count the daily course of the sun about the earth during the same period:

99 months	× 29½ days	= 2920½	(1)
8 years	× 365 days	= 2920	

If one doubled the sequence in order to obtain a 16-year cycle, the surplus would be a full lunar day which could be skipped in order to begin again.

In the fifth century before Christ, an alternative system was created by two Athenian scholars. Euctemon and Meton, who recommended the use of a 19-year cycle of alternating full lunar months of 30 days, each with short lunar months of 29 days, to which they added seven full intercalary months of 30 days, resulting in a total of 6940 lunar days.[8] We do not know by what pattern the full, hollow, and intercalary months were alternated in the Metonic cycle in order to maintain accord with solar phenomena; but the overall numbers will be as follows:

118 full months	× 30 days	= 3540	(2)
110 hollow months	× 29	= 3190	
7 intercalary months	× 30	= 210	
235 lunar months	total days	= 6940	

Meton is justly famous for this advance in reckoning the times. But the Metonic lunar cycle is about 2 minutes too long, and the supposed solar cycle would be about 30 minutes too long; continued use will create problems. Other Greek scholars were not at all satisfied with it.

In order to reduce the problems of the Metonic cycle, another fourth century astronomer and pupil of Aristotle, Kallippos of Cyzicus (ca. 370–300 B.C.), reckoned a longer period.[9] He calculated that 912 lunar months would accumulate

[8]*PW* VI/1 (1907, pp. 1060–1061) and XV (1932, pp. 4458–1466); Ginzel II (1911, pp. 391–392); Duhem I (1914, p. 108); *DSB* IV (1971, pp. 459–460).

[9]Geminos, *Elementa astronomiae* VIII, pp. 120–122: the 76-year period of Kallippos; *PW* X (1919, pp. 1662–1664); Heath (1913, pp. 212–213); Duhem, I (1914, pp. 123–128; *DSB* III (1971, pp. 21–22).

26,918 days, with alternating 29- and 30-day periods for which we have not received an explanation. To this he intercalated 29 full lunar months of 30 days each to add 840, for a total of 27,758 days. According to later reports this result was one day short of four Metonic cycles or 76 solar years. That information requires the solar cycle to be 365¼ days. We can only guess at Kallippos' method of calculation and sequence for 940 lunar months, 76 solar years, and 27,758 days whose relationships are not reported in the evidence.

The variety of such Hellenistic calculations was very great, and evidence for them survives only in fragments. But from the time that Julius Caesar went courting Cleopatra on a barge in the Nile River and returned to Rome with an astronomer, Sosigenes, the 19-year cycle with fourth year embolism, more commonly called the extra day of leap year, was accepted in the solar calendar of the Roman Republic and thus came to be used in capital cities of the expanding Roman Empire.[10] Furthermore, the Metonic cycle had been developed into its classic form with mean lunar months of 29½ days each[11]:

$$
\begin{aligned}
\text{12 mean lunar months} \times 29\tfrac{1}{2} &= 354 \text{ days} \qquad (3)\\
\text{19 years} \times 354 &= 6726\\
\text{7 intercalary months} \times 30 &= 210\\
\text{Embolisms during 19 years} &= 4\tfrac{3}{4}\\
\text{total days} &= 6940\tfrac{3}{4}\\
\text{19 solar years} \times 365\tfrac{1}{4} \text{ days} &= 6939\tfrac{3}{4}
\end{aligned}
$$

There was a remainder of one lunar day which they called *Saltus lunae;* that is, if you simply skip over that one extra lunar day, you may repeat the 19-year cycle and remain rather well in accord with the course of the sun. It was in this form that the 19-year cycle was adapted to Christian usage at Alexandria and Milan, later at Carthage, and eventually at Rome.

III

The evidence for the Christian development of computus or calendar reckoning is very complex but quite abundant because it was necessary. A bishop for example must teach the sequence of prayer and praise as well as the meaning of Christian festivals throughout the year. But so much of the meaning and so much of the timing depends upon the Easter festival, that any bishop must certainly know when to do it. As with so many other obvious practical matters, the letters of bishops which survive from early centuries imply that such questions are being dealt with

[10]Suetonius, *Divus Julius,* p. 40: on calendar reform, without mention of Cleopatra or her astrologer; but see Plinii *NH* II 8.39 and XVIII 211; Ginzel II (1911, pp. 274–281); *PW,* Zweite Reihe III/2 (1927, pp. 1153–1157); *DSB* XII (1975, p. 547): Sosigenes.

[11]The classic form of the Metonic cycle is explained by Rühl (1897, pp. 114–116); Schwartz (1905, pp. 15–17); Ginzel II (1911, pp. 91–392); Jones (1943, pp. 20–21).

but do not repeat the details which they tell priests, deacons, or other leaders of their scattered flocks. One of them was quite verbose, however, the famous St. Augustine, and many of his letters and works have survived.

Augustine grew up in Roman Africa. He taught as a *grammaticus* in Carthage and Rome, and then with the influence of his Manichaean friends he obtained the prominent and lucrative post of imperial rhetorican in Milan. After ten years of a successful career in the capital cities of Rome and Milan, he changed his religion and returned to Africa where he served as a Christian priest in Carthage and Thagaste from A.D. 389/390 to 395 and was then dragooned into becoming bishop of Hippo Regis (A.D. 395–430).[12] Augustine wrote several great books which set forth his thoughts on grace, free will, consciousness and memory, the nature of man, and the nature of God. These writings include many philosophical reflections upon history in which the concept of "time" is often mentioned and sometimes discussed.[13] There are aspects of history and time which have apparently seemed to be too mundane for doctrinal history or philosophical theology but which the man himself could not have avoided: when did this bishop observe Easter during the course of a Julian year? What system did he use to set the day of the month for Easter Sunday each year within his diocese in order that all churches could celebrate together?

A letter had come to Augustine from Ianuarius about Easter, and we have the bishop's answer—a long answer: twenty-five Latin pages in the *CSEL* edition and thirty-three pages of English translation.[14] Ianuarius had asked: Why does the annual commemoration of the Lord's Passion not come around on the same day of the year, as does the day on which he is said to be born? If Easter is irregular because of the sabbath and the moon, what does it mean to observe the flux of sabbath and moon?

In his response Augustine emphasized the Easter festival as the great sacrament of grace through the historical Incarnation, and not any Christmas celebration or date. That is to say, he emphasized the Passion of Jesus and not his birth. Christ died and rose again, and thereby something was passed over to us: *Pasch, transitus, Passover*—something to be received. Incidentally he also affirmed the value of

---

[12]There are many good studies of his life, such as Bonner (1963) and Brown (1967), and of his political concepts, such as Deane (1963) and Adams (1971). But the best introduction may still be H.-I. Marrou, *Saint Augustine and His Influence Through the Ages* (1957). A reliable and often enlightening group of surveys of his many works is *A Companion to the Study of St. Augustine* (1954).

[13]In most essays concerning "time and history" in the thought of Augustine, there is not a word about real time in daily life. Only the juxtaposition of large concepts and wide philosophical meanings are found in the oft-cited essay by Marrou (1950). Another example is O'Daly (1981) which has nothing to do with any sort of measurement, much less of time. "Meaning of History" appears not to lead authors to discuss actual time-sequential experience and its structural necessities.

[14]Epistola LV ad Ianuarium, ed. A. Goldbacher, *CSEL* XXXIV/2 (1898, pp. 169–213), Eng. transl. W. Parsons in *Fathers of the Church,* Vol. 12 (1951, pp. 260–293). A large part of this letter was available for medieval schools in the *Excerpta ex operibus s. Augustini,* prepared in the sixth century by Eugippius, Fulgentius of Ruspe, and Paschasius diaconus, ed. P. Knoell, *CSEL* IX/1 (1885, pp. 425–444). Discussion may be found in Jones (1943, pp. 35–37, 87–89, 400 et passim); and Van der Meer (1961, pp. 285–293).

astronomy with its calculations and eclipses of sun and moon, and he warned against Manichaean fables about such things. Furthermore he twice reiterated the regularity of the "councils and fathers" who had found the date of Easter in terms of:

1. The month of new corn.
2. Luna XIV to XXI.
3. Sunday, the eighth and first day.
4. Forty days for Lent; fifty days for Pentecost.

It will help to recall that the Hebrew calendar was lunar, not solar, and it called its first month Nisan or the month of new corn. Commencing with the first sliver of reflected light as the new moon or luna I appeared somewhere in the sky, the moons were counted to the fourteenth or full moon which corresponded with the vernal equinox. Within the eight days and nights from luna XIV to luna XXI—from the fourteenth to the twenty-first day of the new moon—therefore, the day of Jesus' death would be observed on a Friday, and the third day thereafter (counted inclusively) would be Sunday: the time for celebrating his Resurrection (Richardson, 1940; Jones, 1943, pp. 6–11, 18–20). The lunar cycle itself, however, introduced uncertainties as to the fourteenth day, in that the observed first moon may rise during the light of day or at any time during the night. And some people customarily reckoned the beginning of a 24-hour day at dawn, noon, sundown, or midnight. Such regional and cultural variances thus usually required any set of guidelines to be adapted by local scholars to local practices in order to fulfill the intended goals.

Observance of Christian Sunday rather than Jewish Saturday was repeatedly justified by the bishop without speaking ill of anyone. Christians had not always reorganized weekdays to commence with Sunday, and it may have been adopted first in Alexandria by the third century if not earlier. Although it would seem to be an easy adjustment of the Jewish pattern for communities to make, especially if Christians wished to give priority to the day of Resurrection, such social changes do not always find ready acceptance. Thus celebration of the Resurrection of Christ on a Sunday does not appear to have found general usage within all of the Christian communities before the fifth century. In terms of Roman culture, the first day of the week could equally be counted the eighth, beginning and completion in one, by which Augustine gave rhetorical emphasis to celebrating on Sunday. Quadragesima or Lent was reckoned back from Easter Sunday, but intervening Sundays were not counted for this purpose; and Quintagesima was counted forward beyond Easter Sunday to Pentecost in the same manner.

The system which Augustine explained therefore was dominated by liturgical concerns, as one should expect of a bishop, and it is consistent with the fragmentary reports from the Council of Nicaea. It is also quite simple and is similar—perhaps identical—to what I learned from my mother as we walked to church on Sunday mornings:

> Easter is the first Sunday
> after the first full moon
> after the vernal equinox.

Augustine was well known to be attentive to his episcopal duties, careful and diplomatic in fulfilling them. He was responsible for setting the date each year for Easter to be celebrated in every parish within his diocese. He sat in many church councils of the province of Roman Africa, and he consulted with other bishops in more distant provinces. What he explained to Ianuarius, therefore, ought to have been in accord with his actual practice if not complete in detail. We may fairly ask of him: Are his rules sufficient for his diocese, for the province, for the whole church?

That is a difficult question to answer. Many scholars attempted this task in the early Christian centuries and have left considerable evidence for the dates of Christian Easters from the early third through the ninth centuries and various explanations for the sophisticated systems which they used to determine those dates. From their fine works it will be possible now to describe eight sets of evidence for the chronological work of early Christian scholars which bear directly upon the context in which Augustine worked and the rules which he provided; but they will not be easy for us to explain.

1. The octaëteris or eight-year calendar cycle of Babylonians, Chaldeans, and Greeks seems to have been doubled and applied by Hippolytus to guide the choice of Sundays to celebrate Easter for 16 years.[15] He was a presbyter at Ostia in the Roman civitas who lived about A.D. 170–236 and was exceedingly hostile toward Calixtus, contemporary overseer or episcopus of Rome (217–222), earlier convicted of embezzlement and whom he also accused of heresy. Two books by Hippolytus were written in Greek, including a *Synagoge* or chronicle which placed the birth of Christ in the *annus mundi* 5500. The *Cyclus Hippolyti* which began with luna XIV on Saturday, April 13, 222, survives on two marble tables together with his third century bust at the public entrance to the Lateran´Museum of the Città del Vaticano, Rome. His parallel series of dates were only for the full moon or luna XIV and for the following Sunday of Easter; the latter evidently could range between limits of luna XVI to XXII. Later the cycle was applied for the years 241–256 by adjusting the dates (Krusch, 1880, pp. 189–192; Richard, 1950, p. 243). It was partially explained by an *Expositio bissexti* for use during 64 years, and then it was extended for application during 7 × 16 or a total of 112 years.[16] The evidence

---

[15]The Easter tables and other works of Hippolytus have been edited and discussed by Isaac Argyrus [= Jacob Christmann], *De correctione paschatis* (1595), rpr by J. J. Scaliger, *Hippolyti episcopi Canon paschalis cum commentario* (Leiden, 1595, pp. 2–24), with long excerpts from Argyrus appended in this and other of Scaliger's books; the comments were collected and translated into Latin by Denis Petau, *Uranologion* (Paris, 1630, pp. 359–392). The cycle of Hippolytus was printed and its author discussed also by Aegidius Bucherius, *De doctrina temporum commentarius in Victorium Aquitanum* (Antwerp, 1634, pp. 289–312), rpr *PG* X 1857, coll. 877–884); *Monumenta ecclesiae liturgica*, I (1904, coll. 25*–27*); *DACL* IV/2 (1924, coll. 2419–2434); Jones (1943, pp. 11–13); Richard (1950).

[16]Ms London British Library Cotton Caligula A.XV (s.VIII 2) f.97v–105v; and a manuscript once in the Bibliothèque Saint-Remy of Reims, transcribed by Jean Mabillon into ms Vat. Regin.lat.324 (s.XVII 2) and edited by J. Wallis in *Sancti Caecilii Cypriani opera*, ed J. Fell (Oxford, 1682), rpr *PL* IV (1844,

deriving from Hippolytus shows that third-century Roman practice allowed Easter Sunday to be observed as early as March 18th or even the vigil of that day which would have been March 17th, a full week or more earlier than the Julian equinox of March 25th, and as late as April 21st. But the Passion of Jesus was primary for these Christians: the *Cyclus Hippolyti* avoided celebrating Easter Sunday on either luna XIV or luna XV which it considered to be the moons of Crucifixion and improper for days of fasting.

2. Bishops Demetrios (d.A.D. 232) and Dionysios (fl. 247) of Alexandria also used an 8-year cycle to set limits of luna XIV to XXI for observance of the Easter full moon, from which they then counted forward to the first day of their seven-day week in order to determine Easter Sunday. But in Alexandria they corrected the vernal equinox from the Julian date of March 25th to March 21st, and they would not celebrate during the hours of light on that day but only in the evening which they called luna XV or March 22nd. Their rules allowed Easter Sunday to range as late as April 26th.[17]

3. Anatolios was a teacher in Alexandria before he became bishop of Laodicaea about A.D. 269 to 280. Some of the dates for Easter Sunday used by him have been preserved by Eusebios, and they show that Anatolios set limits of luna XIV to XX for the Easter full moon and presumably accepted the Alexandrian vernal equinox of March 21st (*HE* VII, pp. 14–19, 32).[18] Surviving evidence unfortunately does not make it clear whether he also may have used the 16-year cycle or the Metonic 19-year cycle, but modern attempts to fit his data to the latter have not been convincing.[19]

4. Many changes took place in setting the dates of Easter for the Roman civitas on the Italian peninsula. Presidents, presbyters, and eventually bishops of its dominant basilica tried at various times to use luna XIV to XX, or at other times

---

coll. pp. 939–967); ed, G. Hartel, *CSEL* III/3 (1871, pp. 248–271); *Monumenta ecclesiae liturgica*, II (1913, coll. 82–87); Richard (1950, pp. 237–247); Ogg (1950); the lost Reims manuscript apparently attributed its text of *Expositio bissexti* to Cyprianus, and the editors have thought that this meant the bishop of Carthage (s.III med), to whom a *Chronicon* was later attributed by Paul the Deacon in the late eighth century. The *Expositio bissexti* in ms Ivrea Biblioteca Capitolare XLII (A.D. 813) f.50v–51v is also accompanied by a cycle of 16 × 4 = 64 years which may represent the Hippolytan tradition, and it should be published.

[17]Demetrios wrote paschal letters announcing the date of Easter to the bishops of Antioch, Jerusalem, and Rome according to the annals of Eutyches, ed *PG* CXI (1863, col. p. 989); but doubtless others also received these letters. Dionysios did the same, at least for the churches of Egypt, and one of his paschal letters for A.D. 247 included an octaëteris mentioned by Eusebii *HE* VII, pp. 20–22.

Within the same area of influence Epiphanios, episcopos of Constantia or Salamis on Cyprus (367–403), was still using such an eight-year cycle at the end of the fourth century; Krusch (1880, p. 23); Jones (1943, p. 14).

[18]The reference is VIII 28 in the Latin translation edited by T. Mommsen, "Die lateinische Ubersetzung des Rufinus," in *Eusebius Werke* II/2 (1908, pp. 723–725). Ms Paris Bibliothèque Nationale Lat. 10318 (s.VIII 2), p. 189: Cyclus Paschalis Grecorum seu Macedonum Post Annos XCV (extending from 258 to 352), ed Van de Vijver (1957, p. 11).

[19]Krusch (1880, pp. 311–316); Anscombe (1895); Turner (1895, pp. 699–710); Rühl (1897, pp. 114–116); MacCarthy (1901, pp. lxv–lxvi); Schwartz (1905, pp. 15–17, 138–139); Jones (1943, pp. 20–21).

luna XVI to XXII, as limits for determining the moon which Easter Sunday would follow. Once they even celebrated Easter as late as May 15th (*idus Maii*).[20] But eventually the Roman bishops gave up the attempt to limit the lunar sequence within which to determine the Easter moon. At least by the years A.D. 412 to 414, they began trying to regulate Easter Sunday itself without reference to the equinox and instead set arbitrary limits of XI *Kalends Aprilis* to XI *Kalends Maii* (that is, March 22nd to April 21st) in order to meet a peculiarly local need. They considered it important to avoid the carnival of *natalis urbis dies Romae*, the wild times around an annual commemoration of the founding of the city.[21]

5. Augustine's conversion and his transition, from an ambitious imperial rhetor and aspiring neo-Platonic philosopher to the more moderate life of a Christian theologian, occurred in Milan. There he probably learned the technical terms for Christian festivals from Ambrose, bishop of Milan (A.D. 374–397). Those terms were the Alexandrian limits of luna XIV to XXI, which by the end of the fourth century were found not only in northern Italy and in Roman Gaul (Provence) but also in most of southern Italy (including the heavily populated Neapolitan civitas) and Sicily. In Roman Africa to which Augustine returned, however, the Julian calendar still gave the vernal equinox as March 25th, not the Alexandrian correction to March 21st. Use of Alexandrian tables there with a different equinox would require new calculations and would doubtless cause considerable confusion. Fortunately there is evidence of how various scholars faced their computistical problems in Roman Africa.

A conflict over the calendar had arisen earlier in Carthage. It seems that Augustalis, a third-century bishop, used an 84-year table which extended from A.D. 213 to 297 and that he extended it to A.D. 312 for the convenience of rounding it off to an even 100 years. This Easter table has been reconstituted by Krusch (1880, pp. 17–19). In an attempt to improve upon the Hippolytan tradition by local changes in criteria, it applied the *Saltus lunae* every fourteenth year so that a successive cycle of 84 years could begin again with precisely the same data in all columns. According to this computist the data should include an *epact* of one on the first of January, and this may be the first computus which specified the age of the moon from one to thirty on the first of January each year as a number of *epactae*.

The *Laterculus Augustalis* was continued in Africa by Agriustia of Thimidia Regis who created more tables in A.D. 412, by which he supposed that he could

---

[20]Tertullian (ca. A.D. 160–240) argued that the Romans were wrong to allow luna XIV to fall as early as March 18th, or the evening of March 17th, as had the Hippolytan tables; and he also thought that they had erred by allowing Easter Sunday to be delayed as late as May 15th. His occasional remarks on these matters were conveniently given in *The Calendar of St Willibrord* (1918, pp. 24–25), by H. A. Wilson. Discussions of Tertullian and the most detailed analyses of his writings (Barnes, 1971) have ignored all reference to the date of Easter and to Easter tables, even when citing Hippolytus.

[21]The earliest indication of this practice is the letter of Innocentius, bishop of Rome (A.D. 401/2–417), to Aurelius of Carthage advising him to observe Easter Sunday on luna XVI in the year 414, rather than on luna XXIII. Thus he would set aside the vernal equinox as an early limit for the Easter moon, rather than let it wander into the confusions of carnival. This letter is found in all the collections of early canons in the Roman area of influence; for example, *PL* LXXXIV (1850, pp. 657–658).

maintain the older practices of the Roman civitas and correct the errors of others. He seems to have believed that Easter practices long since abandoned in Rome would be better than the new practices from Alexandria. Agriustia wanted to be more Roman than the Romans (Krusch, 1880, p. 29; Jones, 1943, pp. 15–17).

Such difficulties were furthered also by a mid-fifth century *Ratio paschae* which has also been called *Computus Carthaginensis*.[22] This table commenced with A.D. 439, near the end of an 84-year cycle which culminated 420 years (84 × 5) after the crucifixion in *annus Passionis* XXIX, thus A.D. 449; and all data were presumed to recur thereafter. It attributed many faults to both Augustalis and Agriustia, especially rejecting their use of *Saltus lunae* after the fourteenth year (e.g., 449) and favoring the twelfth in an attempt to make the 84-year sequence truly cyclical; but its 12-year *Saltus* actually appeared following A.D. 449 as well, with potential confusion. It also rejected attempts to reconcile Alexandrian and Milanese limits of luna XIV to XXI or XV to XXI for the paschal full moon, as well as new Roman limits of March 22nd to April 21st for Easter Sunday. The purpose of the Carthagenian *Ratio paschae* was to reassert earlier Roman practices, including those which allowed Easters on or prior to the vernal equinox, as in the Hippolytan tables, or which delayed Easter Sundays as late as the Ides of May. This new computus, however, would show how those rules were valid and meet the difficulties anticipated in the year 455 by recognizing luna XIV on March 22nd and setting Easter Sunday on April 17th.

This controversy in the Roman Africa of Augustine apparently provided occasion also for a peace-keeping effort called the *Acta Synodi Caesareae*. According to Eusebios (*HE* V, p. 24) there was a council at Caesarea presided over by Theophilos in A.D. 180 which was concerned with the tradition of the Passover and with the sending of letters "so that in the same manner and at the same time we keep the sacred day" with Alexandria. These extant *Acta* cite that place and name the bishop but do not address the same questions.[23] The document responded to a different local need during the early fifth century, which was to reconcile those Christian communities who used Alexandrian limits of luna XIV to XXI and those who used one or another of the more recent Roman limits for Easter Sunday. Thus the *Acta* were written after Romanists of Carthage who had rejected both of the current Roman usages and had tried to go back to the earliest and impossible Roman dates. As we shall see, Roman bishops did not fall back on any of these confusing and contradictory ideas of the various "Romanists" in the fifth century.

6. The Romans themselves had objected to using the Alexandrian Easter in A.D. 382 when it would have required that the Resurrection be celebrated on April

[22]Fragments of this computus survive in ms Lucca Biblioteca Capitolare Felinana 490 (A.D. 796–816), f.282–286v; Schiaparelli (1924, pp. 17–19). It seems to have been noticed first by Etienne Baluze (1630–1718), librarian for Colbert during A D 1667–1700, and was published in *PL* LIX (1847, pp. 545–560). A later edition is by Krusch (1880, pp. 279–297), and see further pp. 28, 138–140, 170; Jones (1943, pp. 44, 46).

[23]Four versions of the *Acta* will be sorted out from thirty-two manuscripts and twelve editions in Stevens' *Catalogue* (in preparation). The best modern editions have mixed them by collation of two or more versions in the attempt to create a single critical text: Krusch (1880, pp. 306–310), and Wilmart (1933); Jones (1937; 1939, pp. 44–45; 1943, pp. 87–89).

I

22nd. About A.D. 395 another Theophilos, bishop of Alexandria (A.D. 385–412), sent an Easter Table of 100 years to Rome, so that those bishops would know the correct dates for Easter Sunday.[24] But Easter Sunday fell on April 22nd in the year 417, again at the beginning of the carnival. Either Cyril, bishop of Alexandria (436–444), or his successor, Dioscores (444–451), gave assistance to Rome by sending their paschal tables for years 437–531 with rubrics and explanations in order to make certain that they could be understood there by bishops and curial officials as they already were understood in Milan. A serious effort was made in the Roman churches to follow this Alexandrian table; but it soon faced the same problems of local practice.

In the year 444 Easter Sunday was expected to fall on April 23rd according to the Alexandrian cycle, but at Rome that was two days into carnival. Therefore Leo, bishop of Rome (440–461), enquired of Pascasinus, bishop of Lilybaeum (Sicily), whether this could be avoided. Pascasinus' response[25] noted that the Romans had difficulties in the past partly because they were still using the outdated Julian equinox of March 25th for luna XIV and that they seemed ignorant of the better tables from the bishops of Alexandria. He courteously but firmly advised acceptance of Easter Sunday on April 23rd, 444. A few years later Leo wrote again to Pascasinus concerning an even later Easter Sunday anticipated for April 24th, 455. This time he also approached the emperor Marcianus (450–457), Julianus, bishop of Kios (Bithynia Pontica, ca. 454/5), and finally Proterios, bishop of Alexandria (451–457), to whom Marcianus had referred the question.[26] By that time the Alexandrians had absorbed their octaëteris into the first eight years of a nineteen-year cycle. They still insisted upon the lunar limits of XIV to XXI but, assuming that a full day extended from noon until noon, they excluded the daylight hours for luna XIV. However, people in the Western regions had a different notion of what constituted a day (midnight to midnight), so that Alexandrian luna XIV to XXI would usually have been expressed in Latin as luna XV to XXI if the moon did not rise before night (N.B.: lacking a rule, if the moon did rise between nightfall and midnight, the problem could still exist). Proterios' answer to Leo in A.D. 454

[24]Theophilos prepared a table of Easter data for 100 years from A.D. 380 to 479, commencing with the first consular year of Emperor Theodosius I (378–395). It no longer exists, but there are manuscripts and editions for an *Epistola ad Theodosium* and a *Prologus,* including *questiones* which provided the classic rules in both Greek and Latin for Christian use of the 19-year cycle. The prologue is known in two Latin versions, ed. Krusch (1880, pp. 220–226), and Jones (1943, pp. 29–33); see discussion by Krusch, pp. 84–88. Jones (1943, pp. 26–27) also mentioned several early compromises for the sake of common celebrations of Easter.

[25]Leo's first letter has not survived, but Pascasinus' reply is found in eighteen manuscripts. The earliest is ms London British Museum Cotton Caligula A.XV (s.VIII 2, f.93v–95), *Brittanniae ecclesiarium antiquitates* (Ussher, 1639), p. 926 and rpr (London, 1687), p. 480. The Ballerini edition from ms Vat. Regin.lat.586 (s.X 2) f.115v–117, was reprinted in *PL* LIV (1846, pp. 606–610); Krusch (1880, pp. 245–250); Stevens, *Catalogue.*

[26]The Ballerini edition of these letters (1753) has been reprinted in *PL* LIV (1846): ep.88, 121, 122, 127, 133, 137, 142. Some of them have been edited from other manuscripts by Krusch (1880, pp. 212–217, 255–265, 269–278); they are also discussed by Jones (1943, pp. 56–61). Details of the texts, manuscripts, editions, and bibliography will be found in Stevens, *Catalogue* (in preparation).

upheld Alexandrian practice of course, but he went on to demonstrate the validity of its extremes: that is, Easter Sunday had been celebrated in Egypt as late as April 26th, 387, and this would recur in 482 and again in 550—salt in the wounds of the Romans!

Ultimately, Leo always yielded to the more learned Alexandrians, and in Epistle 138 he announced their date for Easter Sunday on April 24th, 455, within his provinces of central Italy, Gaul, and Spain for the sake of unity with the apostolic sees. It appears from Leo's letters and answers to them that at this time the Roman bishop's influence did not extend to Magna Graecia (southern Italy and Sicily) or to Roman Africa and other African provinces.

7. One result of Leo's concerns was that his archdeacon Hilarus commissioned the *calculator scrupulosus,* Victurius of Aquitaine, to create new Easter tables which would fulfill the goal common to both Alexandria and Rome, that is, to celebrate the principal Christian festival in unity. Victurius worked from Alexandrian tables but did so in a new way and stretched the classic rules for a 19-year cycle quite a bit.[27] He allowed the Easter moon to be recognized as early as March 18th, and he approved of Easter Sundays as late as April 24th. But his primary point of reference was the Passion of Christ on luna XIV of annus mundi 5229 which he reckoned to have been the consular year of the Gemini; this he described as *annus Passionis* I.[28] With reference to the 19-year cycle his *a.P.*I was to be the fourth year, but this had the result that *Saltus lunae* followed *a.P.*XVI. With these shifts of lunar and solar cycles Victurius' year sixteen fell unhappily upon year six of the standard Alexandrian cycle whose *Saltus* followed the ninteenth year. In consequence the Easter terms would be applied differently in the two cycles during 13 years (VII–XVIIII) of the Alexandrian cycle. This made it quite difficult for the churches to celebrate Easter on the same Sundays unless Alexandria agreed to change its usage—a most unlikely prospect which even Victurius did not have the temerity to suggest. In order to obviate this problem he supplied alternate dates, *Graeci* and *Latini,* so that each bishop could use his own cycle; apparently he did not realize that some of the particular Greek dates which he proposed would not be tolerated in Alexandria.

The *Cyclus Paschalis* of Victurius of Aquitaine provided a mechanism for Romans to adapt their local practices to the Alexandrian 19-year cycle. But on many dates the apostolic sees would not be celebrating the Resurrection on the same Sundays. It may be for this reason that Leo and other bishops of Rome, Ravenna, Milan, and Aquileia did not accept them. Some bishops gained approval for them in parts of Gaul, and they seemed to have found favor temporarily in

---

[27]The prologue and tables of Victurius were published by Denis Petau (1627), A. Bucherius (1634), Th. Mommsen (1892), and E. Schwartz (1905); all were superseded by the edition by Krusch (1938, pp. 17–25), rpr *PL Supplementum* III (1963, pp. 381–426). Thus far, fourteen manuscripts of the prologue and five manuscripts of the tables have been verified for Stevens, *Catalogue* (in preparation). Other manuscripts of authentic Victurian tables may yet be found.

[28]Later, the Victurian *annus Passionis* I might have corresponded with *annus Domini* XXVIII in the Dionysian cycle but for the fact that in his actual reckoning Victurius used *annus mundi* 5230 [= *annus Passionis* I].

Ireland. Occasionally the Romans celebrated Easter on one of the Sundays which appears in the Victurian options, but there seems to be no direct evidence that this *cyclus* was actually used in Rome or other parts of Italy and North Africa.

8. The Roman ecclesiastical leaders did not abandon the project of finding some way for observing the Easter celebration together with other Christians according to some predictive system. Yet another attempt to solve the problems was made about A.D. 523 when Bonifatius, *primicarius notariorum* or (as one might say) papal chancellor, asked Dionysius Exiguus to translate Alexandrian rules into Roman practice. Dionysius was a Scythian scholar whose work in the Roman curia was primarily with codification of conciliar decisions. When his Easter tables were ready Bonifatius and his colleague Bonus recommended them to bishop Iohannes (A.D. 523–526). This was a very important development and is well documented.[29] The tables and explanations created by Dionysius Exiguus provided the pattern which eventually prevailed in the Latin churches. He applied the classic rules for the Alexandrian 19-year cycle without significant alteration, and he extended the earlier Alexandrian data for 95 years from DXXXII to DCXXVI. Although he did not suppose that a perpetual calendar had thereby been achieved, Dionysius nevertheless supplied later scholars with the mechanisms by which a nearly perpetual 532-year cycle could be used until the changes of A.D. 1582 brought further adjustments. Nevertheless there is no clear evidence to show that churches of the Roman *civitas* or of the other *civitates* in Italy did use Dionysian Easter tables until the tenth century. By that time they prevailed in schools and cathedrals throughout the Carolingian and Germanic kingdoms, Hispanic regions, Ireland, Cornwall, Wales, Scotland, and England. But apparently Rome and its provinces were the last Western areas to accept them (Krusch, 1938, pp. 4–15; Jones, 1943, pp. 61–67; Stevens, 1981, pp. 92–93).[30]

These eight sets of evidence show that from the second to fifth centuries much scholarly labor was dedicated to discovering the relation of the crucifixion, death, and resurrection of Christ Jesus to the sabbath and to the changeable moon. But

---

[29]Ms Würzburg Univ.-Bibliothek M.p.th.f.46 (ca. A.D. 800) f.lv–6 and Berlin Deutsche Staatsbibliothek Phillipps 1830 (s.IX 2) f.4–10v are the earliest manuscripts containing unaltered Dionysian tables; an earlier fragment is ms Paris Bibliothèque Nationale Lat.9527 (s.VIII) f.20.

An *Exemplum Bonifatii* recommended the tables as meeting both the criteria of the Council of Nicaea and the need to avoid carnival in Rome; it is extant in five manuscripts. Apparently Iohannes died before he had accepted them. There are also twenty-one manuscripts of Dionysius' *Epistola ad Petronium*, twenty-six manuscripts of his *Epistola ad Bonifatium et Bonum*, and at least six manuscripts of his *Argumenta paschalia* explaining the rules for his table. Many other argumenta for Easter tables were modeled on those of Dionysius. There are many editions of these works by Dionysius; the best are by Krusch (1938, pp. 69–86).

Interesting lectures on the contributions of Dionysius by B. Hoffman were printed only in *PL* LXVII (1848, pp. 453–492). See also Jones (1934; 1938, pp. 204–205; 1943, pp. 68–75).

[30]The presence of Alexandrian indictions in a Latin document have often been taken to mean that Dionysian tables were then being used, as for example Rühl (1897, p. 171): Roman documents of A.D. 584. But the Roman civitas and diocese often did business with the imperial chancery, for which indictions provided official dating. Dating by indiction therefore does not imply the use of Dionysian tables.

for a bishop the results were still uncertain. What was Augustine to say to Ianuarius about these questions? His own concern for the date of Easter was quite limited, and he was far less precise than were many Christian scholars during his era. He only proposed that one should observe it on the Sunday which falls within luna XIV to XXI of the month of new corn. That is, everyone should celebrate Easter on the first Sunday after the first full moon after the vernal equinox. Beyond that: not to worry!

IV

These examples of different sets of evidence generated through the fourth, fifth, and sixth centuries can be traced by their uses and elaborations in the many Christian schools of Spain, Gaul, Italy, Germany, Ireland, and England through the sixth, seventh, eighth, and ninth centuries. They display very intense and usually competent use of arithmetic and astronomy for the purpose of determining the proper date of Easter and predicting it for the future. And this would not exhaust the evidence by any means. Many more complications would underline the recognition that calendar and computus engaged the minds of experts at technical levels of arithmetic and astronomy and that the results interacted significantly with the lives of early Christians at worship.

It was a period during which Roman controls, courts, and taxes were disappearing in the West. During the late Roman Empire the well-to-do leaders of Roman society were worried about invaders, potential or actual. The movement to build high and sturdy walls around cities in many parts of the Empire testify to a sense of threat.[31] Furthermore from the third century onwards, Italian and Gallo-Roman patricians practiced economic restraint in education, gymnasium, and the baths. Not only did they enclose their cities with larger walls, but often they simply withdrew their produce from the markets, leaving lesser castes to bear the brunt of the economic chaos which such actions induced. Except for military employment, aristocrats usually refused to associate with the new Asiatic and many Germanic groups which moved in a few at a time with different notions of raising children and perhaps a fear of emasculation through literacy (Musset, Bachrach).[32] One could therefore suppose a break in the tradition of every school with masters and students which had existed up to that period (Roger).[33]

[31]Not only military and political problems but also the sense of crisis as it increased from A.D. 161 to 518 is discussed by Remondon (1964), with emphasis on the third century as a period of reaction which had major consequences especially for cities in the Western regions of the Empire. See also Bayless (1972).

[32]Bachrach (1972, 1973) identified several Asiatic as well as Germanic peoples who were invited and settled within the Western Empire by Roman generals specifically for the sake of using their skills. They did not mix with local leaders or peoples for several centuries.

[33]For this point of view, the fundamental work was M. Roger, *L'enseigement des lettres classiques d'Ausone à Alcuin* (1905). Marrou (1948) also completed his great study of Hellenic and Hellenistic education with insistence that the tradition was lost in the subsequent period: chapter 9 emphasized the diappearance of classical schools after the third century; chapter 10 outlined the appearance of Christian

Doubtless there was great loss, yet the evidence shows also considerable continuity. When did aristocrats fail to tutor their own children at home, even as patrons of small enclaves of students on their estates? Some of the Roman aristocrats were also Christian bishops who looked after the training of children and of priests in their dioceses for pastoral duties (Riché).[34] The Latin classical studies were continued in the households of bishops, encouraged by the positive example of Gregory the Great as bishop of Rome (590–604) and the influence of his highly literate writings, as well as by the Cassiodorean *artes et scientiae vel disciplinae* which extended the tradition of the seven liberal arts.[35] Thus there was no time when schools failed to be found in various regions of the late Roman Empire.

On the other hand if we look ahead several centuries, we find that by the late eighth century endowments and personnel once more had become available for schools on a larger scale, both East and West of the Rhine river, both North and South of the Alps, in Ireland and the British isles. Costly books were supplied again; instructors were trained. But it was not the same. These later schools were not Roman and rhetorical in a Hellenistic sense but were Christian and pietistic, serving quite different functions in monasteries and cathedrals. The disciplines of advanced learning were taking new forms.[36]

Because historians have found an abundance of piety in the new centers of sacred study, and little taste for Horace's nuances of language in the bulk of surviving Latin manuscripts, they have been too sure that there was also no science in the early Christian schools. But they have simply overlooked the demands which the new development of Christian worship and thus the Christian calendar placed upon masters and students for complex calculations of dates and all the exactitude possible in astronomy.

It was not the trivium and quadrivium that made up the new curriculum of monastic schools; it was *Grammatica, Computistica,* and *Cantica:* language, reckoning, and singing—without which neither a cathedral nor a monastery could meet its obligations to God and to man (Jones, 1947; Stevens, 1979).

Thus it is necessary to turn again to the evidence of *Computistica*. What did the masters do with the computus in monastic schools? As illustrated above, it is

---

schools "of the medieval type"; and his Epilogue affirmed "The end of the school of antiquity." Beyond the "Silver Age" of Latin literature however, Marrou relied too much upon Roger, and he did not apply the same careful analysis to the evidence for education in the later periods that he had given to the earlier.

Nevertheless by 1972 Marrou had accepted and emphasized continuity in his "L'école de l'antiquité tardive"; and see also the discussions of his paper on pp. 203–211, especially remarks of F. Prinz.

[34]In section one, "Survivance de l'école antique et organisations des écoles chrêtiennes dans les royaumes barbares méditerranéens (480–533)," Riché provided abundant evidence for continuity of classical forms of education well into the sixth century.

[35]This development was also analyzed by Riché in the first part of his section two, "La fin de l'éducation antique et le développement des écoles chrêtiennes en Italie, Gaule, Espagne (533—premier tiers du VIIe siècle)."

[36]This is the principal theme of Riché's third section, "Les débuts de l'éducation médiévale," challenging the view that the new Irish, Anglo-Saxon, or Carolingian schools were recovering a lost classical curriculum.

not at all a simple matter to use a 19-year cycle, much less to repeat it in successive periods of nineteen years. But it is only from the Christian literature of the Latin middle ages that one may learn the complex details both of its theory and of its practice. The following examples of calculations are summarized from Bedae *De temporum ratione liber* (A.D. 725) and have been explained more fully elsewhere by Stevens (1992).

Although lunar months may be reckoned as alternately 29 or 30 days, twelve such months account for only 354 days, leaving more than 11 days to deal with in terms of a solar year. Three years like that and you have lost more than a month. One keeps track of such things in two ways: first of all, some years were designated common and other embolismic in order to coordinate the lunar with the solar cycles. Thus an embolismic month of 30 days was added to the lunar cycle following years 3, 6, 8, 11, 14, 17, and 19. The eighth year in this series was chosen because of the early tradition of octaëteris, the 8-year cycle. (It is remarkable that the ninth year might have served better than the eighth but was never used.) An extra adjustment was therefore required to maintain the 8-year cycle, so that an embolismic month of 30 days is counted at the end of year 8 in order to keep track of the *epactae*. But in order to do that, the computist must look ahead and borrow two days, one from year 9 and another from year 10.

Second, after twelve lunar months of 354 days there still should be eleven days to make up the 365 days of a solar year, three times out of four. The eleven days which are lacking are reckoned as lunar *epactae,* and a running account is kept of them: after two years one has counted 22 lunar epacts; after three years 33 lunar epacts. At this point the count of epacts has surpassed the days of one full month; therefore an embolismic month had to be noted and its 30 days subtracted from 33, leaving three epacts to carry forward. After four years therefore one will count $3 + 11 = 14$ lunar epacts, after five years $14 + 11 = 25$ lunar epacts, after six years $25 + 11 = 36$ when another embolismic month must be noted and its 30 days subtracted from the count of 36 lunar epacts, leaving six epacts to carry forward; and so on for 19 years—keeping in mind that after eight years the running count of epacts will be rounded up to 30 by borrowing. In summary it looks like this:

$$
\begin{array}{lll}
8 \quad \text{years} \times 11 \text{ lunar epacts} & = 88 + 2 = 90 & (4) \\
\quad \text{less } 30 \text{ days} \times 3 \text{ embolisms} & = 90 \\
\\
11 \quad \text{years} \times 11 \text{ lunar epacts} & = 121 - 2 = 119 \\
\quad \text{less } 30 \text{ days} \times 4 \text{ embolisms} & = 120 \\
\quad\quad\quad \text{Remainder} & = -1 \\
\\
\text{or, } 19 \quad \text{years} \times 11 \text{ lunar epacts} & = 209 \\
\quad \text{less } 30 \text{ days} \times 7 \text{ embolisms} & = 210 - 1 = 209
\end{array}
$$

As a remainder, one day of the embolismic months has been subtracted more than the running count of *epactae*. In order to begin a new 19-year cycle properly, that extra day is skipped over and is called *Saltus lunae*.

I

Cycles of Time   45

Nevertheless the overall count of lunar and solar days does not come out right:

19 years	× 354 days	= 6726	(5)
7 embolisms	× 30 days	= 210	
	Total	= 6936 lunar days	
19 years	× 365¼ days	= 6939¾ solar days	

From equation 3 (page 32 above) it was evident that the 19-year cycle should have 6940¾ lunar days, from which the *Saltus lunae* could be subtracted in order to bring sun and moon together again; it should not have 6936. Where is the error?

Imagine for yourselves the situation of masters, students, and scribes redoing their figures, comparing their tables, looking up each running count of epacts to balance their figures for the 19-year Easter tables. The manuscripts show that this was in fact the single most repeated problem in monastic schoolwork; and its solution led to the greatest number of *argumenta* to be found in computistical handbooks for fourteen centuries. The error stems from that fraction ¼. The day count of 365¼ is not applied in the computus by quarters but only by integers: counting 365 every year and adding one more day every fourth year, or leap year. It is a simple procedure, but one in which the use of a fraction will lead to error! Use of a fraction in that way allowed the computisti to forget that for each day there is not only a sun to reckon but also a moon. Therefore if one adds a day to the solar cycle, one must add the same day to the lunar cycle; thus

Solar Cycle	19 × 365 days	= 6935	(6)
	4 × 1 extra day	= 4	
	Total	= 6939 days	
Lunar Cycle	19 × 354 days	= 6726	
	7 × 30 days	= 210	
	4 × 1 extra day	= 4	
	Total	= 6940 days	

This count of days (6939, 6940) will occur only one time out of four successive 19-year cycles, one in which there would be four leap years. But for the other three 19-year cycles in a sequence of four, there would be five leap years in each cycle; then in each of those three 19-year cycles the solar and lunar totals would both increase by one day (6940, 6941). In each case there will be an excess moon, and that remainder represents the *Saltus lunae* which is skipped in order to begin the next 19-year cycle with sun and moon in accord. By use of the fraction the computist could easily lose track of a full lunar day in counting 19 × 4 = 76 years; but despite their awkwardness at some points, by use of integers the reckonings of lunar and solar cycles will correlate without fail, and the date of Easter Sunday may be predicted reliably.

V

By no means do these reckonings exhaust the problems or the hidden complications to be met in applying the nineteen-year cycle itself, and they only begin to touch the technical and scientific riches of the broader computistical literature. When Frankish scholars in ninth- and tenth-century monastic and cathedral schools noticed a slippage in the position of the equinoxes—the calculations undertaken became increasingly complex and the concepts became steadily more sophisticated. On this basis schoolbooks of arithmetic, of astronomy, and even of geometry were generated—with contents both theoretical and practical (Stevens, 1979, 1981, 1992). It is also on this basis that there was an audience for the more nearly complete books of Euclid and Ptolemy when these later became available in Latin, an audience eager and informed.[37]

The modern philologists' search for the classical quadrivium faltered and partly failed when it was made to depend upon literature that cited and discussed only the classic four fields. Those discussions in late Roman and early medieval letters and encyclopedias were significant and had some influence upon schoolmasters. But during hard times the real need and pressure for studies was stimulated in other terms and by other needs which were not Hellenistic. Oddly enough, it was in the schools of Christian piety that masters and students thought it necessary to drill, to learn, to understand the computus and thus not only to use but also to comprehend the calendar and improve it. In order to do that, basic arithmetic and astronomy were required at a modest level. It has been found that masters and students in monastic and cathedral schools went well beyond those basic elements of computistical science and added planetary theory and plane geometry which were not at all necessary. The best and thus far the least known evidence for studies in early medieval schools may still be found in those 9000 or more Latin manuscripts which include the tracts, tables, and argumenta of computus wherein are calculated repeatedly and diversely the cycles of time.

References

Adams, J. (1971), *The Populus of Augustine and Jerome*. New Haven: Yale University Press.
Anscombe, A. (1895), The paschal canon attributed to Anatolius of Laodicea. *Eng. Hist. Rev.*, 10:515–535.
Aujac, G. (1974), "L'image du globe terrestre dans la Grèce ancienne." *Revue d'hist. sci.*, 27:193–210.
Bachrach, B. (1972), *Merovingian Military Organization, 481–751*. Minneapolis: University of Minnesota Press.
——— (1973), *The History of the Alans in the West*. Minneapolis: University of Minnesota Press.

[37]M. Folkerts, *Euclid in Medieval Europe* (Questio II de rerum natura, 1989); and F. Benjamin, "Campanus of Novara," in Benjamin and Toomer (1971, pp. 3–24). Campanus was a computist and calculator whose commentary on Euclid's *Elements* (1255–1259) and *Theorica planetarum* (1261–1264) were based upon Ptolemaic solar, lunar, and planetary models; both of his works became widely known from the thirteenth to the seventeenth centuries.

Barnes, T. (1971), *Tertullian, a Historical and Literary Study*. Cambridge, U.K.: Clarendon Press.

Bayless, W. N. (1972), *The Political Unity of the Roman Empire During the Disintegration of the West*, A.D. *395–457*. Unpublished doctoral dissertation. Providence, RI: Brown University.

Benjamin, F., & Toomer, G. J. (1971), *Campanus of Novara and Medieval Planetary Theory*. Madison: University of Wisconsin Press.

Boncompagni, B. (1868–1887), *Bulletino de Bibliographica e di Storia delle Scienze matematiche e fisiche*, 20 volumes. Rome.

Bonner, G. (1963), *St. Augustine of Hippo: Life and Controversies*. London: SPCK.

Borst, A. (1988), Computus: Zeit und Zahl im Mittelalter. *Deutsches Archiv*, 44/1:1–82.

Brown, P. (1967), *Augustine of Hippo*. Berkeley: University of California Press.

Bucherius, A. *De doctina temporum commentarius in Victorium Aquitanum*. Antwerp: 1964.

Bullough, D. A. (1965), *The Age of Charlemagne*. London: Elek.

Cabrol, F., & Leclercq, H., eds. (1900–1904), *Reliquae Liturgicae Vetustissimae*, Vols. 1–2. In: *Monumenta ecclesiae liturgica*. Paris: Didot.

The Calendar of St Willibrord, H. A. Wilson, ed. (1918), Henry Bradshaw Society Publications, LV. London.

Christ, W. (1983), "Victorii Calculi," Sitzungsberichte der königliche Akademie. *Philos.-hist.-Klasse*, 1:132–136 (München).

A Companion to the Study of St Augustine (1954), ed. R. W. Battenhouse. New York: Oxford University Press.

CSEL (1866 seq), *Corpus scriptorum ecclesiasticorum latinorum*, multivolume series. Vienna: C. Geroldus.

Cummian's letter De controversia paschali and the De ratione conputandi, ed. M. Walsh, & D. ó Cróinín. Toronto: Pontifical Institute of Mediaeval Studies, 1988.

Deane, H. A. (1963), *The Political and Social Ideas of St. Augustine*. New York: Columbia University Press.

DACL (1907–1953), *Dictionnaire d'archéologie chrétienne et de liturgie*, 15 vols., ed. F. Cabrol, H. Leclercq et H.-I. Marrou. Paris: Librairie Letouzey et Ané.

DSB (1970–1980), *Dictionary for Scientific Biography*, 16 vols., ed. C. C. Gillispie. New York: Charles Scribner's Sons.

Duhem, P. (1914–1959), *Le Système du monde*, 10 vols. Paris: Hermann.

Eastwood, B. S. (1983a), Mss Madrid 9605, Munich 6364, and the evolution of two Pinian astronomical diagrams in the tenth century. *Dynamis: Acta Hispanica*, 3:272–276.

———— (1983b), Origins and contents of the Leiden planetary configuration (ms Voss.Q.79, f.93v). *Viator*, 14:1–23.

———— (1987), Plinian astronomical diagrams in the early middle ages. In: *Mathematics and Its Applications to Science and Natural Philosophy in the Middle Ages: Essays in Honor of Marshall Clagett*, ed. E. Grant & J. Murdoch. Cambridge, U.K.: Cambridge University Press.

Eusebius Werke. In: *Griechische christliche Schriftsteller*, ed. E. Schwartz. Vol. 9. Leipzig: J. C. Hinrichs, 1903–1904.

Folkerts, M. (1989), *Euclid in Medieval Europe*. Questio II de rerum natura, ed. W. M. Stevens. Winnipeg: The Benjamin Catalogue for History of Science.

Fraser, J. T., & Lawrence, N., eds. (1975), *The Study of Time*. Berlin: Springer-Verlag.

———— (1987), *Time, the Familiar Stranger*. Amherst: University of Massachusetts Press.

Friedlein, G. (1871), Der Calculus des Victorius. *Zeitschr. für Mathemat. und Physik*, 16:42–79.

Geminos, *Elementa astronomiae*, Vol. 8, ed. C. Manitius. Leipzig: Teubner, 1898.

Ginzel, F. K. (1906–1914), *Handbuch der mathematischen und technischen Chronologie*, 3 vols. Leipzig: J. C. Hinrichs.

Heath, T. C. (1913), *Aristarchos*. Cambridge, U.K.: Oxford at the Clarendon Press.

Herwagen, J. (1563–1564), *Opera Bedae venerabilis omnia*, 8 vols. Basel: Verlag Herwagen.

Jones, C. W. (1934), The Victurian and Dionysiac paschal tables in the West. *Speculum*, 9:408–421.

———— (1937), The lost Sirmond manuscript of Bede's computus. *Eng. Hist. Rev.*, 52:204–219.

———— (1939), *Bedae pseudepigrapha: Scientific Writings Falsely Attributed to Bede*. Ithaca, NY: Cornell University Press.

———— (1943), *Bedae opera de temporibus*. Cambridge, MA: Medieval Academy of America.

I

48

——— (1947), *Saints' Lives and Chronicles in Early England*. Ithaca, NY: Cornell University Press.

King, M. H., & Stevens, W. M., eds. (1979), *Saints, Scholars & Heroes: Studies in Medieval Culture in Honour of Charles W. Jones*, 2 vols. Collegeville, MN: Hill Monastic Manuscript Library, Saint John's Abbey and University.

King, V. H. (1969), *An Investigation of Some Astronomical Excerpts from Pliny's Natural History Found in Manuscripts of the Earlier Middle Ages*. University of Oxford Bachelor of Letters Thesis, unpublished.

Krusch, B. (1880), *Studien zur christlich-mittelalterlichen Chronologie: Der 84 jährige Ostercyclus und seine Quellen*. Leipzig: Verlag Von Veit. Abbreviated *Studien* I.

——— (1938), *Studien zur christlich-mittelalterlichen Chronologie: Der Entstehung unserer heutigen Zeitrechung*, Abhandlungen der Preussische Akademie zu Berlin, Jahrgang 1937, Philol.-hist.-Klasse Nr. 8. Berlin. Abbreviated *Studien* II.

MacCarthy, B. (1901), Introduction. *The Annals of Ulster*, Vol. 4. Dublin: Irish Record Publications, 3.

McGurk, P. (1973), Germanici Caesaris Aratea cum scholiis, a new illustrated witness from Wales. *Nat. Lib. Wales J.*, 18:197–216.

Marrou, H.-I. (1948), *A History of Education in Antiquity*. New York: Sheed and Ward, 1956.

——— (1950), *L'ambivalence du temps de l'historie chez S. Augustine*. Montréal: Institut des Etudes médiévales.

——— (1957), *Saint Augustine and His Influence Through the Ages*. London: Longmans.

——— (1972), L'école de l'antiquité tardive. *Settimane di Studio* (Spoleto), 19/1:127–143.

Mommsen, Th. *Chronica minora*, 3 vol. in *Monumenta Germaniae Historica*, Vols. IX, XI, XIII. Berlin: 1892–1898.

Monumenta cartographica vetustioris aevi, A.D. *1200–1500*, ed. R. Almagia & M. Destombes. Amsterdam: N. Israel, 1964.

Musset, L. (1965), *The Germanic Invasions: The Making of Europe*, A D. *400–600*, trans. E. and C. James. London: Elek, 1975.

North, J. D. (1975), Monasticism and the first mechanical clocks. In: *The Study of Time*, vol. 2, ed. J. T. Fraser & N. Lawrence. Berlin: Springer-Verlag, pp. 381–398.

O'Daly, G. J. P. (1981), Augustine on the measurement of time: Some comparisons with Aristotelian and Stoic texts. In: *Neoplatonism and Early Christian Thought*, ed. H. J. Blumenthal & R. A. Markus. London: Variorum, pp. 171–179.

Ogg, M. (1950), *The Pseudo Cyprianic De pascha Computus*. London: SPCK.

Parsons, W. (1951), English translation of Augustini *Epistola LV ad Ianuarium*. In: *Fathers of the Church XII*, ed. A. Goldbacher. Washington: Catholic University of America Press, pp. 260–293.

PG (1857–1886), *Patrologiae cursus completus, Series graeca*, 161 vols., ed. J.-P. Migne. Paris: J.-P. Migne.

PL (1844–1882), *Patrologiae cursus completus, Series latina*, 221 vols., ed. J.-P. Migne. Paris: J.-P. Migne.

Pedersen, O., & Pihl, M. (1974), *Early Physics and Astronomy: A Historical Introduction*. New York: Neale Watson.

Petau, D. *Opus de doctrina temporum*, 2 vols. Paris: 1627.

Plinius Secundus, Gaius (A.D. 73–77), *Naturalis historia*, 37 vols., ed. J. Jan & C. Mayhoff. Leipzig: Teubner, 1906.

PW (1893–1968), *Real-Encyklopädie der klassischen Altertumswissenschaft*, multivol, series, ed. A. Pauly, G. Wissowa et al. Stuttgart: A. Druckenmüller.

Remondon, R. (1964), *La Crise de l'empire romain de Marc-Aurele à Anastase*. Paris: Presses Universitaires de France.

Richard, M. (1950), Comput et chronographie chez Saint Hippolyte. *Mélanges de science religieuse*, 7:237–257.

Richardson, C. C. (1940), The quartodecimans and the synoptic chronology. *Harvard Theolog. Rev.*, 33:177–190.

——— (1973), A new solution to the quartodeciman riddle. *J. Theolog. Studies*, 24/1:74–84.

Riché, P. (1962), *Education et culture dans l'Occident barbare, VIe-VIIIe siècle*. Paris: Editions du Seuil. Second edition 1967; third edition 1972. [*Education and Culture in the Barbarian West, Sixth Through Eighth Centuries*, trans. J. Contreni. Columbia: University of South Carolina Press, 1976.]

Roger, M. (1905), *L'enseigement des lettres classiques d'Ausone à Alcuin*. Paris: A. Picard et Fils.

Rühl, F. (1897), *Chronologie des Mittelalters und der Neuzeit*. Berlin: Reuther & Reichard.

Schiaparelli, L. (1924), *Il Codice 490 della Biblioteca Capitolare de Lucca*. Rome: Pressa La Biblioteca Vaticana.

Schwartz, E. (1905), *Christliche und jüdische Ostertafeln*. Abhandlungen der Königlichen Gesellschaft der Wissenschaften zu Göttingen, *Philol.-hist.-Klasse*, Neue Folge 8/6. Berlin.

Stevens, W. M. (1972), Walahfrid Strabo—a student at Fulda. In: *Historical Papers 1971*, ed. J. Atherton. Ottawa: Canadian Historical Association, pp. 13–20.

———— (1979), Compotistica et astronomica in the Fulda school. In: *Saints, Scholars & Heroes*, vol. 2, ed. M. H. King & W. M. Stevens. Collegeville, MN: Hill Monastic Manuscript Library, Saint John's Abbey and University, pp. 27–63.

———— (1980), The figure of the earth in Isidore's *De natura rerum*. *ISIS*, 71:268–277.

———— (1981), Scientific instruction in early insular schools. In: *Insular Latin Studies*, ed M. Herren. Toronto: Pontifical Institute of Mediaeval Studies.

———— (1992), Sidereal time in Anglo-Saxon England. In: *Sutton Hoo: Voyage to the Other World*, ed. C. B. Kendall & P. S. Wells. Medieval Studies at Minnesota, no. 4. Minneapolis: University of Minnesota Press.

———— (in preparation), *A Catalogue of Computistical Tracts*, A.D. *200–1200*.

Strobel, A. (1977), *Ursprung und Geschichte des frühchristlichen Osterkalenders*. Texte und Untersuchungen, 121. Berlin: Akademie-Verlag.

———— (1984), *Texte zur Geschichte des frühchristlichen Osterkalenders*. Liturgiewissenschaftliche Quellen und Forschungen, 64. Münster in Westfalen: Aschendorffsche Verlagsbuchhandlung.

Suetonius Tranquillus, Gaius (ca. A.D. 69–125?), *Divus Julius*, ed. H. E. Butler & M. Cary, 1927. Introduction and notes G. B. Townend. Bristol: Classical Press, 1982.

Turner, C. H. (1895), The paschal canon of Anatolius of Laodicea. *Eng. Hist. Rev.*, 10:699–710.

Ussher, J. *Britannia ecclesiastium antiquitates*. Dublin: 1639. Rpr. London: 1687.

Van de Vijver, A. (1957), L'évolution du comput Alexandrin et Romain du IIIe au Ve siècle. *Rev. d'hist. ecclésiastique*, 52:5–25.

Van der Meer, F. (1961), *Augustin the Bishop*, trans. B. Battershaw & G. R. Lamb. London: Sheed & Ward.

Wilmart, A. (1933), Un nouveau texte de faux concile de Césarée sur le comput pascal. *Studi e testi*, 59:19–27.

Appendix
Luni–Solar Time Cycles

Within the concepts of Hellenistic science, attempts to coordinate observed cycles of the moon and of the sun were numerous and their tabular periods could extend for 8, 11, 16, 19, 28, 76, 84, 95, 100, 112, or 532 years. With the following selected list of cycles created in the Mediterranean basin and in Western Europe, the reader may be assisted in locating them.

Each cycle is identified by *Author or Cycle* as it is usually cited, by the *date* of first appearance and usage, as well as by place of *origin* in so far as these are known in modern scholarship. Greek forms are given for the names of Greek authors within limits of Roman typeface. Two or more dates and two or more place names are provided when the reappearance of a cycle is historically significant. The Roman province for a place name is sometimes indicated: thus Hippo Regis (Africa) is in the province of Roman Africa. Further information about most of these luni–solar time cycles may be found in the Introduction to C. W. Jones (1943), while more recent studies are cited in annotations of the present essay.

Author or Cycle	Date	Origin
Octaëteris	B.C. ante 500	Babylonia
Euctemon, Meton	500–400	Athens
Kallippos of Cyzicus	370–300	Athens
Sosigenes	50	Alexandria, Rome
G. Plinius Secundus	A.D. 73–77	Italy
Laterculus Romanus	213–297	Rome
	412	Africa
Augustalis	213–297–312	Africa
Hippolytus	222–240	Ostia, Rome
Demetrios	232	Alexandria
Dionysios	247	Alexandria
Cyclus Hippolyti	241–256	Rome, Carthage?
	485	Gaul
	577	Gaul, Spain
Expositio Bissexti	241–304	Africa, Italy?
Anatolios	269–280	Laodicea

I

Appendix

Author or Cycle	Date	Origin
Romana Supputatio	312–342	Rome
Council of Arles	314	Arles, Provence
Eusebios	335	Caesaria
Athanasius	335	Alexandria
Romana Supputatio, revised	343–354	Rome
second revision	355–411	Rome
	367–403	Constantia, Cyprus
Epiphanios		
Ambrosius	374–397	Milan
Theophilos	395–412	Alexandria
Augustinus	395–430	Hippo Regis, Africa
Innocentius	401–417	Rome
Agriustia	412	Thimidia Regis, Africa
Ratio Paschae	439–449	Carthage
Acta Synodi Caesareae	400–450	Africa
Cyril	436–444	Alexandria
Dioscores	444–451	Alexandria
Leo	440–461	Rome
Pascasinus	442	Lilybaeum, Sicily
Zeitz Tafel	447	Italy
Proterios	451–457	Alexandria
Laterculus Romanus, revised	455	Africa
Victurius	455–457	Aquitaine
Dionysius Exiguus	520–525	Rome
Exemplum Bonifatii	523	Rome
Martin of Braga	580	Spain, Portugal
Cummian	633	Southern Ireland
Beda	701–725	Jarrow, Northumbria

BEDE'S SCIENTIFIC ACHIEVEMENT

The Jarrow Lecture 1985
(revised 1995)

The monk and choirmaster Beda encouraged and developed in his students
and readers an attention to the nature of things which surpassed most scholars in the
Christian tradition before him. Beda himself did not discuss every aspect of nature;
he was not an encyclopedist. But he did ask many questions, identify certain facets
of things, distinguish many variables, and provide arithmetical sense to accounts of
phenomena when there were series of data which allowed it. As we assess some of
that work, we shall see that not all of it can be considered adequate, even within the
possibilities of Beda's terms of reference and his cultural context. But some of it
was quite precise and of lasting value. Even more, some of his work stands out as so
remarkable that it seems to surpass the boundaries or his time and place and
culture. For the sake of understanding natural phenomena some of Beda's work
had scientific value, even as natural scientists use that fragile term today.

One of his letters concerned the Bissextile or extra day of every fourth year
(leap-year) which is only the simplest of many calculations and corrections required
to keep a Julian calendar in close accord with the actual course of the sun and the
moon.[1] Another letter explained how the circle of the sun itself moves through the
heavens, tilting 23½ degrees to the north and 23½ degrees to the south of the

---

[1] Bedae *Epistola ad Helmwaldum*, ed. J.A. Giles, *Anecdota Bedae, Lanfranci et aliorum* (London:
Caxton Society, 1851), p.1-6, from ms Oxford Merton College 49 (s.XV) f.289v-291; the introductory
portion was reprinted from Giles by C. Plummer, *Venerabilis Bedae Opera Historica* I (Oxford: at the
Clarendon Press, 1896), p.xxxvii. Beda included most of his letter within the text of *Liber De
Temporibus Ratione* XVIII-XXXIX, ed. C. W. Jones, *Bedae Opera De Temporibus* (Cambridge, Mass.:
Medieval Academy of America, 1943), p.250-253, rpr. *CSEL* = Corpus Christianorum, Series Latina
CXXIII B (Turnholt, Belgium: Brepols, 1977), p.399-404: Jones also added the letter's address and
introductory remarks in CCSL CXXIII C (1980), p.629, from ms Vat. Regin.lat. 123 (A.D.1056) f.40v-
42; in that manuscript most of the text had been transcribed from Beda's later work, but the opening
was inserted between the two halves (f.41) from an exemplar of the earlier letter. It must be dated
between *De Temporibus* (703) which it was intended to supplement and *De Temporibus Ratione* (725),
in which it was quoted; but Helmwald has not been identified, and no other clues to a precise date are
known.

celestial equator, twice annually passing a mid-point at which it may be observed simultaneously anywhere on the face of the earth that the hours of light and the hours of darkness are the same: the equinox.[2] This kind of astronomy had been formalized by Eudoxos of Cnidos (died ca.347 B.C.) and was badly explained by Plato (d.348/47 B.C.). It provided basic concepts for Hipparchos of Nicaea (d.post 127 B.C.), Ptolemaios (d.ca.170 A.D.), and all Hellenistic scientific writers.[3] The model and its concepts were assumed by all Latin writers, though their explanations could sometimes become garbled. Before we attempt assess Beda's scientific labours within this framework however, we should first consider some aspects of his piety as a Christian.

I

In your church and in mine the praise of God very often includes celebration of his creation, the earth and the skies, all of nature, the entire universe. The *Book of Genesis* tells of God setting stars in the heavens, sun and moon, signs and seasons, days and years, with light to separate day and night (I 14-18) -- an orderly cosmology, for which the Jews often praised the Creator in Psalms. For example, with *Psalm* XIX the worshipper praised God for the heavens in general and for the sun in particular;

1) The heavens are telling the glory of God;

...

5) In them he has set a tent for the sun, ...

---

[2] Bedae *Epistola ad Wicthedum*, ed. Jones, *Bedae Opera* (1943), p.319-325, rpr. Corpus Christianorum, Series Latina CXXIII C (1980), p.635-642. For the date of writing, A.D.725-731, see Jones (1943), p.138-139. A paragraph of later origin with an example for A.D.776 became attached to this letter in some manuscripts and several editions, discussed by Plummer, *op. cit.*, p.cliv, and Jones *Bedae Pseudepigrapha* (Ithaca, N.Y.: Cornell University Press, 1939), p.41-44.

[3] For Eudoxos, Plato, Hipparchos, Ptolemaios, and other Hellenic and Hellenistic scholars, reliable information is available from the *DSB* = *Dictionary of Scientific Biography*, ed. Charles Gillespie et alii (New York: Charles Scribner's Sons, 1970-1980), 16 vols. One of Plato's attempts to put Eudoxian astronomy into words is *Timaios* 31b-40d, concerning which see E. Maula, "Plato's Agalma of the Eternal Gods," *Ajatus* XXXI (Helsinki 1969) 7-36; idem, "Studies in Plato's theory of form in the Timaeus", *Annales Academiae Scientiarum Fennicae*, series B, 169/1 (Helsinki 1970), p.3-31; idem, "Plato's cosmic computer (Tm. 35a-39c)," *Ajatus* XXXII (Helsinki 1970); and several more recent studies.

> 6) Its rising is from the end of the heavens,
> and its circuit to the end of them;
> and there is nothing hid from its heat.

These thoughts have been extended somewhat by the English dissenting congregationalist, the Reverend Isaac Watts, and published in his *Psalms of David Imitated* (1719):

> 1) The heavens declare thy glory, Lord,
> in every star thy wisdom shine, ...
> 2) The rolling sun, the changing light,
> and night and day thy power confess; ...
> 3) Sun, moon, and stars convey thy praise
> round the whole earth, and never stand;
> so when thy truth began its race,
> it touched and glanced on every lad.[4]

This sentiment was expressed in the more exacting fashion by the *Wisdom of Solomon* which praises "the Lord of all the earth who set all things in measure and number and weight" (II 21). We recall Augustine's break with the Manichaeans because they were unable to give an account of natural phenomena, reason, and human experience which were in accord with each other. Commenting on the passage in *Genesis*, bishop Augustine reminded his canons and parishioners and readers many times of the regular movements of sun and moon through the signs of the zodiac and through the course of years and seasons and days. Manichaean religious assertions implied a certain cosmology to be accepted. The trust which these assertions required should have led to greater understanding, but it did not. Augustine was turned away by the failure of Faustus, their best teacher and bishop, to provide a rational account of the heavens in support of Manichaean dualistic cosmology. He appreciated the rhetoric of Faustus; but he also tested the content and sought a coherence which was not available in that religion. He therefore turned elsewhere for an account of natural phenomena, reason, and human experience which were in accord with each other.[5]

---

[4] These lines are quoted from *The Hymn Book of the Anglican Church of Canada and the United Church of Canada* (1971), no.91.

[5] See his *Confessions* V.6-7. The Latin text with French translation was edited by P.C. de Labriolle, *Saint Augustin: Confessions* (Paris: Les Belles Lettres, 1925; fifth edition corrected 1950); by far the best English translation is by Albert C. Outler, *Augustine: Confessions and Enchiridion* (Philadelphia: Westminster Press, 1955).

4

Ambrosius and Basil had done the same; so did Isidorus of Seville, Boetius [Boethius], and Cassiodorus, Hraban of Fulda, Abbo of Fleury, and many more to this day. Recently in a Winnipeg service of worship we sang in celebration of the creation: "Welcome, happy morning! age to age shall say, ..." But the hymn was not always so general; soon we were praising God for categories of time derived from lunar and solar cycles:

> Months in due succession. days of lengthening light,
> hours and passing moments, praise thee in their flight; ...

The tune was a German melody from the fourteenth century, and the thought from the sixth century poet of Merovingian Gaul, Venantius Fortunatus.[6]

The monk Beda found coherence in the same tradition of nature, reason, and experience as did Augustine; but he showed a greater interest in natural experience. The list of his works and resumé of his life provided by the author himself[7] named first his four books *On the Beginning of Genesis* and continued with 23 biblical commentaries and studies,[8] three saints' lives, as well as accounts of Cuthbert and three abbots of his own monastery. There is his church history "of our island and people" and his book "of the feast days of the holy martyrs." The list concludes with

---

[6] *The Hymn Book, op. cit.*, no. 462: Laus tibi, Christe, ...

[7] Beda provided the list of his own works in *Historia Ecclesiastica* V.24, together with a brief summary of his life. Editions of his collected works were evaluated in 1933 by Bernhard Bischoff, "Zur Kritik der Heerwagenschen Ausgabe von Bedas Werken (Basel 1563)," rpr. and expanded in *Mittelalterliche Studien*, vol.I (Stuttgart: Hiersemann, 1966), p.112-117, concerning biblical and literary materials; and by C.W. Jones, *Bedae Pseudepigrapha* (1939), p.1-19, rpr. in Charles W. Jones: Bede, the Schools and the Computus, Collected works ed. W.M. Stevens (Aldershot: Variorum, 1994), concerning the scientific works. Only four works are authentic of the eleven attributed to Beda on grammar, rhetoric, and various philosophical subjects in volume XC of the *Patrologia Latina* (mostly derived from the 1563 edition of Johannes Heerwagen). Jones also identified six more short tracts of scientific interest either from Beda or from his school, edited in *Bedae Venerabilis Opera Didascalica*, *CCSL* CXXIII C (1980), p.649-672.

[8] Useful reviews are Roger Ray, "What do we know about Bede's commentaries?" *Recherches de théologie ancienne et médiévale* XLIX (1982) 5-20; and B.P. Robinson, "The venerable Bede as exegete," *The Downside Review* CXII/388 (July 1994) 201-226, although the latter did not always use the best editions of texts he discussed.

books of *Hymns* and of *Epigrams* in heroic or elegiac verse,[9] *Orthography*, *The Art of Poetry*, and *Figures of Speech* in Scripture. Thus Beda nurtured that tradition in singing the praise of God, study of Holy Scripture, care for language, the writing of history. But he also developed that tradition further in ways that Augustine had not contemplated and in terms that Isidore had outlined but not carried through. Toward the end of his list of writings Beda included a work *On the Nature of Things* and another on *Times*, two of his earliest schoolbooks written about A.D. 701 and 703; later he added a larger *Account of Times*, accumulating many years of research and completed A.D. 725. These works *De Natura Rerum*, *De Temporibus*, and *De Temporibus Ratione* have never been translated into a modern language, save for brief excerpts or an occasional chapter used to explain some other author. They occupied a major part of his life and provide the primary sources for consideration here.[10] The last and larger account of the times is a difficult book dealing with a complex subject; it is also a contribution to knowledge which earned Beda a high place among scientists of all ages.

Beda's accounts of natural phenomena and of the times have attracted the attention of many scholars to separate topics. All of those works have been edited

---

[9] Michael Lapidge, *Bede the Poet*, Jarrow Lecture 1993 (Jarrow-upon-Tyne: Parish of St Paul's Church, 1994); idem, "Some remnants of Bede's lost *Liber Epigrammatum*," *English Historical Review* XC (1975) 798-820.

[10] The following abbreviations will be used:
HE = *Historia Ecclesiastica*, ed. C. Plummer (Oxford: at Clarendon Press, 1896), 2 vols.;
DNR = *Liber De Natura Rerum*, ed. C.W. Jones, Corpus Christianorum, Series Latina CXXIII A (1975), p.190-235;
DT = *Liber De Temporibus* I - XVI, ed. Jones, *Bedae Opera De Temporibus* (Cambridge: The Medieval Academy of America, 1943), p.295-303, rpr. *CCSL* CXXIII C (1980), p.585-601; to the *CCSL* edition has been added *DT* XVII-XXII: *Chronica minora*, ed. Th. Mommsen (1919);
DTR = *Liber De Temporibus Ratione* I-LXV, ed. Jones, *Bedae Opera* (1943), p.175-303, rpr. *CCSL* CXXIII B (1977), p.263-460, to which has been added *DTR* LXVI-LXXI: *Chronica maiora*, ed. Theodore Mommsen, Chronica Minora, vol.III (Monumenta Germaniae Historica, Auctores antiquissimi XIII; Berlin 1899);
Plinii *NH* = *Plinii Secundi Naturalis Historiae Libri XXXVII*, ed. C. Mayhoff (Leipzig: B.G. Teubner, 1892-1906), 5 vols.;
CCSL = Corpus Christianorum, Series Latina (Turnhout, Belgium: Brepols, 1954 et seq.);
CSEL = Corpus scriptorum ecclesiasticorum latinorum (Wien 1886 et seq.);
Jones, *Bede* = Charles W. Jones, *Bede, the Schools and the Computers*: Collected Studies (Aldershot: Ashgate, 1994);
TKr = Lynn Thorndike and Pearl Kibre, *A Catalogue of Incipits of Mediaeval Scientific Writings in Latin* (2 ed. London: The Mediaeval Academy of America, 1963).

6

and have been considered to our benefit by Charles W. Jones whose introduction to *Bedae Opera De Temporibus* (1943), p.3-122 discussed Beda's use of the Dionysian 19-year table and its extension to a 532-year perpetual calendar, application of the Dionysian *aera Incarnationis* to annals and chronicles, identification and evaluation of the multifarious earlier sources upon which Beda drew and his improvements upon them, as well as discussions of solar, lunar, and celestial cycles. But further understanding and evaluation may require a somewhat different approach. In this lecture I shall not review the Dionysian and Bedan tables or his discussions of how they work. Rather they will be drawn upon as sources in order to discover how Beda explained and used certain aspects of Hellenistic cosmology, and to what extent his work may have contributed to its further development and improvement. Parts of his work will be sampled in translation, and particular attention is directed to his concern for the land and the sea, that is, the shape of the earth and the theory of tides.

II

What did Beda teach about the nature of the universe, early in his career?[11] In his book *On the Nature of Things* Beda affirmed that the universe was created by God, as is told in *Genesis:* it was not a capricious act and the results were orderly.

---

[11] Many studies of Beda have omitted this major part of his work, but we may cite useful studies by Karl Werner, *Beda der Ehrwürdige und seine Zeit* (Wien 1881), p.107-145; Max Manitius, *Geschichte der lateinischen Literatur* I (Munchen 1911), p.70-87 with Nachträge in II (1923), p.794-795; Franz Strunz, "Beda venerabilis in der Geschichte der Naturbetrachtung und Naturforschung," *Zeitschrift für deutsche Geschichte* I (1935) 311-321, rpr. *Scientia* LXVI (Bologna 1939) 57-70, with French translation *ibidem*, Supplementum (1939), p.37-49; Beda Thum, "Beda venerabilis in der Geschichte der Naturwissenschaften," *Studia Anselmiana* VI (1936) 51-71; C.W. Jones, *Bedae Opera* (1943), p.125-129 and notes to the text of *DTR*; idem "Beda," in *DSB* I (1970) 564-566; P. Hunter Blair, *The World of Bede* (New York: St Martin's Press, 1970), ch.24 "Number and Time"; T.B. Eckenrode "Venerable Bede as a scientist," *American Benedictine Review* XXI (1971) 496-507; idem, "The growth of a scientific mind: Bede's early and late scientific writings," *Downside Review* no.316 (1976) 197-212; idem, "The Venerable Bede: a bibliographical essay, 1970-81," *American Benedictine Review* XXXVI (1985) 172-194, which also listed earlier bibliographies; K. Harrison, *The Framework of Anglo-Saxon History to A.D. 900* (Cambridge University Press, 1976), and his many other essays on the usefulness of computistical texts for analysis of annals and chronicles; G. Bonner, "The Christian Life in the thought of the Venerable Bede," *Durham University Journal* LXIII (1971) 39-55; idem, "Bede and medieval civilization," *Anglo-Saxon England* II (1973) 71-90, esp. p.82-83, 89; A.A.M. Duncan, "Bede, Iona, and the Picts," in *The Writing of History in the Middle Ages: Essays presented to Richard William Southern*, ed. R.H.C. Davis et alii (Oxford: at the Clarendon Press, 1981), p.1-42, esp. p.20-41.

Although he accepted the patristic notion that the firmament was composed of two-fold created things, both spiritual and corporeal, he did not speculate about seven heavens above the earth.[12] He taught that phenomena of the skies and phenomena of the earth could be explained rationally.[13] Certain diagrammatic schemata for such rational explanations had become conventional in Hellenistic science. Thus, the earth could be conceived as made of four basic elements - earth, air, fire, and water - mixing in various combinations to compose the dirt and stones and streams, mountains, metals, ice, animals, and every sort of thing we know by sense experience (ch.III, IV). All things of our earth and atmosphere move naturally towards the centre of the cosmos (mundus, universe) which is of course the Aristotelian concept of gravity. The earth lies in the centre of the whole universe with stars of the heavens circling around it (ch.III, V-XI).

The earth itself could be described over-all as a globe banded by five circular zones; it was extremely hot around its middle (torridus) but extremely cold in the north and in the south (frigore), while the central zones both north and south were temperate and habitable (ch.IX). One of Beda's favourite authors was Augustine who had roundly affirmed that the earth is a sphere[14] but that no evidence existed for people called Antipodes, referring to inhabitants of the southwest quarter of the

---

[12] Speculation about seven heavens above the earth was attributed to Bede by Jones, *Bedae Opera* (1943), p.125-126, and P. Hunter Blair, *The World of Bede* (1970), p.262; but this was corrected by Jones in his review of Hunter Blair's book in *Speculum* XLVII (1972) 288.

[13] Bedae *DNR*, ch.I. Operatio divina, quae secula creavit et gubernat, quadriformi ratione distinguitur: ... Tertio, quod in materia materias, secundum causas simul creatas non iam simul, sed distinctione sex primorum dierum in caelestam terestremque creaturum formatur. ...
Ch.II. Die vero prima lux facta est, et ipsa de nihilo. ... Septimo Dominus requievit, non a creaturae gubernatione, cum in ipso vivamus et moveamur et simus, sed a novae substantiae creatione.
His sources for these two chapters are the Christian scriptures and the ideas of Augustine, for example *Confessiones* XII.7-8; but closest to his words in ch.II may be Iunilius II.ii, ed. *Patrologia Latina* LXVIII. 25-28. See also *Bedae Libri Quatuor In Principium Genesis*, ed. Jones, *CCSL* CXVIII A (1967), especially the hexaemeral section (I.495-582) which discusses aspects of the natural world and includes some chronological calculations for the sake of paschal typology in commentary upon *Genesis* I.17-19. This first part of Book I has been dated A.D. 703-709 by Jones (p.vi-x, Introduction); see also idem, "Some introductory remarks on Bede's Commentary on Genesis," *Sacris Erudiri* XIX (1969-1970) 115-198, esp. p.119-121, 174-176, and 189-192, rpr. Jones, *Bede* (1994): item IV.

[14] *De Genesi ad litteram* I.10, ed. *CSEL* XXVIII (1894): 15.6; *Qaestiones evangelicarum* II.14, ed. *Patrologia Latina* XXXV (1841) 1339.

orbis quadratus.[15] Isidore on the other hand used the five banded schema of the globe and affirmed that both temperate latitudes could be inhabited, adding that in the southern band the Ethiopians were scorched from living too near the central torrid zone.[16] Beda's information was not limited to inhabitants of the northern temperate zone, but he passed over such questions at first and returned to them later, as we shall see.

The stars are at different distances from the earth, according to Beda (ch.VI), and they are made of various sorts of matter, such as vapours, rain, hail, and snow that affected weather (ch.VII, XI) but exercised no influence upon man's character and determined no human activities (ch.XIII). All the stars reflected light from the sun (ch.VI, VII), especially the moon whose phases and eclipses were carefully described. Eclipses of both sun and moon were clearly presented in terms of the relative positions of sun and moon observed from the earth (ch.XX, XXII).

Beda's sources thus far were a small schoolbook *On the Nature of Things* by Isidore of Sevilla which he used extensively and that *opus pulcherrimum*, the large encyclopedia of *Natural History* by Pliny the Elder, of which only books II through VII may have been available in Northumbria. He saw no difficulty in quoting this pagan author about lunar and solar eclipses in conjunction with information from

---

[15] *De civitate dei* XVI.9-17, ed. B. Dombart and R.A. Kalb, *CCSL* XLVII-XLVIII (1955). His references to parts of the globe and assertions of philosophers about them are numerous, but his details and assumptions would fit better with the *orbis quadratus* model of the earth than with the tripartite *rota terrarum* (T-O map), so often cited by historians of cartography. At XVI.9 Augustine points out that the logic used about the global shape of the earth by those philosophers often do not result in knowledge of lands or waters or peoples such as *Antipodae* on the opposite side, without some evidence. But the consequence of such knowledge or ignorance for religious questions of salvation is nil. This section XVI.9 was repeated as a gloss to Bedae *DTR* XXXIV in the manuscripts of "B-glosses" attributed to Bridfert in ed. *PL* XC (1850) 453-454, cited by C.W. Jones, *Bedae Pseudepigrapha* (1939), p.33, rpr. Jones, *Bede* (1994).

[16] Augustini *De Civitate Dei* XVI.9; Isidori *Liber De Natura Rerum* X.38-39, ed. J. Fontaine, *Isidore de Seville, Traité de la nature* (Bordeaux: Feret et fils, 1960), whose source was Hygini *Astronomica* I.8. Beda *DTR* XXXIV.59-64 agreed with them. Ethiopians were often cited as evidence for the existence of *Antoikoi*, people living near the sources of the Nile in the southeast quarter of orbis quadratus. See H. J. Mette, *Sphairopodia* (Munich: Beck, 1936), p.66-78, and W. Wolska, *La topographie chrétienne de Cosmas Indicopleustes* (Paris: Presses Universitaires de France, 1962), p.258-259, 267-269. Unfortunately, many valuable discussions of medieval literature are still vitiated by confused references to *Antipodes*, as if the word indicated a region or a place opposite North, rather than those people who live opposite to the *Sunoikoi*, the latter being "our sort of people."

Ambrose, Vegetius, Isidore, and Basil to explain the phenomena (ch.XXVII, XXVIII, XXXI).[17] And his sources also included Isidore's *Origines*, as well as an anonymous tract *De ordine creaturarum*.[18]

In no instance can he be found to have improved upon his sources at this time, about 701; and they were not such tracts as had been stepping stones to the physical sciences at Rhodes or Alexandria. It was simply the case that Beda and his contemporaries had available no works of sophistication, not even the works of Boetius. There was a Latin literary tradition however that would convey quite a lot of data about nature if one searched it out. Some of that tradition was about exceptional phenomena which Beda included: clouds were great conglomerates of moisture and air which could bump into each other, and when they did so there would be thunder and lightning like the sparks from striking stones (ch.XXVIII-XXX). But why then do we not experience these phenomena simultaneously? The reason was given that the lightning is seen more quickly by the eye than the thunder is heard by the ear (ch.XXIX) -- an observation of some importance but one which was not pursued to a detailed explanation. Much more information was given which is of interest, some of it rather specific, some rather general, and not all of it can be thought to be worthwhile. But it was the regularities of things which were considered with more precision, as Beda's studies developed.

Towards the end of his work *On the Nature of Things* Beda again took up aspects of cosmology which he had once assumed were adequately clear and implicit. Earlier he had simply stated that the earth was a globe (ch.VI: *globo terrarum*; XXII: *globo terrae*; XXIII: *globo terrarum*), but he returned to discuss again the central location of the earth in the cosmos (ch.XLV) and its shape as a globe (ch.XLVI. *Terram Globo Similem*). Once more he described its concep-

---

[17] Plinius Secundus in opere pulcherrimo naturalis historiae ita describit ... (*DTR* XXVII.2-3). Sed et Plinius, saecularibus literis sed non contemnendis, ... (ch.XXXI.20).

[18] These sources have been identified by C.W. Jones in his edition of *DNR*. The Pliny codex which Beda used may have partly survived in ms Leiden Voss F.4 (Northumbrian miniscule s.VIII[1]) f.4-33; see E.A. Lowe, *Codices Latini Antiquiores* X (1963) 1578. This manuscript was not mentioned in the Jarrow Lecture for 1982 by M.B. Parkes, "The scriptorium of Wearmouth-Jarrow," and may have been written in some other school in the region. Many questions about Beda's use of Pliny are unsettled, but see now Arno Borst, *Das Buch der Naturgeschichte. Plinius und seine Leser im Zeitalter des Pergaments* (Academie der Wissenschaften, Heidelberg, 1994), p.98-120.

tualization in five bands of *klimata* (ch.XLVII) and how to use a gnomon or long pole to form a right angle triangle as the simple but effective instrument by which data could be accumulated for differentiating the five parallel zones of the zonal model (ch.XLVlII). He also added a final chapter LI on divisions of the earth into three huge continents (Asia, Africa, Europa), a longstanding Hellenistic image of the globe. The tripartite *rota terrarum* had a very long life as a schema representing the globe, at least from the fifth century B.C. when Aeschylos referred to the continents. It has often been cited as the "T-Rota" by historians of cartography.[19] It may be that these items were written in response to questions raised by Beda's Northumbrian students, for whom he elaborated some of the basic concepts of Hellenistic science.

The shape of the earth for example is not everyone's most urgent problem; we do not need to think about it in going about everyday affairs. Perhaps this is the context which allowed some modern writers and school teachers to attribute rather banal notions of the earth to medieval people. However, the only late Greek or Latin writer who described the whole earth as being flat rather than spherical was a sixth century author, Cosmas Indicopleustes; and there were no Roman or medieval writers who did so. Cosmas' ideas about the Christian faith were also out of the ordinary and certainly not those which were common or orthodox in either Greek or Latin regions of the Hellenistic world. According to the only careful analysis of that work,[20] he must have been a heretic under the influence of Syrian Nestorians, though it remains uncertain whether he was a merchant or a monk. Fortunately for art historians, Cosmas' work was illustrated; but the surviving copies are rare and late; it had no evident influence on Western thought. What did the intelligent person believe during those difficult times about the whole earth?

Quite often in the course of ordinary life there is need to draw together general observations and general concepts about the way lands and seas meet at

---

[19] Facsimiles and a valuable list of manuscripts displaying the tripartite *rota terrarum* are in *Monumenta cartographica vetustioris aevi, A.D. 1200-1500. Mappamondes*, ed. R. Almagià and M. Destombes (Amsterdam: N. Israel, 1964). Chapter two is devoted to examples prior to A.D.1200, but the facsimiles do not display the earliest known examples of the various types. Here, as well as elsewhere, *orbis quadratus* was unrecognised.

[20] W. Wolska, *op.cit.* (note 16).

shores, about crossing either lands or seas for great distances, or about measuring those distances. In order to reach a destination and to come home again, a traveller must somehow get his bearings from East and West, North and South. For these purposes one may use the projected northern pole of the sphere of stars, or he may observe the varying course of the sun from rising to meridian to setting as its position on the horizon shifts day by day, season by season. Most people who have travelled England more than once, who have lived by the seashore, or who have ever crossed a great sea in any direction would develop some concept of the dimensions and shape of the surface that they are covering; quite a few will draw implications from their experience for the shape of the earth as a whole, and others will accept someone else's concept if it is brought to their attention. Such people probably include every writer whose works survive from all lands bordering the Mediterranean, the North Sea, and the Atlantic Ocean, as well as many of those anywhere within the continent of Europe who have travelled in a bishop's or a count's retinue, who have transported goods from one market to another, or who who have made a pilgrimage of any consequence.[21] Ordinary human experience has not always been taken into account by modern writers. Yet what does a farmer see across a great wheat field near Tours but his barn rising to him in the distance as he comes in for dinner? What does a sailor see but the port gradually sinking below the horizon as he leaves Streanæschalch (Whitby) or Yarmouth for herring and cod, or Bristol in search of whale? The various ways of experiencing a curvilinear surface could be expected to imply a sphere.

But what alternative concepts are possible? As I have discussed elsewhere,[22] evidence for ideas of the notion of a flat earth among early peoples has been often asserted but rarely found: there is a single clay tablet of Sumerian origin whose interpretation is uncertain; among the Hebrews the notion of a flat earth probably did not exist; no Egyptian evidence has survived for any concept of the earth as a whole; but it is possible that some pre-Socratic Greek philosophers may have conceived of the earth in the shape of the cylindrical section of a stone column. On

---

[21] *Travel and Travellers in the Middle Ages*, ed. A.P. Newton (London: Routledge & Kegan Paul, 1926), esp.p.1-18 (Newton) and p.19-38 (M.L.W. Laistner).

[22] W.M. Stevens, "The Figure of the earth in Isidore's De Natura rerum," *Isis* LXXI (1980) 268-277, rpr. as item III of this volume (1995).

12

the other hand a Pythagorean notion of the sphere of the stars led by analogy to the concept of a sphere of the earth, accepted and taught by Plato and Aristotle in the fourth century B.C. From that time, all Hellenistic scholars taught that the earth was a sphere, and this was accepted and taught without question by all the Christian writers who commented on it.

Yet it is one thing to receive a tradition from the Mediterranean world on the continent or in Anglo-Saxon England, quite another to understand it. In his later and more sophisticated *Account of Times*, Beda composed chapters XXXI and XXXII in order to explain the different lengths of days and nights and the variation of shadow lengths cast on the earth's surface at different times during the year. He began by simply stating that the reason for unequal length of days is the globular shape of the earth. The term used in Scripture was *orb*, as in *Isaiah* XL.22, *Job* XXII.14, and *Proverbs* VIII.27, meaning a figure of enormous range.[23] The word *orb* could also be narrowed to mean a two dimensional circle of the earth's horizon or the circular path of the sun's circuit, as in *Psalm* XIX. Beda was concerned that the more restricted usages of the term not be taken as descriptive of the earth and firmly asserted that both the Holy Scriptures and secular letters speak of the earth as *orb* meaning round, *rotunditas est*.[24] It is a fact, he said,

> that the earth is placed in the centre of the universe not only in latitude so that it is round like a shield but rather in every direction like a ball, no matter which way it is turned.[25]

---

[23] Bedae *DNR* XLVI.2-5; Orbem terrae dicimus, non quod absoluti orbis sit forma, in tanta montium camporumque disparilitate, sed cuius amplexus si cuncta linearum comprehendantur ambitu, figuram absoluti orbis efficiat. See also Stevens, *op. cit.* (1980), p.272.

[24] *DTR* XXXII.2: Causa autem inequalitatis eorundem dierum terrae rotunditas est. Among Beda's writings are at least twelve commentaries on thirteen books of the Old Testament (including *Tobit*) requiring perhaps 27 codices; six commentaries on twenty-five books of the New Testament requiring perhaps 23 codices; plus two books on the Jewish Tabernacle and the Temple of Jerusalem in 5 codices, as well as at least three collections of readings from the *Pentateuch, Joshua, Judges, Psalms*, and the New Testament other than Gospels. M.L.W. Laistner and H.H. King, *A Hand-List of Bede Manuscripts* (Ithaca: Cornell University Press, 1943), has often been supplemented during the past four-and-a-half decades and could well be revised.

[25] *DTR* XXXII.3-6: ... neque enim frustra et in scripturae divinae et in communium literarum paginis orbis terrae vocatur. Est enim re vera orbis idem in medio totius mundi positus, non in latitudinis solum giro quasi instar scuti rotundus sed instar potius pilae undique versum aequali rotunditate persimilis; ...

These three-dimensional thoughts are his own, and no source has been found for them. He went on to say that the earth must be a globe because its seasons are the same for all those who live on the same latitude, because the hours of the day are simultaneous for those who live along any north-south line of longitude, and because when travelling North or South one can see certain stars appear to rise over the horizon or to disappear below it. It is the bulge of the earth that causes this experience. The earth is round, and he suggests that the reader climb a hill and look around to see what he means.[26]

One consequence is that Beda thought of the globe as populated on all sides, and he spoke of the sun passing over those who inhabit the southern side of it.[27] This amplified his thoughts of twenty-five years earlier. About A.D. 701 Beda had read and learned from Pliny the description of five latitudinal bands of the earth, two of which were temperate and habitable. This was essential Eudoxian astronomy as it was customariy applied to description of the earth.[28]

Another consequence is that latitude on the surface of the sphere may be determined rather well by either of two methods, and Beda was soon describing the earth in eight bands of *klimata* from South to North beginning with India, the Red Sea, and the northern coast of Africa.[29] From various authorities Pliny had collected names of many regions, cities, and prominent places both within and without the Roman Empire in parallel zones of *klimata*, sometimes adding hours of daylight for certain places or the height of some tall object together with its shadow length. Beda used some of this material for *DNR XLVII-XLVIII*. Pliny had complained about the diversity of data and worried about whether these reports were useful: the frame of reference for his eight parallels is lacking, save for the

---

[26] *DTR* XXXII.40-41: Quae cuncta de monte quolibet pergrandi undique circum habitato valent facillime probari.

[27] *DTR* XXXII.21-23: ... cum plagam austri circumiens hiberno tempore pervehitur eos qui meridianum terrae latus inhabitant ante oriens adit sed serius dimittit occidens, ...;   40-41: ... terrae moles opposita spatium praecludit aspectandi. Quae cuncta ... undique circum habitato...

[28] *DNR* IX, quoting Plinii *NH* II.177.

[29] In *DNR* XLVII Beda described eight bands of *klimata* about the earth, quoting Plinii *NH* VI.211-220.

five-banded zonal rota of the globe, and the other kinds of data seem too simple. But first impressions may be misleading; hours and shadow lengths could actually provide enough information for quite accurate calculations of what we now might call latitudes.[30] One good method was to observe the longest period of daylight at summer solstice, and reciprocally the shortest period of daylight at winter solstice. Pliny had named regions from India and Persia to the Gates of Hercules within each of his parallels, and Beda selected some of these for *dies longissimus*: for example Meroe was said to have twelve and a half hours of daylight at noon on the summer solstice, Syene less than thirteen hours, a line from southern India through the Red Sea to Gibraltar fourteen hours, Crete and the upper coast of North Africa fourteen hours and perhaps two-fifths of an hour, Syracuse and Cadiz fourteen hours and just over a half hour, northern Sicily and Narbonese Gaul fourteen hours and two-thirds, from Macedonia through Tarentum and the Balearic Islands to central Spain fifteen hours. Marseille fifteen hours and a ninth, Venice and Cremona fifteen hours and three-fifteenths, northern Germany sixteen hours. The reports gathered by Pliny and repeated by Beda are not sufficient for advanced cartography by later standards. The hours of daylight given for Meroe and for Syene would place them too far south, the extent of Sicily would have to increase hundreds of miles, while Narbonne is much too far north, as the calculation of degree equivalents will make clear.

Most of the text of Bedae *DNR* XLVII is presented here in order that one may compare the shadow lengths for each of eight parallel bands of *klimata* circling the earth, and the hours and portions of hours which were usually expressed in fifths or fifteenths, approximately between 30° and 45° North Latitude. In addition modern equivalents have been given in degrees and minutes, but they are placed on the right in brackets, in order to make it plain that no such precision had been intended by

---

[30] The circles of the heavens were lines on the Hellenistic model of Kosmos or mundus, but when applied to the earth they did not serve as lines of latitude in our sense. There was a more general term, *klima-klimata*, which could be used for broad areas such as *plaga mundi* (a quarter of the kosmos) but could also be applied to parallel bands of heaven or earth. Klima of earth could refer to the zone between two parallel circles or the zone on both sides of a parallel, but it did not refer to a parallel circle as a line of latitude. *Latitudo* is a term referring to the breadth of the zodiacal band, with parallel circles of latitude above and below the ecliptic. That is still Beda's use of *latitudo* in *DNR* XVI.5-7, *DTR* VII.31-32 and XXVI.30-31. See further Otto Neugebauer, *A History of Ancient Mathematical Astronomy* (Berlin: Springer, 1975), vol.II, p.725-736.

users.[31] It should be pointed out also that the obliquity of the ecliptic is assumed to be 23°30' for the sake of calculating those degree equivalents which are approximate.[32]

Primus, ab Indiae parte australi per Rubri Maris accolas et Africae Martima ad columnas Herculis pervenit;...	XIIII	[30°46']
Secundus, ab occasu Indiae per Medos vadit et Persas, Arabiam, Syriam, Cyprum, Cretam, Lylibaeum, et septentrionalia Africae contingens;...	XIIII et bis quinta parte	[35°24']
Tertius oritur ab Indis Imauo proximis, tendit per Caspias portas, Taurum, Pamphiliam, Rhodum, Cicladas, Syracusas, Catinam, Gades;...	XIIII atque dimidiae cum tricesima	[36°50']
Quartus ab altero latere Imaui per Ephesum, mare Cicladum, septentrionalia Siciliae, Narbonensis Galliae exortiva;...	XIIII et tertias duas	[38°11']
Quinto circulo continentur ab introitu Caspii maris Bactrii, Armenia, Macedonia, Tarentum, Tuscum mare, Baleares, Hispania media;...	XV	[41°21']
Sextus amplectitur Caspias gentes, Caucasum, Samothraciam, Illyricos, Campaniam, Etruriam, Masiliam, Hispaniam, Terraconensem mediam, et inde per Lusianam;...	XV addita nona parte	[42°20']

---

[31] Calculation of modern degree equivalents from shadow lengths or hours of daylight require spherical trigonometry, explained by D.R. Dicks, *Early Greek Astronomy to Aristotle* (London: Thames & Hudson, 1970), p.19-23; for this I am pleased to acknowledge the assistance of my son, Wesley A. Stevens (Toronto). Concerning Ptolemaios' approach to these problems, see O. Neugebauer, *Exact Sciences in Antiquity* (Providence: Brown University Press, 1957; 2 ed. New York: Dover, 1969), p.214-218.

[32] The origin of Pliny's data quoted by Beda is not known, and the correct obliquity therefore cannot be ascertained. It may be noted by comparison that Hipparchos of Nicaea (ca.190-120 B.C.) seems to have used the equivalent of 23°51' in his calculations, whereas modern astronomers project 23°43' for his period. See *The Geographical Fragments of Hipparchus*, ed. D.R. Dicks (London: Athlone Press, 1960), Fragment 41, p.90-91, with notes on p.167-169 and 176-194, esp.p.193-4 with summary table of positions.

Reports of *klimata* by Pliny have also been discussed by George Sarton, *A History of Science*, vol.I (Cambridge University Press, 1959), p.174-175, who did not take into account hours of daylight if they did not accord with gnomon ratios expressed by shadow lengths; but both types of data were discussed by E. Maula, "Ancient shadows and hours," *Annales Universitatis Turkuensis*, series B, no.126 (Turku, Finland, 1973).

16

Septimus ab altera Caspii maris ora incipit, vaditque per Thraciae aversa, Venetiam, Cremonem, Ravennam, transalpinam Galliam, Pyreneum, Celtiberiam;...	XV et quintarum partium trium	[43°5']
Octavus a Tanai per Maeotim Lacum et Sarmatas, Dacos partemque Germaniae, Gallias ingreditur; ...	XVI	[49°]

Further Beda selected more information from Pliny to add two southerly and two northerly parallels:

His circulis antiqui duos praeponunt: unum per insulam Meroen et Ptolemaidam Rubri Maris urbem,...	XII et dimidia hora	[8°33']
Alterum per Syenem Aegypti, ...	XIII duosque subiciunt	[14°36']
Primum per Hyperboreas et Britaniam	XVII	[54°28']
alterum Scythicum, a Ripheis iugis in Thulen,	in quo dies continuantur nocteque per vices.	[66°30']

These shadow lengths do provide a means for estimating the north/south relationships of places and regions in parallel bands about the earth. Beda thought them useful.

A year or two after making up his first list, in *DT* VII he added Alexandria with fourteen hours and Italy with fifteen hours in a brief summary, illustrating the significance of the solstice. Then in *DTR* XXX-XXXIII, he provided Pliny's material more extensively within a rather full explanation of the system.[33] About A.D.731 in *Historia Ecclesiastica* I.1, he commented upon his own northerly position in Britain where

> the winter nights are also of great length, namely XVIII hours... In summer too the nights are extremely short; so are the days in winter, each consisting of six standard equinoctial hours,

---

[33] Most of the data was not available from works of Solinus or Isidore. Bedae *DT* VII added bits of information from Plinii *NH* II.186-187. Bedae *DTR* XXXI.47-65 is partly dependent upon Plinii *NH* II.186-187 and IV.104; *DTR* XXXII.34-40 used *NH* II.178; *DTR* XXXIII.7-80 cited and then quoted *NH* VI.211-220 extensively. Plinii *NH* VI.211-220 has been newly translated by O.A.W. Dilke, *Greek and Roman Maps* (London: Thames and Hudson, 1985), p.185-187, with discussion on p.67-71.

adding by comparison *longissima dies XV, brevissima VIIII* for Armenia, Macedonia, and Italia. Of course variable hours would not do for this purpose, and Beda often emphasized that one must use a standard hour-length: those twelve equal hours of daylight from sunrise to sunset at equinox when they are the same for all seasons and anywhere in the world.[34]

Another method for determining latitude is also deceptively simple: one may take a stick of a man's height and measure the length of its shadow at noon on the day of equinox. The ratio between the height of the stick (gnomon, umbilicus) and the shadow-length for each place may be taken as sides of a right-angle triangle, or the gnomon ratio. By use of carpenter's square, a geometer's square, or later the shadow square on the quadrant or the dorsal side of an astrolabe, that ratio could be resolved into proportions of the earth's surface which are the same for all places equidistant from the equator.

This type of ratio was commonly used by astronomers to find the altitude of sun, stars, and especially the pole (near the "pole star"). Eudoxos and Eratosthenes understood that daylight hours, as determined by the sun's rising and setting on the horizons, determined the observer's place on the earth's north/south circumference. Hipparchos expressed this in a table of chords, equivalent to a table of sines in later trigonometry.[35]

Hipparchos and Ptolemaios had applied this method of gnomon ratios to elaborate a system of 360° for the surface of the earth's globe as well as for the

---

[34] Sed aequinoctialis dies omni mundo aequalis est. Vario autem lucis incremento in meroe longissimus dies XII horas aequinoctialis et octo partes unius horae colligit, ... (Bedae *DT* VII.4-6; see also *DNR* XLVII.6 and *DTR* XXXI.48). Beda was rejecting the tradition of seasonal hours for which there is quite widespread evidence: Neugebauer, *History* (1975), II, p.980-982.

[35]See Dicks, *Geographical Fragments* (1960), p.154-164; Neugebauer, *History* (1975), I, p.43-45 (on Hipparchos) and vol.II, p.934-940 (on the *Geographia* of Ptolemaios).

zodiac as a scale for observing planetary motions.[36] But engineers and surveyors used the same method to determine altitude of buildings or mountains long before one could speak of the tangents and cotangents of the angle of a triangle. Roman *mensores* also had made this much simpler for using the height/base ratio to determine relative latitude; Latin *mensores* means literally measurers, that is, engineers and surveyors as well as city planners and city clerks for keeping records of their public activities. The surveyors took a common pole length with six divisions, as one side of the triangle. The height of the gnomon or umbilicus for this purpose was assumed to be the height of a man, probably six Roman feet. If one Roman foot were 29.57 cm or about 11 5/8 inches English measure, that would be a gnomon of just over 5 feet 9 ½ inches. Thus, a series of shadow lengths could serve to indicate relative position between the equator and the north pole, and so that for each position the shadow length alone, umbra corporis humani, need be recorded.[37]

Pliny provided much information for this purpose and Beda passed it on to his students and readers in the same passages cited above.[38] The gnomon and the

---

[36] These uses of the gnomon are discussed by Dicks, *Early Greek Astronomy* (1970), p.19-23, 154; O. Pedersen and M. Pihl, *Early Physics and Astronomy* (Amsterdam: Elsevier, 1974), p.42-45, 50-53.

Concerning the shadow square and other set-squares: D.J. Price, "Precision instruments to 1500," in *A History of Technology*, ed. C. Singer et alii, III (Oxford University Press, 1957), p.608 and figures 329 and 329a (on p.528-529); E. Maula, "Ancient shadows and hours," *op.cit.* (note 32); N.L. Hahn, *Medieval mensurarion*: Quadrans vetus and Geometrie due sunt partes principales ... (Philadelphia: American Philosophical Society, Transactions LXXVII/8, 1982), p.lxxi-lxxiii, with early explanations from *Quadrans vetus* 17-20 (p.166-168).

The earliest surviving Latin astrolabe has a shadow square on its dorsal side, dating from about A.D.975, possibly from the region of Barcelona: see M. Destombes, "Un astrolabe carolingien et l'origine de nos chiffres arabes," *Archives internationales d'histoire des sciences* LVIII/LIX (1962) 3-45, plate II. That instrument was designated a fraud by Derek de Solla Price in *A Computerised Checklist of Astrolabes*, ed. S.L. Gibbs, J.A. Henderson, and D.S. Price (New Haven 1973), no.3042; nevertheless, so far as I could judge from personal inspection, its Latin inscriptions are authentic Visigothic script of late tenth century. Its origin and date have not been agreed upon by the thirteen authors of *El màs antiguo astrolabio latino*, ed. G. Beaujouan, A.J. Turner, W.M. Stevens, to be published in 1995; but its authenticity has been upheld by eleven of them.

[37] Umbilicus umbrae corporis: ms Bern Burgerbibliothek 250 (s.X/XI) f.14ᵛ. O.A.W. Dilke, "Illustrations from Roman surveyors' manuals," *Imago Mundi* XXI (1967) 9-29; idem, *Greek and Roman Maps* (London: Thames & Hudson, 1985), p.89. A rich account of the education of *mensores* and their work is by Dilke, *The Roman Land Surveyors* (Newton Abbot: David & Charles, 1971), based upon some of the surviving manuals in seventh and ninth century manuscripts.

[38] Bedae *DNR* XLVII, from Plinii *NH* VI.211-220. For calculating degree equivalents which follow, the advice of my son, Mark M. Stevens, has been invaluable.

parabolic course of its shadow were used with quite elementary procedures to find the local meridian precisely and thus true North/South, the moment of noon, the solstices, and the length of the tropical year. The ratio of heights of various sorts of gnomen or umbilici and their shadow lengths at noon on an equinoctial day for seven of the eight circles are reported to be (with degree equivalents added):

Ab Indiae parte australi ad columnas Herculis	7/4	[cotangent 29°45']
Creta, sept.Africae	35/23	[33°19']
Syracusa, Gades	gnomonis cunctae umbram XXXVIII unciarium	[36°?]
	21/16	[37°18']
Narbonensis Galliae Baleares, Hispania media	7/6	[40°36']
Massilia	9/8	[41°38']
Venetia, Cremona	35/36	[45°42']

Southern India is much farther south than here supposed, the Pillars of Hercules farther north, and both cannot be within the same circular band. But the others are surprisingly well placed. Beda added similar information for which the ratios are not given directly:[39]

Aegyptus [Alexandria]	Umbilici umbra paulo plus quam dimidiam gnomonis mensuram efficit	[ca.30°]
Urbs Romae	Nona pars gnominis deest umbrae	[ca.42°]
Ancona, Venetia	Superest xxxv	[ca.45°+]

---

[39] Bedae *DNR* XLVIII. Gnomica De Hisdem, from Plinii *NH* II.182-183.

Neither Pliny nor Beda explained the methods by which common terms could be found in order to standardize this data, even though the ratios given seem to imply that the reader knows the height or at least the scale of the gnomon being used. Yet the ratios themselves in series locate each region and all places within one circular band to the North or to the South of those in another band.

Beda's work stimulated others to explain the techniques for both of these methods of estimating geographical latitude, and many scholars and students soon began to add explanatory notes, tables, or *rotae horarum* which elaborated the data for Beda's locale or for their own positions on the curved face of the earth. Although it is the hours of daylight and darkness at the summer or the winter solstice that should give terrestrial latitude, users would sometimes work out the hours of daylight also for each of the four seasons and then for all twelve months of the year; such hours were often entered at the foot of each month in calendars. Twelve month calendars often added the hours of the night and hours of day at the foot of each month. There are hundreds of examples from these early centuries.[40] Although it is shadow lengths at equinox which give relative latitude, tables for both equinoxes and both solstices survive in many manuscripts. Both types of data were summarized in tables or in *rotae* for convenience. Shadow lengths for pairs of months through eleven hours of the day are given in numerous manuscripts at all periods and deserve further attention for their utility.[41]

---

[40] For example the calendar of ms Oxford BL Digby 63 (A.D.862-892) f.40-45ᵛ (December: nox XVIII, die VI); similar hours are given in Fleury ms Berlin Staatsbibliothek Phillipps 1833 [Cat.138] (s.X ex) f.23-28ᵛ with English influence from the time of Abbo, and in the Fleury ms Bern, Burgerbibliothek 250 (s.X/XI) f.15-17ᵛ.

[41] An example was printed by Johannes Bronchorst (Noviomagus), *Opuscula complura de temporum ratione* (Köln 1537), fol. lviii recto as a gloss to Bedae *DTR* XXXIII: Concordia duodecim mensium per umbram cuius gnomon erit humani corporis longitudo, rpr. *PL* XC (1850) 447-448. This type of information was set out in a fan-shaped horologium in Thorney ms Oxford St. John's College 17 (A.D.1102-1110) f.37 with the text of Isidori *DNR* V. 1 glossing Bedae *DTR* XXXI; the shadow length is given for each of eleven hours at both equinoxes and both solstices in *rotae*, with the same data repeated in a different horologium below, again with four *rotae*. This type of data is found as well in the Fleury ms Paris BN Lat.5543 (A.D 847) f.137 and 140ᵛ-141. See Faith Wallis, "Ms Oxford St John's College 17: a mediaeval manuscript in its context" (University of Toronto dissertation 1985), p.400-402.

A traveller heading North or South therefore should take along not only a tall marked hiking pole but also a scrap of parchment on which he had copied a table or a rota of shadow lengths or hours of daylight, quite useful tools which would be nonsensical, were not the earth a sphere, as Beda repeatedly emphasized.

III

The Effective Power of the Moon is discussed in Bedae *DTR* XXVII, in which he relates several stories about the moon's forces on human, animal and physical life. Does dew change on the grass with waxing or waning of the moon? Could there be a marvelous stone which shines in response to the correct Easter moon but not to a mistaken one? There is even a bit of healing suggested; and it would seem to a modern reader that Beda is opening the way for astrology which he not only avoids in other contexts but warns against. Actually he is only attempting to shift our attention to a subject which has few proofs but many correlations. Today, the police captain or the telephone operator does not have to be told, when work picks up, that the moon is full -- for they know, like Beda, that at full moon "the crazies come out!" There is a correlation in ordinary experience, even though it remains unexplained. Beda's reason for introducing such matters appears in the following and longer chapter XXIX. *De Concordia Maris Et Lunae*, in which he had something quite new to say to the serious student for an understanding of the larger correlation of the moon's effective power and the great rhythms of the sea: that is, the tides. Tyd or Tide was an English term meaning season, time, hour, but also the time between flux and reflux of the sea or even the movement itself. The corresponding Latin terms for season, time, and hour were not adapted in that way. Rather, sea tide was indicated in Latin by the terms: *aestus, aestus maritimi mutuo accedentes et recedentes,* or *marinorum aestuum accessus et recessus.*

Hellenistic scholars and church fathers, both Greek and Latin, expressed very diverse views about tides and their causes, mostly erroneous but some valid. A twice daily ebb and flow was mentioned only by a few writers, though it had to be common knowledge. Poseidonios of Rhodes (ca. 135-50 B.C.) learned also of the semimonthly variations, especially strong at equinox. Many remarked that the tides somehow followed the phases of the moon, or they noticed ebb-tide before a new moon. Basil (d.379) explained in his morning sermon to workers on the way to their jobs that "the water of the straits flows from one side to the other during different phases of the moon, but at the time of its birth [new moon] they do not remain quiet for the briefest instant but they are constantly tossing and swaying backwards and forwards until the moon, again appearing, furnishes a certain order for their reflux." Ambrose followed Basil in his own Latin sermon.[42]

An attempt to correlate tides with an eight-year calendar was reported in the first century A.D. by Plinii *NH* II.215. This octaëteris is the earliest attempt we know to coordinate lunar and solar cycles. Eight years of 365 solar days come to 2920 days, whereas 99 lunar months (alternating 30 with 29 day months) also come to 2920 days, but with twelve hours left over. After two such periods of eight years there would be 24 hours of the moon left over and one lunar day could be skipped in order to bring the sun and moon together and begin again. In the long run this was not going to work very well as a calendar, but then none is ever quite right. Not only this eight year calendar and also the nineteen-year luni/solar cycle ascribed to Euctemon and Meton of Athens ca. 430 B.C., but every calendar ever devised has had to skip a moon or add a measure to the sun's course,[43] in order to correlate two cycles. Unhappily, no whole number multiples of days or hours or parts of hours comprising the periods of the lunar cycle and the solar cycle will ever allow a combination which is complete for each cycle series; that is, they are ultimately incommensurable. But their correlation could be improved.

Usually one should expect more knowledge and understanding of the nature of things from cultural centres within the Roman Empire than from those outside. It was certainly useful for the first century encyclopedist to report an earlier attempt to account for tidal movements with an eight-year luni/solar calendar. It may also

---

[42] Basilius of Caesarea, *Hexameron* 6-11; Ambrosius of Milano, *Hexameron* IV.7. A useful survey of early literary references to the tides remains unpublished by Thomas R. Eckenrode, "Original aspects in Venerable Bede's tidal theories with relation to prior tidal observations" (St. Louis University dissertation, 1970), chs.4-6.

[43] C.W. Jones, *Bedae Opera* (1943), p.11-15.

have been reasonable for both an Irish author and a young school teacher on the North Sea coast to assume that the report was good.[44] But waters of Mare Nostrum have insignificant tides, averaging about four feet for the range from ebbtide to hightide during a year, but only about one-and-a-half feet at Alexandria. Poseidonios had commented on data which came not from Rhodes but from the Atlantic coasts of the Spanish peninsula, and most useful information about tidal phenomena must derive from observations either in the western Mediterranian where the range increases somewhat or from the extremities of oikumene where tidal movements of the western and northern coast of France range up to 50 feet. Thus, we find improved accounts of tidal action by Priscian of Lydia in the sixth century and by two anonymous seventh century tracts written in the North, perhaps in Ireland.

Prisciani *Solutionum ad Chosroem* was written in Greek but survives only in Latin translation in a single late manuscript, though deriving from a Carolingian exemplar.[45] He described tidal action on the coasts of both the western Mediterranian and the Atlantic ocean emphasizing straits and including four-day tidal bore on the Thames and Rhine. He recognized that the water level changes daily with two ebbtides and two hightides and that the moon has an attractive force which draws water after as it moves to the west but with a definite lunitidal delay, and he also noticed that the higher tides each month coordinate with conjunction of moon and sun at new moon as well as with full moon when they are in opposition. He has combined scattered remarks from Aristotle, Poseidonios, Seneca, and Pliny to advance tidal theory explicitly towards recognition of Spring-tides and Neap-tides and of the very high tides at equinoxes.

Further advances in tidal theory were made in mid-seventh century Ireland by the author of *De mirabilibus Sacrae Scripturae liber* whose preface gave his name as bishop Augustine. He had been stimulated to write the work by Manchianus (d.

---

[44] Plinii *NH* II.215; Bedae *DNR* XXIX; and see "Liber De Ordine Creaturarum" IX, discussed below.

[45] Ms Vatican Urbin.lat.1412 (s.XV) f.1-71; ed. I. Bywater, "Priscianus Lydus," in *Commentaria in Aristotelem Graeca*, Supplement I, Part II (Berlin 1885), p.41-104, esp.p.70-73. Incipit of Priscian's valuable "Questions to Chosroes" is Cum sint multe varie in questione propositiones... (TKr 341). The best analysis is by Eckenrode, "Original aspects" (1970), p.196-209.

652) who may have been the centre of a very active circle of scholars in southern Ireland.[46] In part I.7 (written ca.655), he affirmed that God had ceased creating on the seventh day and started to govern; any new wonder was henceforth based upon a principle laid down in the depths of nature. The book discussed many wonders of God's creation in three divisions of Holy Scripture, including the Deluge and other types of flooding. Here he brought up the tides of Ireland, affirming their regularity but denying any understanding of why tides acted the way they did or what source they had. Nevertheless, he asserted that moon and tides were related, in more detail than any previous writer.

The Irish Augustine described *ledo* and *malina*, terms having appeared in extant literature previously only by the fourth century physician Marcellus of Bordeaux whose remedies must often be found *die Jovis liduna*. In late Latin there is only one other reference to *malina*, a term which in the context meant neap-tide (the least of the high tides), as *ledo* meant spring-tide (the highest of the high tides).[47] The author recognised that neap-tides flow and recede evenly for six hours each way, whereas spring-tides build up faster in only five hours but take seven hours to withdraw. Such new knowledge must derive from observation. Analysis of concordance between moon and tides led to more detail: he thought that the hebdomada required for either neap-tide or spring-tide was itself divided by the quarter of the moon into periods of three days and twelve hours; this accumulated to XXIV *ledones* and XXIV *malinae* in each common year of XII lunar months, or XXVI of each in an embolismic year of XIII lunar months. He also thought that

---

[46] *De Mirabilibus Sacrae Scripturae*, ed. *PL* XXXV (1845) 2149-2200; it was placed among works by Augustine of Hippo due to similarity of name. Ireland is mentioned twice therein at I.7 and II.4; the latter refers to tides. On pseudo-Augustine see J.F. Kenney, *Sources for the early history of Ireland* (New York: Columbia University Press, 1929), p.275-277 and D. O Cróinín, "The Irish Provenance of Beda's Computus," *Peritia* II (1983) 229-247, esp.p.238-242, who also clarified several confusing references to Manchianus. A new edition of this work by Gerard McGinty (Dublin: University College dissertation 1971) is not yet available. ó Cróinín pointed out that the author was working with Victorian Easter tables in the region where Cummian had been promoting the Dionysian; his article emphasizes that it was from this region that the examplar for the "Sirmond group of manuscripts" had its origin, that is, the collection of tracts in Beda's computistical library; see C.W. Jones, "The lost Sirmond manuscript of Beda's computus," *English Historical Review* LII (1937) 204-219, rpr. Jones, *Bede* (1994): item X; idem, *Bedae Opera* (1943), p.106-107.

[47] Marcellus Empiricus, *De Medicamentis*, ed. G. Helmreich (Leipzig 1889); see Eckenrode, "Original Aspects" (1970), p.217-220.

malinae or spring-tides were highest at both the two equinoxes and the two solstices, though observation partly failed him at this point (that is, springtides are now seen to dissipate at solstice and not necessarily to occur on that day). These advanced teachings were recognised to some extent also in another work *De ordine creaturarum* of late seventh century whose ch.IX also may have derived from work within the southern Irish milieu.[48] Beda used information common to both *De mirabilibus Sacrae Scripturae* and *De ordine creaturarum* but appears to have drawn more directly upon the latter's section IX.[49]

Despite some errors and inadequacies in all his sources, Beda could combine the information received from Pliny with the very much improved knowledge of these northern seventh century tracts and evaluate them to enrich anyone's understanding of tides. Pliny had mentioned a daily variation of tidal inundations and separately noted a daily retardation of the moon's course relative to the sun which Beda could verify from his computistical studies of lunar and solar cycles. Many others had remarked these phenomena, especially the Irish Augustine; but Beda's work undertook to define the relationship of luni/solar retardation to tidal variations and to quantify the time function as nearly as possible. For this purpose he initially accepted Pliny's period of retardation but later applied a corrected

---

[48] *Liber de ordine creaturarum: un anónimo irlandés de siglo VII*, ed. Manuel C. Díaz y Díaz (Santiago de Compostella: Monografias 10, 1972); idem, "Isidoriana: Sobra el libro de ordine creaturarum," *Sacris Erudiri* V (1953) 147-166; an older edition was rpr. *PL* LXXXIII (1850) 913-945. Attribution of the work to Isidore in some late manuscripts was false, but its Irish origin is not altogether certain.

Ordine Creaturarum survives in Freising ms München Staatsbibliothek CLM 6302 from the last third of the eighth century in a Carolingian hand but with many signs of Irish background in script and contents. See *CLA* IX (1959) 1267; B. Bischoff, "Wendepunkte in der Geschichte der lateinischen Exegese im Frühmittelalter" (1954), rpr. *Mittelalterliche Studien* I (Stuttgart: Hiersemann, 1966), p.205-273, esp. p.226, 230, 236, 245, 255, 257; idem, *Die südostdeutschen Schreibschulen und Bibliotheken in der Karolingerzeit* I (rev. ed. Wiesbaden 1960), p.81-82. An early copy is ms Paris, Bibliothèque Nationale Lat.9561 [olim Suppl.Lat. 254.2] (s.VIII med) f.1-14^v: Anglo-Saxon Uncial; see *CLA* V (1950) 590; E.A. Lowe, *English Uncial* (Oxford 1960), p.23, reviewed by B. Bischoff, *Gnomon* XXXIV (1962), rpr. *Mittelalterliche Studien* II (1967) 333. A third early copy is ms Basel, Universitätsbibliothek F.III.15b (s.VIII^2) f.1-19: Anglo-Saxon minuscule perhaps from the North, according to *CLA* VII (1956) 844; but the two insular hands have begun to take on some Carolingian characteriscs by f.10-12^v and may have been working in Fulda where this manuscript was known at least by s.X. See also H. Gneuss, "A preliminary list of manuscripts written or owned in England up to 1100", *Anglo-Saxon England* IX (1981): nos.785 and 894.

[49] Bedae *DTR* XXIX.38-44.

period; he also accepted initially the Augustinian hebdomadal coordination of ledones and malinae with lunar phases but later adjusted the weekly intervals, too.

Before the specifics of an improved tidal theory could be developed and rewritten, they must have been refined in many ways during twenty-two years. When published in A.D. 725 Beda's *De temporum ratione* had encompassed new knowledge which could not have been obtained without observations and measurements over long periods of time from quite distant places which Beda himself could not have carried out. During those years he must have enlisted the cooperation of many friends, as we know he did for preparation of his *Historia Ecclesiastica*. Indeed three of the places from which he learned of the tides were there named: Lindisfarne, Whithorn, and the Isle of Wight whence he had correspondents. Concerning the *insula Lindisfarnensi* he could report its foundation as an episcopal see together with information about each abbot-bishop, their connexions with other sees, and the interest taken by each Northumbrian king in the community and diocese. He was well informed not only about the construction of a church and distinctive liturgical usage but also about recent activities of the brothers and priests there. Lindisfarne is "enveloped by sea twice a day like an island, and twice a day the sand dries and joins it to the mainland." He knew at least four priests: the sacrist Guthfrith, Herefrith from Mailros who was later abbot of Lindisfarne, Baldhelm, and Cynemund who crossed over to the mainland often in their pastoral duties; Lindisfarne was the administrative centre of an extensive region.[50] Visitors could wait on the tides before crossing or make local enquiries about reaching the island, but supplies of goods and services for the bishop and abbot and their communities could lose any profit by expensive delays of the loaded carts and pack animals and workers.

The Holy Island lies at about 54°50' North Latitude. If one steps across the stones into the North Sea towards the end of ebbtide about three-and-a-half days

---

[50] Beda reported activities of abbot-bishops and kings: *HE* Praef,; III.2,3,12,17,25-27; IV.4,26-31; V.1,17,23; about building a church: III.23; about Lindisfarne usage at Laestingaeu/Lastingham: III.25; and about activities of the general community: Praef.; III.22; IV.4,12,27,29,31; V.1,19. He also named four priests who supplied him with information about Cuthbert in the *Vita sancti Cuthberti*, ed. B. Colgrave, *Two Lives of Saint Cuthbert* (Cambridge University Press, 1940), p.141-307; see D. Whitelock, "Bede and his teachers and friends," in *Famulus Christi*, ed. G. Bonner (London: SPCK, 1976), p.19-39, esp.p.31.

after New Moon, the Spring-tide may turn and pursue the observer relentlessly back to higher ground. A schedule of tides would be of great aid, and it is difficult to conceive of Lindisfarne without a tide-table posted or someone on duty with the information. These are data of use to Beda, who probably visited the community to review his prose *Life of St. Cuthbert* with the monks before it was issued. Improvement of the tidal data over a reasonable period of time can be expected by some monk of Holy Island who would cooperate by supplying more specific details under Beda's guidance.[51]

From the Western shores Beda often received news from both Iona and Whithorn. Far to the North at 56°17' North Latitude, the insula Hii was a prime source of Christian piety in the Northumbrian as well as the Pict kingdoms, and the Hienses monachi supplied much information for him, both historical and contemporary. But he also kept in touch with the see whose foundation was attributed to Ninian in the sixth century at Hwit aern or Hwiterne, meaning Candida casa but today Whithorn rather than White House. This is a town near the southern tip of a large Galloway peninsula on a meridian with the Isle of Man and at 54°38' North Latitude, almost due West of Jarrow. Along this way since megalithic times has run a major trade route from the Garonne and Loire river basins on the Atlantic coast Gaul northwards either around or across Bretagne, past Cornwall and Wales into the Irish Sea. Archeologists have demonstrated that this route continued North to the Hebredes and the shores of Pictland. Yet a tidal current passes from North to South to produce strong tidal flows past the Island of Islay on the North and between Rathlin Island and the Mull of Kinthyre on the South to produce strong tidal flows, with eddies and whirlpools. Before the Spring-tide current weakens in the broad Irish Sea it is still flowing at six knots off the Mull of Galloway, sufficient to deter some merchants who used an alternative route on the inner side of the Mull into Luce Bay and across a few miles of flat land to Loch Ryan for easier sailing to

---

[51] Bedae *HE* III.3: Insula Lindisfarnensi... qui videlicet locus accedente ac recedente reumate bis cotide instar insulae maris circumluitur undis bis renudato litore contiguus terrae redditur. The term reuma was rare, but see Jones, "Beda and Vegetius," *The Classical Review* XLVI (1932) 248-249, rpr. Jones, *Bede* (1994): item II.

the North.[52] Closely related with Hwit aern/Whithorn is St. Ninian's Cave about four miles south west on Luce Bay, while the Isle of Whithorn, perhaps with a monastic community, is southeast by the same distance. Anyone from the region would know of the great current thwarting traders who supplied the bishop, abbot, or town. One of Beda's contacts with Aldhelm was the latter's deacon and monk Pecthelm[53] who probably was a source about the West Saxons. He may have been active for some time in recovering the Galloway and Dumfriesshire regions for Northumbrian influence before he became bishop of Hwit aern (ca.A.D.730-734),[54] and he was named as an active informant. From the western shores, Pecthelm could be expected to cooperate in observations of tidal currents, bore, and other phenomena at Beda's request.

Vecta insula or the Isle of Wight is described as lying three miles off the southern coast of England at the mouth of river Homelia/Hamble. He told us this in order to explain the unusual meeting of two great tidal currents flowing around England from North to South and causing great turbulence in the strait called the Solent, forcing water up the river and then letting it flow back into the sea. Wilfrid had worked in this area with the South Saxons; Acca and others in Wilfrid's party had told Beda of their experiences in the region, and he knew of Bosanhamm near the south shore as well as of the peninsula Selaescu or Selsey used by Wilfrid as a retreat, east of the Isle of Wight. These places were in the diocese of Daniel, bishop

---

[52] Insula Hii: Bedae *HE* III.3, 4, 17, 21; V.9, 15(twice), 24; Hienses monachi: V.21, 22, 24; Hiensium dominum: V.15.

Hwit aern. Bedae *HE* III.4; V.13, 23. Tidal currents, eddies, and whirlpools concerning Whithorn: see M. Davies, "The diffusion and distribution pattern of the Megalithic monuments of the Irish Seas and North Channel coastlands," *Antiquaries Journal* XXVI (1946) 38-60, esp.p.40-42, 50, and Figure 1 displaying trade routes, currents, and speeds; Charles Thomas, *Britain and Ireland in Early Christian Times, A.D.400-800* (London: Thames & Hudson, 1971), p.78-82; E.G. Bowen, *Saints, Seaways and Settlements* (Cardiff: University of Wales Press, 1969), p.18, 23-27.

[53] Bedae *HE* V.13, 18, 23. The name is given as Pehthelm or Pechthelm in ms London, BL Cotton Tiberius C.II (s.VIII^2), insular minuscule, according to E.A. Lowe, *Codices Latini Antiquiores* II (rev. 1972), p.191. His successor at Whithorn in A.D.735 was Frithuwold, named in the *HE* continuation of ms Cambridge, University Library Kk.5.16 (the Moore Bede), date and origin of which is uncertain; see M.B. Parkes, *The Scriptorium of Wearmouth-Jarrow* (The Jarrow Lecture 1982), p.26, n.35.

[54] N.K. Chadwick, "Early culture and learning in North Wales," *Studies in the Early British Church*, ed. Chadwick et alii (Cambridge University Press, 1958), p.60-62; C. Thomas, *The Early Christian Archaeology of North Britain* (Oxford University Press, 1971), p.14-18 et passim.

of the West Saxons, who also sent him accounts of the region and island.[55] Reports from one or more of these informants about tidal actions around the Isle of Wight contributed to his theory of tides which accounted for such irregularities.

Along with Lindisfarne, Jarrow, and Wearmouth, there are a surprising number of ports and monastries on or near North Sea which were the locations for historical and hagiographical accounts in the *Historia Ecclesiastica*. Some of those stories from those many seaside places must derive from local persons who were informed about the coastal life upon which they all depended, persons who augmented the author's communications network during several decades.[56] Beda reported events from twenty-three places along the eastern sea coast, including several peninsulas or islands:

Aebbercurnig/Abercorn   55°55' North Latitude

> *HE* I.12; IV.26: Northumbrian monastery a few miles west of Edinburgh on the Firth of Forth (high tidal bore); Trumwine worked from there as a bishop of the Picts (IV.28).

Colodesburh/St. Abb's Head   55°49'

> *HE* IV.19,25: double monastery on the coast near Coldingham deriving its name from the abbess Aebbe, sister of Osuiu and aunt of Ecgfrith. The priest Edgisl informed Beda about irregular life and carelessness there (IV.25).

---

[55] Tidal currents and tidal bore: Bedae *HE* Praef.; IV.16,23; other references to Wight: I.3,15; IV.13,19; geographic centre of the island lies about 50°37' North Latitude. Bosanhamm/Bosham (IV.13) lies at 50°46' North Latitude, near the South Saxon shore to the north of Selsey and northeast of Wight. Selaescu/Selsey (IV.13; V.18) lies at 50°40' North Latitude, a tidal island of 87 hides east of Wight but now submerged. Concerning information from Acca, Nothelm, and Pecthelm about the Isle of Wight, the Solent, and Wessex, see Jones, *Saints' Lives and Chronicles in Early England* (Ithaca, N.Y.: Cornell University Press, 1947), p.48-49, 190-194, and note 46.

[56] The latitudes for these locations are approximate, as read from the Ordinance Survey one-inch maps of *Ancient Britain, North Sheet* and *South Sheet* (2 ed., London 1964). See also the maps and studies in C. Thomas, *Britain and Ireland* (1971) and *The Early Christian Archaeology of North Britain* (1971); *The Saxon Shore*, ed. D.E. Johnston (Council of British Archaeology Research Report no.18, 1977); M. Gelling, *Signposts to the past, Place-names and the history of England* (London: Dent, 1978); *The Anglo-Saxons*, ed. J. Campbell, E. John, P. Wormald (Oxford: Phaidon, 1982); D.A. Bullough, "The missions to the English and Picts and their heritage (to c.800)," in *Die Iren und Europa im früheren Mittelalter*, ed.H. Löwe, vol.I (Stuttgart: Klett-Cotta, 1982), p.80-98.

30

Insula Lindisfarnensi   55°35'

> *HE* III.3: monastery on a small peninsula North of *urbs regia* Bebbanburh, the Holy Island was the seat of the bishops of Northumbria, five miles north of Bebbanburh by sea, twice the distance by land. Beda had frequent communications with the monks there; his numerous references are cited in note 50 above.

Insula Farne   55°32'

> *HE* III.16; IV.27,29-30: used by both Aidan and Cuthbert as a retreat about 7 miles from Insula Lindisfarnensi but less than 2 miles from the *urbs regia* at Bebbanburh. Oidiluald/Oethelwald remained after Cuthbert's death.

Bebbanburh/Bamburgh   55°32'

> *HE* III.6,12,16-17: from the time of Oswald and Osuiu, this was an *urbs regia* for Northumbrian royal families; Aidan died A.D. 651 at a site on the royal estate which may be the present-day St. Aidan's Church.

Adtuifyrdi/Alnwick?   55°20'

> *HE* IV.28: on river Alne, site of a council with king Ecgfrith and archbishop Theodore of Canterbury when Cuthbert was elected bishop of Hexham and Lindisfarne and was fetched there from Farne Island (685); he declined Hexham.

Ostium Tini/Tynemouth   54°56'

> *HE* V.6: monastery on north bank of the river Tyne as it meets the North Sea downstream from Jarrow.

Ingyruum/Jarrow   54°55'

> *HE* V.21,24: monastery on north bank of river Tyne before it opens to the North Sea, but just beside a secondary stream.

Flumen Uiuri/South Shields?   54°51'

> *HE* IV.23: monastery between rivers Tyne and Wear but on north bank of river Wear, founded by Hild, sister of Osuiu, who soon moved to Hartlepool and then to Streanæshalch; Verca was later abbess (Bedae *Vita sancti Cuthberti* 35).

Uiuraemuda/Wearmouth   54°50'

> *HE* IV. 18; V.21,24: monastery on north bank of river Uiura/Wear just as it bends north to enter the North Sea.

Heruteu/Hartlepool   54°47'

> *HE* III.24; IV.23: monastery for women on an insula Cerui in the sea, a natural harbour at northern mouth of river Tees, 16 miles south of Uiuraemuda; founded by Heiu after the death of Aidan.

Streanæshalch/Whitby   54°26'

   HE III.24-25; IV.23,26: double monastery on a promontory with shore facing North
   and a good tidal harbour. Beda called it *sinus fari*, Bay of the Watchtower (or
   Lighthouse?).

Hacanos/Hackness   54°15'

   HE IV.23: monastery built in A.D. 680 by Hild about 13 miles South of Streanæshalch
   and a few miles inland.

Wetadun/Watton   53°53'

   HE V.3: monastery for women 7 miles north of Beverley and about 10 miles from the
   sea.

Inderauuda/Beverley   53°42'

   HE V.2,6: monastery for men on river Hull in silva derorum, 12 miles from North Sea
   coast and 9 miles north of the Humber estuary, on which were the ports for primary
   traffic with the continent. The Humber has a great tidal bore running over 40 miles
   from the North Sea.

Peartaneu/Partney   53°08'

   HE II.16; III.11: monastery south of the Humber shore and ten miles from the sea;
   on river Limine/Steeping. Beda received stories from the priest and abbot, Deda.

Cnobheresburh   52°33'

   HE III. I9: the castrum at urbs Cnobheri on a deep bay was used by Fursa as a base
   for his work among the East Angles; perhaps identified with the Saxon shore fort,
   now called Burgh Castle.

Dommoc/Dunwich?   52°16'

   HE II.15: civitas with a bishop for work with the East Angles; on the coast 24 miles
   south of Cnobheresburh and now submerged. Abbot Esi was Beda's informant.

Ythancaestir/Bradwell-on-Sea   51°41'

   HE III.22: Saxon shore fort and church where river Penta (Blackwater) meets the
   North Sea; for ten years Cedd's base of work among the East Saxons.

Tilaburh/Tilbury   51°35'

   HE III.22: monastery-base for Cedd's work among the East Saxons on river Thames
   northwest of Rochester and on the road to Barking. The river Thames has a great
   tidal bore and high Spring-tides.

Lundonienses/London   51°27'

   HE II.3; IV.6: a trading centre on the river Thames.

Hrofaes caestrae/Rochester   51°21'

> *HE* II.3,6; III.14; IV.2,5,12; V.23 (twice), 24: civitas Dorubrivis was a walled town at a crossing on river Medway which opens into a large bay and then into the broader estuary of the river Thames; 19 miles west of Canterbury.

Racuulfe/Reculver   51°21'

> *HE* I.15; V.8: monastery at a Roman fort on Isle of Thanet, north of river Wantsum [Yant] or Genlada which separated the Isle from the mainland of Kent.

On the western shore he reported events from four places, about three of which he seems to be less well informed. In addition to insula Hii/Iona and Hwit aern/Whithorn discussed above, there are

Lugubalia/Carlisle   54°56'

> *HE* IV.29: Luel or Luguvalium was a walled civitas on river Edan as it nears the Solway Firth. Old Carlisle is a Roman fortress 10 miles southwest.

Caerlegion/Chester   54°04'

> *HE* II.2: Legacaestir or civitas Legionum was a Roman fort on river Dee before it empties into the bay.

As well, one ought to add the well-known teachers and friends of Beda who were great travellers on land and sea: Ceolfrith who was his master and then abbot at Jarrow; Hwaetberht, his fellow teacher and then abbot, to whom he dedicated *De temporum ratione* in 725; and Acca, his friend and fellow singer. Ceolfrith travelled far and often. As abbot he wrote to king Naitan or Nechtan IV of the Picts about A.D. 706 or later to explain the new Easter reckoning.[57] Although it is not improbable that Beda drafted that letter for his abbot, Ceolfrith had been his teacher for many years and presumably introduced him to computus, as Trumberht taught him the Scriptures. Hwaetbehrt may have been one of those who brought Beda a report from their trip to Roma that the *annus Passionis* DCLXVIII was used there at Christmas, A.D.701. Beda referred to his friend Hwaetberht three times as "Eusebius," and it was probably he who wrote the *Aenigmata Eusebii* in forty sets of verses, four lines each, which include five on aspects of nature pertinent to the computus (V.*De celo*; VI.*De terra*; X.*De sole*; XL.*De luna*; XLVIII.*De die et nocte*)

---

[57] Bedae *HE* V.21.

and two on important adjustments necessary for using the 19-year cycle (XXVI.*De die bissextili; XXIX.De aetate et saltu lunae*).[58] Acca had been trained in York and travelled with Wilfrid to Rome, Frisia, and many parts of England, succeeding him as bishop of Hexham (709/10 - 731/2). Through Acca, Beda had easy access not only to York and Hexham but also to the double monastery at Streanæshalch on the sea.[59]

From such learned friends, it is most likely that Beda received competent assistance which greatly broadened his range of data. There were probably many other associates whose names are not recorded as mathematicians but who are among the crowd of calculators, Coelfrith said, who "were able to continue the Easter cycles" and could keep "the sequence of sun, moon, month, and week in the same order as before."[60] Good information from different places allowed Beda to verify the tidal regularities which he formulated and to disallow those earlier reports; it also allowed him to specify certain aspects of tidal action that could only be understood locally but not generalized even along a contiguous coastline.

What would he have asked his learned associates to do in their various coastal observation posts? Essential steps for astronomical or cartographical orientation are simple procedures. The North/South meridian can be determined with precision on a clear night by setting a vertical stick (gnomon) in the ground and sighting the northern pole on the circling heavens; the resulting three point

---

[58] Bedae *DTR* XLVII.61-65; note that the date 668 was misprinted 688 in P. Hunter Blair, *The World of Bede* (1970), p.269. *Aenigmata Eusebii*, ed. M. De Marco, *CCSL* CXXXIII (1968), p.209-271.

[59] See P. Hunter Blair, *The World of Bede* (1970), p.161-169, 269-270; idem, "The letters of Pope Boniface V," in *England before the Conquest*: Studies in primary sources presented to Dorothy Whitelock, ed. P. Clemoes and D. Whitelock (Cambridge University Press, 1971), p.12-13; D.P. Kirby, "Northumbria in the time of Wilfrid," in *Saint Wilfrid at Hexham*, ed. Kirby (Newcastle upon Tyne, 1974), p.24; D. Whitelock, "Bede and his teachers and friends," in *Famulus Christi* (1976), p.19-39, esp.p.24-28.

[60] *Epistola Ceolfridi*: Tanta hodie calculatorum exuberat copia, ut etiam in nostris per Brittaniam ecclesiis plures sint, qui mandatis memoriae veteribus illis Aegyptiorum argumentis facillime possint in quotlibet spatia temporum paschales protendere circulos, etiam si ad quingentos usque et XXX duos voluerint annos; quibus expletis, omnia quae ad solis et lunae, mensis et septimanae consequentiam spectant, eodem quo prius ordine recurrunt. Quoted by Bedae *HE* V.21, ed. Plummer (1896), p.341. See Judith McLure, "Bede and the life of Ceolfrid," *Peritia* III (1984) 71-84; G.H. Brown, *Bede the Venerable* (Boston: Twayne, 1987), p.144.

alignment of eye, stick and the pole is the North/South line. The same meridian may be determined during a clear day by taking a cord and drawing a circle about a gnomon: as the sun moves from dawn to dusk, the shadow will pass from West to East, and the parabolic path of the shadow's end will cut the circle twice; a calculator may bisect the cord determined by those two intersections; the mid-point of the cord will form a line with the base of the gnomon which is a segment of the true North/South meridian, and a right angle will give East and West.

Measurement of the longest and shortest shadow should show the days on which solstices occur, summer and winter, but the end of the shadow is usually fuzzy, and its length may be uncertain from minute to minute. A better method is to follow the curve daily as it crosses the meridian and measure the greatest distance from the gnomon which should be coterminous with its length on the meridian. But one may also place a small disque on the head of the gnomon and use the centre of the disque as the extremity. Even if all edges are fuzzy, the centre of its oval shadow is then easier to determine than the indefinite end of a shadow cast by a straight stick. Another way to make the end of the shadow clear is to use an aperture in the disque and to follow the spot of light as its curve crosses the meridian.[61]

Finding the equinox is also not a difficult procedure, though the results will be somewhat less exact. Having determined the East/West line (above), one may set out two stakes and await the day of perfect alignment of them with sunrise and sunset. This is quite impractical however, as a cloud on either horizon on that day will void the observation, practical results may be achieved only if this is repeated for several days before and after the anticipated date. By use of two long poles for sighting across open space to the East and West horizons and a plumbline to keep the poles vertical, it is possible to align them on the rising sun, adjust to the setting sun, and eventually find a day on which they form a four point alignment from sunrise to sunset. That could be a day of twelve hours daylight, defined as the period of the sun's visibility, and it could also be reasonably near the day of

---

[61] Bruin, F. and M. "The limits of accuracy of aperture-gnomons," in *PRISMATA. Naturwissenschaftsgeschichtliche Studien*. Festschrift für Willy Hartner, eds. Y. Maeyama and W.G. Saltzer (Wiesbaden: Franz Steiner Verlag, 1977), p.21-25; David C. Lindberg, "The theory of pinhole images from Antiquity to the thirteenth century." *Archive for the History of the Exact Sciences* 5 (1968) 154-176.

astronomical as well as terrestrial equinox. This may have been the only non-trigonometrical method feasible for determining the equinox, but its results must be understood as having a tolerance of plus or minus three or possibly two days in England after three years of observations for both vernal and autumnal equinoxes. Beda accepted the twelfth Kalends of April [XII Kal.Aprilis = 21 March] as vernal equinox and pointed out the need to adjust the older use of the eighth Kalends [VIII Kal.Aprilis = 25 March], due to *horologica consideratione*. That qualification probably means that he used a plane, horizontal sundial with a vertical gnomon. However, modern estimates would indicate that three to four more days' shift forward to 18/17 March would have been more correct for the astronomical vernal equinox at that time than 21 March.[62] This may indicate a severe limitation in Beda's work, or it could suggest a modern difficulty in assessing it.

Because of the sun's distance from the earth the equinox will occur simultaneously at all latitudes along the same meridian. But observing it in England is difficult because of many restrictive viewing conditions. Logically it might seem that the equinox could be determined by unaided observations within one day tolerance by the method described. In practice however there are many indeterminate factors, such as whether there are hills or houses or trees on either horizon and whether a place with good visibility is within easy walking distance, as one will have to go there twice a day for a minimum of five days. As Harrison pointed out, there is also a question of what part of the sun is to be taken as the key; the sun does not appear as a point of light but as a rather large disque, and one must key into its leading edge, following edge, or centre. of disque, using the same key every day, morning and evening. The observer also discovers that the disque moves across the horizon very rapidly at an acute angle with the horizon; in case anything happens to distract him, the moment will have passed and the experiment cannot be repeated until after a year's delay. Some other conditions about this scientific enterprise included the discovery may be that poles used for such observations have a fascination for passers-by, and they are likely to be vandalized. It is also crucial that the sky be clear and the weather cooperates. My own city of Winnipeg on the Canadian prairies lies near the parallel of 50° North Latitude and is well known for

---

[62] K. Harrison, "Easter cycles and the equinox in the British Isles," *Anglo-Saxon England* VII (1978) 1-8.

its great number of sunny days each year. Yet, I have not found clear enough horizons on five successive days around the times of vernal or autumnal equinoxes for effective observation.

Greater precision could be achieved nearer the equator where there is more likelihood of clear morning and evening weather and perhaps less atmospheric distortion on the horizons. Yet, there will be significant differences between direct observation, instrumental observation, and calculation. From the data published annually by *The Astronomical Almanac* or by *Whitaker's Almanack*, it will be clear that one must allow a tolerance of several days between an observed day of twelve hours from sunrise to sunset and the supposedly corresponding astronomical day of equinox. A similar disjuncture of several days will be found for a day of solstice.

What could we expect from a man of Beda's interests and superior abilities? Anyone who has attempted the procedure at a latitude of 55º North will realize that Beda may have been wise to accept a broad degree of imprecision in the date of the vernal equinox. Such observations could be sufficient to reject the Julian date but insufficient to overthrow the equinoctial date received from Alexandrian scientists by way of the Council of Nicaea and the works of Dionysius Exiguus.

Measuring and timing tides themselves is no more difficult than these preliminaries, and the resulting data can be used with fewer complications. Only one sighting pole is required, with bits of cloth and a sundial. Measure off the pole in cubits or handbreadths and digits with your knife; set it firmly in the water near the mouth of a stream but not on the coast itself (since the pole may not stay secure against waves), so that both low and high water may be observed. Tie pieces of cloth at those extremes and at several median points for observing tidal stand and slack water, as well as commencement of ebb and flow. Use a sundial from a building nearby on the north bank of your stream. The sundial on the South face of Bewcastle Cross is perhaps early s.VIII, and the one on the South wall of St. John's Church, Escombe, is contemporary with St. Paul's Church Jarrow (A.D.685). They could be used not only for display but also for work periods and for the liturgical schedule, as the vertical planar sundials with horizontal gnomen show tidal marks at mid-morning, noon, and mid-afternoon; the Bewcastle sundial has two further divisions between tides, with twelve periods from dawn to dark (equivalent to twelve variable hours of daylight).[63] True noon and true equinox are the only stable elements with this type of sundial, while the hour lengths vary by season.[64] But there are several types, and you might bring a portable one.[65] Sundials alone however would not be sufficient to meet Beda's observational need for standarized hours. Therefore you should also take along an hourglass for greater precision.

Beda's studies specify the use of standard hours, 24 in the daily solar cycle; and he cited horologia so often and so firmly that it is certain that he used them and expected readers to use their own without further explanation. He did not explain what type of horologium he used. However his references to the horologium could usually mean a parchement rota displaying not hours of the day but changing hour patterns of day length and night length by season or by month (as described above). Such instruments often appear in manuscripts, either for solar cycles by day or for lunar cycles by night.[66]

The longer you continue the observations, the less your data series will be distorted by the uncertainties of hours and parts of hours. Thus prepared, you could easily set down the times and heights of low water and high water daily, the phase of the moon as it changes or, better, the lunar days I to LIX over two continuous synodical lunar months. From your record would soon appear the lower or Neap-

---

[63] See H.M. Taylor, *English Architecture in the Time of Bede* (The Jarrow Lecture 1961), p.17-18; H.M. and J. Taylor, *Anglo-Saxon Architecture* I (Cambridge 1965), p.234-238; M. Pocock and H. Wheeler, "Excavations at Escombe Church, County Durham," *Journal of the British Archaeological Association*, Third Series, XXXIV (1971) 11-19.

[64] For stone and brass instruments, see D.J. Price, "Portable sundials in antiquity," *Centaurus* XIV (1969) 242-266; F.A. Stebbins, "A Roman sundial," *Journal of the Royal Astronomical Society of Canada* LII (1958) 250-254; and S.L. Gibbs, *Greek and Roman Sundials* (New Haven: Yale University Press, 1976), who provided a catalogue of those known from the Hellenistic period, though others have now been identified. Gibbs explained each type in terms of modern trigonometry but not in terms of the contexts.

[65] Nevertheless, those sundials surviving from Anglo-Saxon England deserve much more attention; see A.R.G. Green, "Anglo-Saxon sundials," *Antiquaries Journal* VIII (1928) 489-516, and the Roman portable example at the Oxford Museum of the History of Science which gives hours according to latitude, including 57° for Britain [?Aberdeen], cited by Stebbins and Price.

[66] An example of a horologium with cursor still attached is seen in ms Bern, Burgerbibliothek 250 (s.X/XI) f.14, for use with the moon. *Horologia nocturna* provide another topic with abundant evidence deserving further attention.

tides and the higher or Spring-tides with their five-day correlations with full moons and new moons, respectively. If you continue to do this over several months or hopefully over a full year or more, your record will likely display the two highest tides at equinox, though it may not show any remarkable change of the lesser tides at solstices. These observations then should be sent in to the central research centre at Jarrow in order for Beda to correlate the data from extreme southern, extreme western and extreme eastern coasts of Britain, as well as at least three of the 23 places he named along the North Sea coast.

The theory derived by Beda from his correspondents and from his own observations at Jarrow and perhaps other places was not a rediscovery of lost science from a glorious past, nor a leap into the Newtonian future; but it was a remarkable reorientation of an entire field of study. He could now set aside the eight-year luni/solar cycle and establish the 19-year cycle which he understood and taught with great detail and precision.[67] In his earliest book of science he had accepted the report from Pliny that the moon is retarded in its course at the rate of dodrantis semiuncias horarum which is 3/4 plus 1/24 hour; these Roman fractions indicate that in our terms the moon would reach the same position in the sky 47 ½ minutes later every 24 hours. Beda had then introduced the new statement that the tides also fall back and rise later each day by the same delay.[68] By A.D.725 however he was able to explain the daily retardation of both moon and tide as more nearly quatuor puncti or 48 minutes.[69] He also accounted for it in terms of 59 solar days of two synodic months during which the moon actually circles the earth only 57 times.

---

[67] Bedae *DTR* XXIX.82-84: Aestus adfluere naturalis ratio cogit, per denos autem et novenos annos, iuxta lunaris circuli ordinem; etiam maris cursus ad principia motus et paria incrementa recurrit. This modified while repeating phrases from Bedae *DNR* XXXIX.12-14 and Plinii *NH* II.215. Surviving tidal tables from s.VIII-XII should be studied.

[68] Bedae *DNR* XXXIX.3.5: Qui cotidie bis adfluere et remeare, unius semper horae dodrante et semiuncia transmissa videtur, ... See also *De ordine creaturarum* IX.

[69] Bedae *DTR* XXIX.9-15: Sicut enim luna, iuxta quod et supra docuinus IIII punctorum spatio quotidie tardius oriri tardius occidere quam pridie orta est vel occiderat solet; ita iam maris aestus uterque, sive diurnus sit et nocturnus seu matutinus et vespertinus, eiusdem pene temporis intervallo tardius quotidie venire tardius redire non desinit.
Then he added for clarity: Punctus autem quinta pars horae est, quinque enim puncti horam faciunt. The difference between the two periods was much discussed in glosses to these chapters in later copies; some have been cited by Jones, *Bedae Pseudepigrapha* (1939), p.44 and *Bedae Opera* (1943), p.363-364.

He further explained the relationship in 29-day and 15-day periods, and enlarged its scope into ordinary and embolismic years with detailed calculations for each.[70] He gave the overall pattern of Spring-tides and Neap-tides every seven or eight days, dividing the month by quarters,[71] and he noted further that the Spring-tide usually starts five days before the new moon or full moon and that the Neap-tide usually begins five days before the first quarter and third quarter of the moon.[72] This led him to discover that the series of Spring-tides reaches a climax sometime around each of two equinoxes and that the tide near summer solstice will be relatively low.[73] But here he also loosened the specificity of other reports with the phenomena were not in accord, especially for Spring-tides in mornings or evenings or in certain months.[74] He challenged anyone who thought that tidal inundations came *uno puncto temporis* in every region and country and offered in contrast the testimony of his informants from the coasts of Britain, particularly to the North and

---

[70] Bedae *DTR* XXIX.15-23: Unde fit ut, quia luna in duobus suis mensibus, id est diebus LVIIII, quinquagies et septies terrae orbem circuit, aestus oceani per tempus idem geminato hoc numero, id est centies et quatuordecim vicibus, exundet ad superiora et tot aeque vicibus suum relabatur in alveum. Quia luna in XXVIIII diebus vicies octies terrae ambitum lustrat et in XII horis, quae ad naturalis usque mensis plenitudinem supersunt, dimidium terrae circuit orbem; ut quae verbi gratia praeterito mense super terram meridie, nunc media nocte sub terra solem accedenda consequatur, per tantumdem temporis geminatis aestus sui vicibus, quinquagies septies.
   DTR XXIX.34-38: Et quoniam luna per annum, id est menses XII suos qui sunt dies CCCLIIII, duodecim vicibus minus, hoc est trecenties quadragies et bis, terrae ambit orbem, aestus oceani tempore eodem DCLXXXIIII vicibus et ipse terras adluit ac resilit.

[71] Bedae *DTR* XXIX.38-44: Imitatur autem lunae cursum mare non solum communi accessu et recessu sed etiam quodam sui status profectu defectuque perenni, ita ut non tardior solum quam pridie, verum etiam maior minorve quotidie redeat aestus. Et crescentes quidem malinas, decrescentes autem placuit appellare ledones. Qui alternante per septenos octonosve dies successu, mensem inter se quemque quadriformi suae mutationis varietate dispertiunt.

[72] Bedae *DTR* XXIX.50-52: Et siquidem aestu vespertino vel novilunio vel plenilunio instante malinam nasci contigerit, idem aestus quotidie per septem malinae dies subsequentes fit maior et violentior aestu matutino.
   DTR XXIX.77-79: Ergo malinam quinque fere ante novam sive plenam lunam diebus, ledonem totidem ante dividuam saepius incipere comperimus.

[73] Bedae *DTR* XXIX.79-80: Et circa aequinoctia duo maiores solito aestus adsurgere, inanes vero bruma et magis solstitio.

[74] Bedae *DTR* XXIX.62-67: Scimus enim nos, qui diversum britannici maris litus incolimus, quod ubi hoc aequor aestuare coeperit ipsa hora aliud incipiat ab aestu defervere, et hinc videtur quibusdam quia recedens aliunde aliorsum unda recurrat, iterumque relictis quos adierat finibus, priores festina repetat. Ideoque se ad tempus maior malina his litoribus abiens amplius abducat ut alibi adveniens amplius exundare sufficat.

to the South of Jarrow, that a quiet sea will begin to flow at one place at the same hour it ebbs at another and vice versa, that it rises and falls earlier to the North of Jarrow but later to the south.[75] Thus he seemed to recognise a tidal movement from North to South, in that region and at Wight, but he was not aware of the great curved wave with which the tidal crest moves across open seas.

The tides certainly did not occur simultaneously around the coast of Britain but at different times in different places on distant shores or those nearby, according to Beda. The force, speed, and height could vary quite a lot from weather conditions, from wind direction and strength, and from conditions of tidal bore in or near the mouth of a river. These variations were so great that he referred them not only to natural force but even to chance (*sorte*) and accident which upset the regularity (*ordine turbata*).[76] There was however a pattern of the moon's relation with the tide (*regula societatis*) which held true for each region.[77] In addition to the nearly 48-minute luni/tidal interval, there could be a different space of time between the moon's passage and the high water that follows but which could be determined only by observation.

Beda's proposal that this interval varied according to latitude, due to the moon's force being weaker in the north because it was farther away, was one of Beda's errors.[78] Early Stoic philosophers such as Seleucus of Babylon and Poseidonios may have discussed general effects of sun and moon upon the ocean, especially

---

[75] Bedae *DTR* XXIX.72-76: Porro aliis in partibus ab ea caeli plaga recessum maris luna qua his signat accessum. Non solum autem sed et in uno eodemque litore que ad boream mei habitant multo me citius aestum maris omnem, qui vero ad austrum multo serius accipere pariter et refundere solent.

[76] Bedae *DTR* XXIX.44-48: Saepe quidem aequa uterque sorte septenis diebus ac dimidio cursum consummantes, saepe vel ventis impellentibus aut repellentibus vel alia qualibet accedente sive naturali vi cogente tardius citiusve venientes aut minus ampliusve solito feventes -- ita ut aliquoties ordine turbato malina plures sibi aestus hoc mense, pauciores vindicet in alio.
Sors, sortis and accedens (accido, accidi) may be understood as referring to events occuring to some extent due to indeterminate contingencies rather than due to a friendly or malevolent Fortuna.

[77] Bedae *DTR* XXIX.76-77: Serviente quibusque in regionibus luna semper regulam societatis ad mare quamcumque semel acceperit.

[78] Bedae *DTR* XXIX.80-82: Semperque luna in aquilonia et a terris longius recedente mitiores quam cum in austro digressa propriore nisu vim suam exercet. This account was retained from *DNR* XXXIX and thence from Plinii *NH* II. 215.

evaporation and storms; but none of them suggested either causal relation or correlation of the sun with Spring-tides or Neap-tides. The speculations of George Sarton and many other historians of science about this have been misleading.[79]

Beda made it clear that the patterns he discovered must be applied locally with variations and exceptions. Recognition of these local variants within the overall regularities is the basis for what is now called "rule of port." He went further to show that need than anyone had done or would do for many centuries. It should be kept in mind that "establishing the port" for any location now requires local mean tides and intervals which would not be so very different from the arithmetical "averaging" done by medieval calculators;[80] but I do not know of evidence pertaining to tides. Beda's contribution was to clarify why a general theory of tides could never determine tides and intervals correctly without observation and adjustment for local conditions.

When all the variations are tabulated and coordinated with the 19-year lunar table of Beda's computus, one may predict movements of tides in any place daily and monthly to a high degree of accuracy. His theory of tides could still be improved upon in several respects: in particular, he did not notice the sun's force on tides. It is now understood that Robert Grosseteste realized not only that the waters of the globe bulged when the moon passed near and its reflected light was strong but also that the waters contracted when it was far away. He also affirmed "the help which the sun gives the moon" to increase the tides, and he speculated about their influence on waters on the opposite side of the earth.[81] Yet the sun's direct

---

[79] Sarton, *A History of Science* I (1959), p.204 was corrected by Eckenrode, "Original aspects" (1970), p.109-112 and 122-127, but not repeated in his subsequent articles.

[80] R.C. Gupta, "The Process of Averaging in Ancient and Medieval Mathematics," *Ganita-Bharati*; Bulletin of the Indian Society of History of Mathematics III (1981) 32-42.

[81] On Robert Grosseteste see R.C. Dales, "The text of Robert Grosseteste's *Questio de fluxu et refluxu maris* with an English translation," *Isis* LVII (1966) 455-479, esp.p.459-468 (text); and A.C. Crombie, *Robert Grosseteste and the Origins of Experimental Science, 1100-1700* (Oxford University Press 1952), p.112, 149. Robert's authorship has been debated, e.g. by S.H. Thomson, *The Writings of Robert Grosseteste, Bishop of Lincoln, 1235-1253* (Cambridge University Press 1940), and D.A. Callus, *Robert Grosseteste, Scholar and Bishop*, (Oxford University Press 1955). But the development of his ideas about tides in several works allows this text to be dated about A.D.1226-1228, according to Dales, "Robert Grosseteste's scientific works," *Isis* LII (1961) 381-402, esp.p.387-391; idem, "The authorship of the *Questio de fluxu et refluxu maris* attributed to Robert Grosseteste," *Speculum* XXXVI (1962) 582-588. In his *Opus maius* IV.iv.6, Roger Bacon also made use of Robert Grosseteste's ideas.

influence on the seas went unrecognized by Bede in the eighth century, by Robert in the thirteenth, and by all others until the late seventeenth century.

Beda's theory of tides was a valuable contribution to knowledge which further research did not fault and could scarcely improve until A.D.1686 when Isaac Newton provided an essentially new framework for thinking about mutually effective power of moon, sun, and earth upon each other.

IV

Beda's preeminence in the field of calender studies has not been disputed, although this has not always been acknowledged as of value for theoretical science in contradistinction to practical work. That distinction between science (theoretical) and application (practical) has been an icon for historians of science in the twentieth century which natural scientists seem to have abandoned long since. In fact Beda's work has both theoretical and practical value for the development of science in the subjects of his research.

In conclusion then, we should turn to the greatest accomplishment of Beda, so often acknowledged even if not very well understood: his creation of the western calendar.

Beda's computus was based upon the 19-year table from Alexandria which had been received by Dionysius Exiguus and reformulated about A.D.525 into terms and limits preferred in the Roman *civitas*, and then extended for five cycles usable for 95 years. This is what Beda developed and applied on a larger scale. Two centuries after Dionysius, most bishops and all masters of computus in monastic and cathedral schools were teaching and using the Dionysian calendar system on the authority of Beda. It may appear from Wilfrid's address to King Osuiu at the Council of Streanæschalch/Whitby that if Peter holds the Keys to the Kingdom of Heaven, surely the Pope holds the keys to the computus. But in fact no pope could understand the problems about finding a proper date for Easter. Beda discussed the

Synod of 664 as an event which determined "the Easter Question" when King Osuiu accepted Wilfrid's claim with the reluctant concurrence by Colman of Lindisfarne that Peter the apostle had a certain authority for entering heaven -- though he said this with a smile according to the *Vita Wilfridi* but not mentioned by Beda.[82] We may conficently assert that the party of *Romani at this event, led by Acgilbert the Frankish bishop of Wessex* at that time, were presenting an Alexandrian computus. On the basis of this evidence however, it is not possible to determine whether it might have been the Dionysian version of that computus. There was no canon and there is no positive evidence for Roman practice of the Dionysian system or Easter tables until later tenth century. Thus it would be improper to mention it as canonical for the Roman bishop or Roman curia or for those central Italian and a few other scattered bishops whom they influenced during this period.

What the extant evidence does show most abundantly throughout is the papal and curial failure to understand the problem of incommensurable lunar and solar cycles, their inability to develope any tables which could be maintained for more than two decades without confusion, and their neglect of those long-term computistical tables which would have been most helpful: the tables created by Victurius of Aquitaine (A.D.455) and Dionysius Exiguus (A.D.525) at the behest of Roman curial officials or by Beda on his own.[83] On the other hand there are letters of several bishops of Rome that ask or that assert which Easter Sunday should be celebrated in Rome or England or Ireland. But there are no paschal tables promulgated by any of them and no canons declaring the necessity of any computistical system.

Beda proceeded to apply and to redevelope the tables and instructions of Dionysius without any papal or canonical authority. He explained natural phenomena on the basis of observation and reason, adjusted in practice to human

---

[82] Bedae *Historia Ecclesiastica* III.25; *Vita Wilfridi*, ed. W. Levison, *M.G.H. Scriptores rerum Merovingicarum* VI (Hannover/Leipzig 1913), p.163-263, esp.p.202-204 (section 10); D.H. Farmer, "Saint Wilfrid," in *Saint Wilfrid at Hexham*, ed. D.P. Kirby (Newcastle: Oriel, 1974), p.43. Authorship of the *Vita* by Eddi Stephanus is contested.

[83] See Jones, *Bedae Opera* (1943), p.55-77, and Stevens, "Scientific Instruction in Early Insular Schools," in *Insular Latin Studies*, ed. Michael Herren (Toronto: Pontifical Institute of Mediaeval Studies, 1981), p.83-111, esp.p.83, 90, 95-98, rpr. in this volume, item IV.

experience. It was Beda's science which convinced rational men in the medieval centuries as to how they should give a dependable time structure to experience and memorable events and which demonstrated how the regular motions and powers of sun, moon, and stars contributed to a better understanding of their experience.

Beda's sources for reasoned understanding of natural phenomena were quite limited. From them he nevertheless selected a respectable body of Hellenistic science. As early as A.D. 701, he was drawing inferences which were better than his sources. After twenty-four years he had developed a richer understanding of some parts of that science than could be found anywhere else in the world. He accepted the basic spherical cosmology of heavens and earth and improved upon the understanding of lunar periods. He was careful in his calendar studies to weigh evidence and make the best choices. He refrained from forcing his data series into untenable precision about astronomical phenomena, such as the equinox, and about the prediction of terrestrial phenomena, such as local tides. We could wish for him more success than he actually had, but it is enough to acknowledge the improvements which he succeeded in making. Finally, he advanced the study of ocean tides to a new level and indeed put it on a basis which deserves every praise from those who appreciate scientific accomplishments.

Beda is best venerated as a monk and choirmaster and schoolmaster who prayed and sang the glory of God daily with mind and heart. It is consistent with this *opus Dei* that his approach to natural phenomena was reasonable and methodical. In some of the matters which he studied, Beda was not able to advance beyond the knowledge he received and may even have been mistaken in what he understood. But in other matters he not only could see more than others but also could explain well what no one had explained before him. This was an advance in human knowledge and a scientific achievement of the highest quality.

Appendix I: BEDA'S SCIENTIFIC WORKS

The following account of the scientific tracts and letters of Beda is organized by title, incipit and explicit, extant manuscripts, and editions. A selective bibliography is provided for the sake of textual analysis and description but not general discussion of contents. Brief comments are added as a guide to some of the principal questions of text and authorship.

Usually a bibliographic reference will be given in full once and thereafter cited by last name of author and year of publication. All citations of manuscripts which were given by the works of Ernst Zinner, *Verzeichnis der astronomischen Handschriften des deutschen Kulturgebietes* (München: Beck, 1925); M.L.W. Laistner and H.H. King, *A Hand-list of Bede Manuscripts* (Ithaca, N.Y.: Cornell University Press, 1943); *Clavis Patrum Latinorum,* ed. E. Dekkers and A. Gaar, *Sacris Erudiri* III (2 ed. Turnhout, Belgium: Brepols, 1961); and L. Thorndike and P. Kibre, *A Catalogue of Incipits of Mediaeval Scientific Writings in Latin* (rev.ed. London: Medieval Academy of America, 1963) have been incorporated without further bibliographical mention. The following abbreviations will be used:

Giles (1843) .... *Venerabilis Bedae Opera quae supersunt omnia, The Complete Works of Venerable Bede,* ed. J.A. Giles (London 1843-1844), 12 volumes.

Heerwagen .... *Opera Bedae Venerabilis Omnia,* ed. Johannes Herwagen [Hervagius] (Basel 1563, rpr. Köln 1612 and 1688), 8 volumes.

Jones, *Bedae Ps* .... C.W. Jones, *Bedae Pseudepigrapha*: Scientific writings falsely attributed to Bede (Ithaca, N.Y., 1939), rpr.*Bede, the Schools and the Computus,* ed. W.M. Stevens (Aldershot: Variorum, 1994).

Jones, *Bedae Opera* .... *Bedae Opera De Temporibus,* ed. C.W. Jones (Cambridge, Mass., 1943).

Jones, A (1975), B (1977), or C (1980) ... *Bedae Venerabilis Opera,* Corpus Christianorum, Series Latina CXXIII: Opera didascalica, parts A, B, C (Turnhout, Belgium: Brepols, 1975, 1917, 1980).

Mommsen, *Chronica minora* .... *Monumenta Germaniae Historica, Auctores Antiquissimi,* vol.IX (Berlin 1892): Chronica minora I, ed. Th. Mommsen; vol.XI (1894); ibidem II; vol.XIII (1898): ibidem III.

46

General abbreviations and usages:

A.D.	annus domini	f.	folio of codex, book
A.M.	annus mundi	in.	initio (earliest years of s.)
ca.	circa	inc.	incipit (first words of text)
cap.	capitulum, chapter	ms.	manuscript
ed.	edition, editor	rpr.	reprint of edition cited
ex.	exeunte (latest years of s.)	s.	seculum, siècle, century
expl.	explicit, final words		

A. De Natura Rerum Liber (ca. A.D.701)

Inc: 1) Naturas rerum varias, labentis et aevi/... [4 verses]

2) De quadrifario dei opere [cap. I]

3) Operatio divina, quae saeculo creavit et gubernat,...

Expl: ...atque inde africa a meridie usque ad occidentem extenditur.

Ms: 134 manuscripts were listed by Jones, A (1975); another is Aberystwyth Nat.Lib. Peniarth 540 (s.XII[1]): two bifolia.

Ed: Iohannes Sichardus, *Bedae...De Natura Rerum Et Temporum Ratione Libri Duo* (Basel: Henricus Petrus, 1529) f.1-6[v]; Noviomagus II (1537) f. 1-23[v], rpr. Heerwagen II (1563) 1-49 with glossae at scholia; rpr. Giles, I (1843), p.99-122, rpr. *PL* XC (1850) 187-278; Jones, A (1975), p.189-234, esp.p.185-186; idem, *Bedae Ps* (1939), p.1-20 concerning the editions.

Analysis and description: Jones, A (1975), p.173-187; this supersedes idem, "The manuscripts of Bede's De natura rerum," *ISIS* XXVII (1937) 430-440.

Comment: This work is often cited but very little studied. Although its structure seems to be that of Isidori *De natura rerum*, Beda was reluctant to depend upon Isidore as a source, according to Jones, *Bedae Opera* (1943), p.131. A contrary view was given by J. Fontaine, *Isidore de Séville, Traité de la nature* (Bordeaux 1960), p.79. Ms Aberystwyth containing cap.XXVII-XLVIII has been discussed by Daniel Huws, "A Welsh manuscript of Bede's De natura rerum," *Bulletin of the Board of Celtic Studies* XXVI (1976-1978) 491-504

B. De Temporibus Liber I-XVI (A.D.703)

Inc: 1) De momentis et horis. (cap I)

2) Tempora momentis horis diebus mensibus annis, saeculis et aetatibus dividuntur...

Expl: ...Sexta, quae nunc agitur, nulla generationum vel temporum serie certa sed, ut aetas decrepita

ipsa, totius saeculi morte finienda. (cap. XVI)

Ms: 93 manuscripts were listed by Jones, C (1980), p.580-583; others are:
Amiens B.mun. 222 (s.IX) f.1-7[v] incomplete
Avranches B.mun. 135 (s.XII/XIII) f.120-121 fragment
Napoli BN V.A.13 (s.IX[2]) f.45-48, cap.I-XI
Paris BN Lat. 7418A (s.XII) f.26-33

Ed: Sichardus (1529) f.709ᵛ; Noviomagus II (1537) f.25-28ᵛ, rpr. Heerwagen II (1563) 205-210ᵛ with scholia, rpr. Giles, I (1843), p.123-132, rpr. *PL* XC (1850) 277-288; Jones *Bedae Opera* (1943), p.295-303, rpr. idem, C (1980), p.585-601.

Analysis and description: Jones, *Bedae Opera* (1943), p.161-167 et passim; idem, C (1980), p.580-583; C.H. Beeson, "The manuscripts of Bede," *Classical Philology* XLII (1947) 73-84, esp.p.75-76.

C. De Temporibus Liber XVII-XXII: Chronica Minora (A.D.703)

Inc: 1) Cursus et ordo temporum (cap. XVI)

2) Prima ergo aetas continet annos iuxta Hebreos I.DCLVI, iuxta LXX interpretes II.CCXLII.

Adam cum esset CXXX annorum,...

Expl: ...Tiberius dehinc quintum agit annum ind. prima. Reliquum sextae aetatis deo soli patet.

Ms: 25 manuscripts were listed by Jones, C (1980), p.580-583.

Ed: Sichardus (1529) f.10-11ᵛ; Noviomagus II (12537) f.28ᵛ-29ᵛ, rpr. Heerwagen II (1563) 211-212, rpr. Giles, I (1843), p.132-138, rpr. *PL* XC (1850) 288-292; Mommsen, *Chronica minora* III (1898), rpr. Jones, C (1980), p.601-611. For other editions see Mommsen, III (1898), p.243; Jones, *Bedae Ps* (1939), p.1-20.

Analysis and description: Mommsen, III (1898), p.224-230; Jones, *Bedae Opera* (1943), p.161-167 et passim; idem, *Saints' Lives and Chronicles in Early England* (Ithaca: Cornell University Press, 1947; rpr. Hamden, Conn.: Archon Books, 1968), p.23-30 *et passim*.

Comment: The Mommsen edition begins with Ch.XVI De mundi aetatibus. He placed Beda's *chronica maiora* (item F, below) at top of the page as the primary text, with parallel lines of these *chronica minora* at bottom the same page.

D. De Temporum Ratione Liber I-LXV (A.D.725)

Inc: 1) De natura rerum et ratione temporum duos quondam stricto sermone libellos discentibus...

2) De computo vel loquela digitorum. (cap.I)

3) De temporum ratione, domino iuvante, dicturi necessarium duximus...

Expl: ...qui aliquando in quaestionem venerant, quando vel quales fuerint, evidentius agnoscant.

Ms: 245 manuscripts were listed by Jones, B (1977), p.242-256; the earliest of those (with new additions to Jones) are given below in Appendix II.

Ed: Sichardus (1529) f.48-64ᵛ; Noviomagus II (1538) f.30 and f.XXXI-LXXIIIᵛ, rpr. Heerwagen II (1563) 49-170 with glossae et scholia, rpr. Giles, VI (1843), p.139-270, rpr. *PL* XC (1850) 293-520; John Smith, *Historiae Ecclesiasticae Gentis Anglorum Libri Quinque, una cum reliquis eius Operibus Historicis* (Cambridge 1722), p.1-33, 653-654; Jones, *Bedae Opera* (1943), p.172-291, rpr. idem, B (1977), p.263-460. Concerning these and other editions see Jones, *Bedae Ps* (1939), p.1-19; idem, B (1977), p.256-257.

Analysis and description: Jones *Bedae Opera* (1943), p.125-161, 175-176. For the recently discovered eighth century insular fragments, see Bernhard Bischoff, *Libri Sancti Kyliani* (Würzburg 1952), p.82, 104; Jürgen Petersohn "Neue Bedafragmente in northumbrischer Unziale saec.

VIII," *Scriptorium* XX (1966) 215-247; idem, "Die Bückerburger Fragmente von Beda De Temporum ratione," *Deutsches Archiv* XXII (1966) 587-597.

Comment: The only critical edition is the text by Jones, *Bedae Opera* (1943), p.175-291; the reissue of that edition in Jones, B (1977) was printed without correction from his proofs, which have also been misplaced. The Bückeburg and Münster fragments are from the same codex in Northumbrian uncial, written between A.D.746 and 750; they were recovered from seventeenth century bookbindings. The Bern, Wien, and Würzburg fragments appear to have been written in Irish minuscule.

It should be noted that ms Salisbury Cathedral 158, f.20-83 Bedae *DTR* complete, is a separate codex written on the continent in ninth century; thus it appears that no manuscript written in England before the Conquest survives, other than the fragments noted above.

E. De Temporum Ratione Liber LXVI-LXXI: Chronica Maiora (A.D.725)

Inc: 1) De sex huius saeculi aetatibus. (cap.LXVI)

2) De sex huius mundi aetatibus ac septima vel octava vitaeque caelestis...

3) Prima est ergo mundi huius aetas ab Adam usque ad Noe, continens annos iuxta Hebraicam veritatem mille DCLVI, iuxta LXX interpretes II.CCXLII,...

Expl: ...ut post temporales caelestium actionum sudores aeternam cuncti caelestium praemiorum mereamur accipere palmam.

Ms: Jones *Bedae Opera Didascalica* B (1977), p.242-256 listed 41 manuscripts.

Ed: Sichardus (1529) f.64^v-67; Noviomagus (1957) f.LXXIIII-LXXXVII and f.XCIIII^v-XCVI^v, rpr. Heerwagen II (1563) 170-173 incomplete, rpr. Giles, I (1834), p.270-342 with additions, rpr. *PL* XC (1850) 520-578; Smith (1722), p.34 only; Mommsen *Chronica minora* III (1898), p.247-327, rpr. Jones C (1980), p.463-544, with variants. Concerning these and other editions see Mommsen, III (1898), p.243, Jones, *Bedae Ps* (1939), p.202-221, and idem, C (1980), p.241-242, 256-257.

Analysis and description: Mommsen III (1898), p.224-230; Jones, *Bedae Ps* (1939), p.20-21; idem, *Bedae Opera* (1943), p.140-144 *et passim*; idem *Saints' Lives* (1947), p.23-30 *et passim*.

Comment: This chronicle was not edited by Jones who recommended the Mommsen edition; it has been reissued with some alteration of the text by editors of *Corpus Christianorum*, Series Latina CXXII B (1977) 461-544.

F. Epistola Ad Helmwaldum (ante A.D.725)

Inc: 1) Dilectissimo in Christo fratri Helmwaldo Beda famulus Christi salutem.

2) Gavisus sum, fateor, multum dilectissimo in Christo frater, quod ubi.../

...per epistolas alloquandam ac de necessariis consulendam putasti.

3) Saltum lunae quem dicunt, locus et hora citior incensionis eius per X et VIIII annos efficet, ita e contrario bissextum non alia causa quam tarditas solaris cursus generat..../...ex ore doctissimi tractatoris intelligas.

Expl: ...quod faciet homo desiderium tuum dilectissimo in Christo frater.

Ms: Oxford Merton College 49 (s.XV) f.289^v-291
       Vat. Regin. Iat. 123 (A.D.1056) f.40^v-42

Ed: J. A. Giles, *Anecdota Bedae, Lanfranci et aliorum* (London: Caxton Society, (1851), p.1-6, from ms
       Merton; Bedae *DTR* XXXVIII-XXXIX, ed. Jones, *Bedae Opera* (1943), p.250-253, rpr.
       idem, B (1977), p.399-404.

Analysis and description: A. Wilmart, *Codices Reginenses Latini* I (Roma 1937), p.289-293; Jones,
       Bedae Opera (1943), p.172, 371-374; idem, C (1980), p.628.

Comment: This letter explained intercalation of the bissextile day every fourth year in the Julian
       calendar as adapted by Dionysius and Beda. It survives intact only in ms Merton College 49
       from which the Giles edition was made. Beda however excerpted the major portion of his
       letter for inclusion in *De Temporum Ratione* XXXVIII and XXXIX, opening with a few
       transitional lines: De ratione bissexti non nova nunc cudere ... Sicut, inquam saltum lunae
       quem dicunt, ... and continuing verbatim. C. Plummer, *Bedae Opera Historica* I (Oxford
       University Press, 1896), p.xxxviii, reprinted the introductory portion of the letter from Giles.
       It is found also in ms Vat. Regin.lat.123, f.41, following *De Temporum Ratione* XXXVIII and
       an argumentum based upon XXXVIII 12-48: Bissextus ex quadrantis ratione, per
       quadriennium conficitur...; it is followed by XXXIX and an *argumentum* on Greek numerals
       which concludes in an epistolary style similar to Beda's letter to Helmwald: ... Per omnia
       tuae sanctae fraternitati quantum congaudens tantum etiam favens [...]dus et profectibus
       bonis. Deus te conservat semper satietque in bonis desiderium tuum vel dilectissimi in
       Christo frater.

       The earlier ms Regin.lat.123 completes the introduction with *putasti*, whereas ms Merton 49
       reads *decrevisti*.

G. Epistola Ad Pleguinam (A.D.708)

Inc: 1) Fratri dilectissimo et in Christi visceribus honorando Pleguinae, Beda in domino salutem.

         2) Venit ad me ante biduum, frater amantissime, nuntius tuae sanctitatis,...

Expl: 1) ...Vere enim dictum est quia si momorderit serpens in silentio, non est habundantia
         incantatori.

         2) ...In cantatorem dominus autem omnipotens frater nitatem tuam sospitem conservare
         dignetur.

Ms: London BL Cotton Vitellius A,XII (x.XI^2) f.83-86^v
         Oxford Merton College 49 (s.XV) f.285-288
                             175 (s.XV) f.28-29^v
         San Marino Huntington Library HM 27486 (ca. A.D.1400) f.91-93^v
         Vat.Regin.lat.123 (A.D.1056) f.48^v-52

Ed: James Ware, *Venerabilis Bedae Epistolae Duae...* (Dublin 1664) 1-20, rpr. (London 1693) 241-251,
       rpr. Giles, I (1843), p.144-154, rpr. *PL* XCIV (1862) 669-675; Jones, *Bedae Opera* (1943),
       p.307-315, rpr. idem, C (1980), p.617-626.

Analysis and description: Jones, *Bedae Ps* (1939), p.94; idem, *Bedae Opera* (1943), p.132-135, 171-172;
       idem, C (1980), p.615-616; W.F. Bolton, *A History of Anglo-Latin Literature, 597-1066*, I

(1967), p.150-152; D. Schaller, "Der verleumdete David. Zum Schlusskapital von Bedas Epistola ad Pleguinam," in *Literatur und Sprache in Europäischen Mittelalter. Festschrift für Karl Langosch*, ed. A. Önnerfors et alii (Darmstadt 1973), p.39-43.

Comment: In the presence of Wilfrid (bishop of Hexham and abbot of Ripon), an unnamed person had accused Beda of heresy for his abuse of Scriptures. This stemmed from *De Temporibus* XVI in which Beda had presented a new *annus mundi* whereby the birth of Jesus had occurred A.M. 3952, differing radically from the reckonings of Jerome, Eusebius, Victurius, and all other chroniclers. Pleguina had been present when the accusation was made and brought it to Beda's attention. The letter refers to an otherwise unknown monk David who was acquainted with Beda's book and whose goodwill should correct that *frater desipiens*. It also appears to expect Wilfrid not to keep silent in the face of reckless words when he was knowledgeable about questions of chronology. The interpretations of Jones and Bolton have been corrected by Schaller.

H. Epistola Ad Wicthedum (A.D.725-731?)

Inc: 1) Reverentissimo ac sanctissimo fratri Wicthedo presbytero, Beda optabilem in domino salutem.

2) Libenter accepi literas tuae benignitatis, amantissime in Christo frater, et capitula quae rogasti promptus describere ac tibi dirigere acceleravi,...

Expl: 1) ...Ipsum quoque pascha quod dominus pridie quam pateretur cum discipulis fecit aut nona kalendarum aprilium die non fuisse aut ante aequinoctium fuisse confirmet.

2) ...Bene vale semper in domino dilectissime frater.

Ms: 34 manuscripts were listed by Jones, C (1980), p.633-634; others are
Evreux B.mun. 60 (s.XII) f.136^v-139^v
Geneva Univ. Lat.50 (s.IX^1) f.38^v-41^v
London BL Royal 12.F.II (s.XII ex) f.112-117^v
Malibu Getty Museum Ludwig XII.2 (ca.A.D.876) f.31^v-33
Ludwig XII.5 (s.XII^1) f.88^v-90
Paris BN Lat.1956 (s.XII) f.81-82^v

Ed: Noviomagus II (1537) f.XCVII-XCIX, rpr. Heerwagen II (1563) 343-345, rpr. Giles, I (1843), p.155-168, rpr. *PL* XC (1850) 599-605; Pierre Chifflet, *Bedae Presbyteri et Fredegarii Scholastici Concordia* (Paris 1631), p.13-18, from ms Geneva Univ. Lat.50 (s.IX^1) f.38^v-41^v; Jones, *Bedae Opera* (1943), p.319-325, rpr. idem, C (1980), p.635-642.

Analysis and description: Jones, *Bedae Ps* (1939), p.41-44; idem, *Bedae Opera* (1943), p.138-139; idem, C (1980), p.633-634.

Comment: This letter was also known as *Epistola De Aequinoctio*, or as *De Pasche Celebratione Liber sive De Aequinoctio Vernali* [to which is sometimes added *iuxta Anatholium*]. It was a further challenge to the Anatolian Canon, respected by the Anglo-Saxons (including Beda) as well as by the Irish. An addition to it was written in A.D.776, *Primo anno circuli decennovenalis...*, version B of a tract usually headed *Calculatio Quomodo* ... This appeared separately in ms Köln Dombibl. CIII (ca. A.D.800) f.51, along with three verses by Alcuin; both additions were then included as part of Beda's letter by Noviomagus, Heerwagen, and Giles. In the ms Oxford BL Digby 56 (s.XIII) f.192^v-194 this letter and other excerpts from Beda were incorporated into the text of the computus of Garlandus (ca. A.D.1038-1057?).

I. De Computo Vel Loquela Digitorum (ante A.D.725)

Inc: Cum ergo dicis unum, minimum in laeva digitum inflectens, in medium palmae artum insiges;...

Expl: ... ambas sibi manus insertis invicem digitis implicabis.

Ms: Bern Burgerbibl. 207 (s.VIII/IX) f.2-2ᵛ
Bern Burgerbibl. 417 (s.IX¹) f.18ᵛ-19
New York Pierpont Morgan M.925 (A.D.1007) f.37ᵛ-39ᵛ
Paris BN Lat.7530 (s.VIII ex) f.280ᵛ-281
Vat. Pal.Iat.235 (s.IX in) f.36ᵛ-37
Further manuscripts are listed by Jones (1939, 1943) and by Cordoliani (1948, 1957), cited in Appendix II, below.

Ed. Heerwagen I (1563) col. 172-174, rpr. *PL* XC (1850) 689-692; A. Cordoliani, "A propos du chapitre premier du temporum ratione de Bède," *Le Moyen Age* LIV (1948) 214-217 with parallel texts from the *Romana Supputatio*.

Analysis and description: Jones, *Bedae Ps* (1939), p.22-23, 53-54; idem, *Bedae Opera* (1943), p.329-331; Cordoliani (1948), p.209-217; idem, "Les plus anciens manuscrits de comput ecclésiastique de la bibliothèque de Berne," *Revue d'Histoire Ecclésiastique Suisse* LI (1957) 109-110.

Comment: This item is an excerpt from Bedae *DTR* I which often appears separately in manuscripts or is absorbed into other computi. There were at least four versions of finger-reckoning before and during Beda's lifetime. One of his sources was *Romana Supputatio*, ed. Jones, C (1980), p.671-672. There are innumerable essays concerning finger-reckoning; for its broader context in the history of science see A.P. Juschkewitsch, *Geschichte der Mathematik in Mittelalter* (Basel 1966), p.336-338.

J. De Bissexti Praeparatione

Inc: Primo igitur anno praeparationis bissexti, prima hora noctis...

Expl: ...que praecedit XV kal. Aprilias Arietem possit ingredi.

Ms: 17 manuscripts were listed by Jones, C (1980), p.648; others are
Paris BN N.a.lat.456 (s.IX/X) f.92-95ᵛ
London BL Harley 3017 (s.IX²) f.121,124

Ed: Heerwagen I (1563), rpr. *PL* XC (1850) 357-361; Jones, C (1980), p.649-651.

Analysis and description: Jones, *Bedae Ps* (1939), p.32; idem, *Bedae Opera* (1943), p.139, 372; idem, C (1980), p.648.

Comment: This argumentum was apparently described by Bedae *DTR* XXXVIII 33-37 as his own, and its calculations are consistent with his teaching (Jones 1943). It appears anonymously in both versions of the *Compilatio DCCCVIIII* under the rubric, Qualiter Bissextus Adrescat, and in other collections with headings, Qualiter Bissextilem Diem Quarto Suo Compleat Anno, or Qualiter Sol Cursu Suo Per Menses Et Signa Bissextilem Diem Quarto Anno Compleatur, or De Cursu Solis Per Menses Et Signa: Qualiter etc. In Heerwagen's edition it appears among the "Bridferti Glossae" to Beda.

52

K. Opuscula Ex Bedae Computo

I.   Inc: Itaque stella Veneris et Mercurii hoc a superioribus tribus planetarum stellis differunt...

Expl: ...et idcirco longiores aetheris metas peragat.

II. De Apsidibus Planetarum

Inc: 1) Circulus zodiacus qui XII signis constat ...

2) Hipparchus in eo opere quo scribitur de secessibus atque intervallis solis et lunae ...

Expl: ...non LXXX sed potius LXX dierum numerum reddat.

III. Argumentum Ad Inveniendam XIIII Lunam.

Inc: Sume regulares XXXVI quas mense martis apponere debes...

Expl: ...reliquium per septenos multiplicare.

Ms: 24 manuscripts were listed by Jones, C (1980), p.656-657, and there are very many more.

Ed: Jones, C (1980), p.658-659.

Analysis and description: Jones, C (1980), p.656-657.

Comment: These short pieces appear in the same order in many manuscripts and often stand between or in sequence with Bedae *DT* and *Epistola Ad Wicthedum*; they may be included after the text of *DT* XVI but before its explicit. Thus it seems that they may have derived from Beda or his school; certainly they were assumed to be his in an English exemplar in which the five items appeared together in the series: De Temporibus, Itaque stella Veneris ..., De Apsidibus Planetarum, Argumentum Ad Inveniendam XIIII Lunam, Epistola Ad Wicthedum.

Appendix II. BEDAE DE TEMPORUM RATIONE LIBER

Manuscripts dating from the eighth and ninth centuries

This list is intended to include all manuscripts known or reported to contain one or more chapters of Bedae *DTR* I-LXV, dated from near the time of composition to the end of the ninth century (that is, s.IX/X). Further manuscripts containing chapters or fragments of cap.I-LXV have been included, but not those reported to have only cap.LXVI-LXXI: *Chronica maiora*. Further details about many of these manuscripts are available in the brief descriptions given by Jones, *Bedae Ps* (1939), p.111-140, *Bedae Opera* (1943), p.140-161, and *Bedae Opera Didascalica* B (1977), p.242-256 which are listed above in Appendix I; each of those lists supplements and corrects but also remains dependent upon the previous ones. I am very much indebted to Professor Charles W. Jones (deceased 1989) for his studies and also for his generous encouragement for many years. I have also drawn upon the files of The Benjamin Catalogue for History of Science, A.D.200 to 1600, formerly known as the Benjamin Data Bank (Dr. N.L. Hahn, Director), and files of the International Computer Catalog of Medieval and Renaissance Manuscripts (Professor M. Folkerts, Director); their assistance is very much appreciated.

Many studies of the manuscripts have been consulted, especially those of E. A. Lowe, Neil R. Ker, Bernhard Bischoff, Patrick McGurk, Bruce S. Eastwood, Donald A. Bullough, Dáibhí Ó Cróinín, and numerous other scholars who take note of computistical and astronomical concerns expressed so often in early Latin manuscripts. New manuscripts or fragments continue to be identified and further study often improves our understanding and evaluation of those which were supposed to be well known. In certain cases diverse codices have been bound together, but they have been indicated separately here if their scripts or other evidence suggest separate origins. So far as possible, the new manuscripts and reevaluations have been incorporated into the information given below.

From the present list it may be seen that, of the earliest twenty manuscripts and fragments dating from mid-eighth century to about A.D.800 (that is, s.VIII/IX and s.IX in), only three (Berlin, Phillipps 1831; Köln CIII; CLM 14725) were available for the Jones edition (1943, rpr 1977). Some of the manuscripts now known had not as yet been identified at that time, but many could not have been studied and collated in European archives during 1940-1945 because of warfare. In the case of each manuscript therefore, I have indicated whether Jones had been able to collate the *DTR* text in full or in part. Those manuscripts which have been inspected by me during the past three decades have been indicated *(asterisk), and I have attempted to assess their scripts as thoroughly as possible. The many manuscripts which I have studied by use of microfilm copies alone have not been indicated as personally inspected because film does not allow palaeographical evaluation of scripts, and often does not allow dependable judgment about textual readings, especially when there have been corrections.

The reader is invited to correct the data of this list in order to improve it as much as possible in the expectation that it may be further elaborated and republished.

s. VIII med

*Bückeburg, Niedersächsische Staatsarchiv Dept.3/1 (A.D.746-750) f.i-viii: eight fragments in Northumbrian Uncial

Münster-in-Westfalen, Staatsarchiv Misc.I.243 (A.D.746-750) f.1-2^v, 11-12^v: two fragments in the same Northumbrian Uncial as the Bückeburg fragments

*Darmstadt, Hessische Landes- und Hochschulbibliothek 4262 (ante A.D.735?): one fragment in Northumbrian Uncial

s. VIII^2

*Bern, Burgerbibliothek 207 (A.D.779-797) f.II-II^v, 1-24: Irish- or Breton-influenced miniscule, early provenance St. Benoît-sur-Loire

*Paris, Bibliothèque Nationale Lat.7530 (s.VIII^2) f.280^v-281: Beneventan minuscule; cap.I within a grammatical miscellany written at Monte Cassino, A.D.779-796

*Paris, BN lat.14088 (s.VIII^2) f.23-25^v: fragment of ms Bern 207; cap.XIII-XVI, XXXV-XXXVI

*Würzburg, Universitätsbibliothek M.p.th.f.183 [olim M.ch.f.206, Fragment 5] (s.VIII^2) f.1-1^v

*Wien Österreichische Nationalbibliothek Lat. 15298 [olim 2269] (s.VIII ex) f.1-4^v: four fragments in Northumbrian minuscule

s.VIII/IX et s.IX in

*Berlin, Staatsbibliothek Phillipps 1831 [cat.128] (Verona ca.A.D.800) f.16-89^v: from Metz; collated

*Berlin, Staatsbibl. Phillipps 1896 [cat.137] (Verona s.VIII/IX) f.l5-88: epitome of DTR

*Besançon, Bibliothèque publique 186 (s.IX in) f.41^v-46: three chapters

*Besançon, Bibl.publ. 186 (s.IX in) f.70^v-161^v

*Köln, Dombibliothek 83^{II} (Köln ca. A.D.805) f.86-125^v

*Köln, Dombibl. CIII (Köln ante A.D.819) f.52^v-184: collated

Leiden, Bibliotheek der Rijksuniversiteit Scaliger 28 (s.IX in) f.2^v, 17-18, 22-22^v: four chapters

Leiden, Bibl.-Univ. Scaliger 28 (Flavigny ca. A.D.800/801) f.43-138

*München, Bayerische Staatsbibliothek CLM 14725 (No.east France s.IX in) f.14^v-16: three chapters; f.25-166^v: collated

*Vaticano (città del), Biblioteca Apostolica Vaticana Pal.lat.1448, f.1-39 (Trier A.D.810) f.5^v: cap.IX

*Würzburg, Universitätsbibliothek M.p.theol.f.46 (s.IX in, A.D.800?) f.23-97^v: Salzburg scribe with early Arno style in St.-Amand; completed by a later Salzburg scribe, f.98-144^v (A.D.821-828)

s.IX^1

*Basel, Universitätsbibliothek F.III.15k (Benediktbeuern A.D.820-840) f.57-60^v: from Fulda; two chapters

*Berlin, Deutsche Staatsbibliothek Phillipps 1869 [cat.131] (Trier? s.IX^1) f.15-139: collated in part

*Bern, Burgerbibliothek 417 (s.IX^1) f.27-31^v: three chapters

Cambrai, Bibliothèque publique 925 (Cambrai Cathedral s.IX^1) f.1^v-63^v

*Geneva, Bibliothèque publique et universitaire Lat.50 [Inv. 122] (Massai s.IX^1) f.45-46, 47-120: collated in part

*Geneva, Bibl.univ. Lat.50 (s.IX^1) f.41^v-44: three chapters

*Ivrea, Biblioteca Capitolare XLII(6) (A.D.820-835) f.2-50^v: cap.VII-LXVI

*Karlsruhe, Badische Landesbibliothek Aug.CLXVII.Perg. (Reichenau s.IX^1) f.23-64: collated

*Karlsruhe, Landesbibl. Aug.CLXXII.Perg., f.1-19 (Reichenau ca. A.D.822) f.1-14

*Karlsruhe, Landesbibl. Aug.CCXXIX.Perg. (Reichenau A.D.806-822) f.25^v-30^v: nine chapters

*Kassel, Landesbibliothek Astron.Fol.2 (Fulda A.D.814?) f.10-82^v: collated

Madrid, Biblioteca Nacional 3307 [L.95] (Metz A.D.820-840) f.24^v-52: scattered chapters

*Melk, Stiftsbibliothek 412 [370; G.32] (Auxerre s.IX, first quarter) p.59-191: collated

*Milano, Biblioteca Ambrosiana D.30.inf. (s.IX^1, ca. A.D.836) f.23-121: from Bobbio; collated in part

*München, Bayerische Staatsbibliothek CLM 210 (NW Austria A.D.818) f.16-16^v, 54^v-112: most chapters scattered in a three-book Compendium of A.D.809.

Napoli Biblioteca Nazionale VI.B.12 (Beneventan A.D.817-35) f.116^v-255

*Paris, Bibliothèque Nationale Lat.7296 (s.IX^1) f.1-111: cap.IX to LXV incomplete; collated

*Paris, BN Lat.13013 (Auxerre ca. A.D.830) f.48-161

Paris, BN Lat.13403 (Corbie s.IX^1) f.1^v-110: collated in part

*Paris, BN N.a.lat.1613 [Libri 88] (s.IX^1) f.1^v-2^v: three chapters

*Paris, BN N.a.lat.1615 [Libri 90] (Fleury? A.D.820) f.19-126^v

*Roma, Biblioteca Vallicelliana E.26 (Lyons A.D.824-840) f.1^v-2, 43^v-71^v, 91-136^v

*Sankt Gallen, Stiftsbibliothek 251 (S.Gallen ca. A.D.810) p.45-181: collated

*Sankt Gallen, Stiftsbibl. 397 (S.Gallen s.IX^1) p.35, 98-102, 121-122: five chapters

*Vaticano (città del), Biblioteca Apostolica Vaticana Pal.lat.834 (ca. A.D.836) f.36-37^v: cap.I

*Vat. Pal.lat.1448, f.40-60 (Mainz s.IX^1) f.40^v-41; cap.IV, I; f.45-59^v: epitome of cap.II-XLVI

*Vat. Pal.lat.1448, f.61-116 (Lorsch/Aachen s.IX^1) f.71, 74^v-75^v, 92-104^v, 112^v-115: selected chapters

*Vat. Pal.lat.1449 (Lorsch s.IX^1) f.27^v-104, 121-145^v

Vat. Regin.lat.838 (s.IX^1) f.84-86: cap.I

*Wien, Österreichische Nationalbibliothek Lat.387 (Salzburg A.D.818) f.16-16^v, 58^v-60, 102^v-114: most chapters scattered in a three-book Compendium of A.D.809.

s.IX med

*London British Museum Cotton Vespasian B.VI (No.France ca.A.D.848) f.1-102: collated

*München, Bayerische Staatsbibliothek CLM 246 (Weltenburg? A.D.840-850) f.1-2^v, 7, 105-113

*Paris Bibliothèque Nationale Lat. 5543 (St.-Benoît-sur-Loire A.D.847) f.25-76

*Paris BN N.a.lat.1612 [Libri 87] (St. Martin Tours s.IX med) f.7-22: cap.LII-LXVI incomplete

*Sankt Gallen, Stiftsbibliothek 184 (S.Gallen s.IX med) p.122-133

Tours, Bibliothèque publique 334 (St.-Martin, Tours s.IX med) f.1-8^v

s.IX^2

Angers, Bibliothèque publique 477 (461) (Landevennec? s.IX^2) f.44^v-86^v

*Berlin, Deutsche Staatsbibliothek Phillipps 1832 [cat.130] (Metz A.D.873) f.16-54^v, 56-78^v

Chartres, Bibliothèque publique 75 (55) f.1-122 (s.IX^2) f.8-122: destroyed

*London, British Library Harley 3017 (St.-Benoît-sur-Loire? A.D.861-864) f.133^v-134^v, 143-143^v, 165-168^v: six chapters

*London, British Library Harley 3091 (s.IX^2 from Nevers) f.19^v-21: cap.XXIX; f.41-128: cap.I-LXVII

*London, British Library Regius 15.B.XIX (Reims s.IX^2) f.38-78^v: cap.I-XXIX; collated

*Milano Biblioteca Ambrosiana M.12 sup. (Corvey s.IX^2) p.47-210: pages confused; Tironian notation

Montpellier, Bibliothèque de la Faculté de Médecine H.157 (Lyons s.IX^2) f.61^v-71: selected chapters

Montpellier, Bibl. Fac. Méd. H.306 (s.IX^2) f.7: cap.IX

*Monza, Biblioteca Capitolare f-9/176 (Lobbes A.D.869) f.52-61: selected chapters scattered in a seven-book Compendium of A.D.809.

*Oxford, Bodleian Library Digby 63 (So. England? A.D.867) f.30^v-33^v, 46-48^v: selected chapters

*Paris, Bibliothèque Nationale Lat.4860 (Mainz s.IX^2) f.77^v-88

*Paris, BN N.a.lat.1632 [Libri 41] (St.-Benoît-sur-Loire s.IX^2) f.10-67^v: collated in part

Rouen, Bibliothèque publique I.49 [cat.524] (St.-Wandrille s.IX^2) f.96-197^v

Salisbury, Cathedral Library 158 (s.IX^2) f.20-83: Carolingian minuscule

58

*Sankt Gallen, Stiftsbibliothek 248, p. 1-98, 149-277 (S.Gallen s.IX²) p.149-212

*Sankt Gallen, Stiftsbibl. 250 (S.Gallen A.D.889) p.104, 109, 164-425

Trier Stadtbibliothek 2500 [olim Ludwig XII.3] (Laon s.IX²) f.36ᵛ-71ᵛ, 74-95

Valenciennes, Bibliothèque publique 174 (St.-Amand? s.IX²) f.1, 42-168

Valenciennes, Bibl.publ. 343 (St.-Amand s.IX²) f.84-179ᵛ

*Vaticano (Città del), Biblioteca Apostolica Vaticana Lat.645 (St. Quentin s.IX²) f.39-41, 51, 76ᵛ-77ᵛ: four chapters

s.IX ex et s.IX/X

*Bern, Burgerbibliothek 610 (s.IX ex) f.81-81ᵛ: remainder lost

*Firenze, Biblioteca Medicea-Laurenziana Ashburnham 10 [54/11] (s.IX ex, A.D.895?) f.153-157

*Paris, Bibliothèque Nationale N.a.lat.1616 (St.-Benoît-sur-Loires s.IX ex) f.6: cap.I incomplete

*Paris, BN N.a.lat.1632 (St.-Benoît-sur-Loire s.IX/X) f.10-67ᵛ

Vaticano (Città del), Biblioteca Apostolica Vaticana Lat.3852 (s.IX/X) f.1-30ᵛ

*Vat. Regin.lat.309, f.4-29, 59-119 (St.-Denis s.IX ex) f.12ᵛ-14, 65, 77ᵛ-89, 104ᵛ: selected chapters

*Vat. Regin.lat.339 (s.IX/X) f.53ᵛ: cap.IV

III

THE FIGURE OF THE EARTH IN ISIDORE'S "DE NATURA RERUM"

During the Hellenistic, Roman, and medieval periods, descriptive literature concerning the cosmos almost uniformly discussed the earth as a globe and the heavens as a sphere.[1] This literature, often intended for school instruction, was sometimes accompanied by stylized diagrams depicting the globe. The diagrams are of three types: the globe divided both horizontally and vertically into four parts by waters and their projections; the globe with parallel lines representing three, five, or more zones of *klima* (or latitude); and the globe with the three partially known continents of the *oikoumenê* extended over the entire surface.[2] Such diagrams may be inferred from early texts in various languages and from various periods, but with the Babylonian exception mentioned later, the diagrams themselves do not survive until examples belonging to the seventh, eighth, and ninth centuries of our own era.

The earliest extant diagram of the earth as a globe occurs in manuscripts of the *De natura rerum* by Isidore of Seville, a schoolbook intended to outline the organized knowledge proper for an educated man in seventh-century Visigothic Spain. It is a diagram of the third type, known as *rota terrarum* or *orbis terrae*. Inexplicably, given its importance, it has suffered comparative neglect, as a brief comparison of its place in the manuscript tradition with its recent publishing history will make clear. The neglect may relate to two problems—the assumption that Isidore did not conceive of the world as a sphere, and the assumption that the *rota* represents a disk and not a sphere. The *rota terrarum* however displays a more sophisticated conception of the world than has often been acknowledged. Occurrences of reversed *rotae* in some of the later English manuscripts of *De natura rerum* and other Latin works have similarly been ascribed to error. A reconsideration of the Isidorean *rota* in the context of Greco-Roman depictions of the globe should correct these misapprehensions.

I

A two dimensional representation of three dimensions is by its very nature problematic. Thus in the absence of primary literary evidence, the clay tablet from the seventh or sixth century B.C. found in ancient Sippar in southern Babylonia—the

My research was undertaken with support from the Social Sciences and Humanities Research Council of Canada (formerly The Canada Council) and the University of Winnipeg.

[1] A. P. Newton, "The Conception of the World in the Middle Ages," *Travel and Travellers of the Middle Ages* (London: Routledge & Kegan Paul, 1926), pp. 1–18. The relevant Latin texts have also been reviewed by F. S. Betten, "Knowledge of the Sphericity of the Earth During the Earlier Middle Ages," *Catholic Historical Review*, 1923, N.S. 3:74–90; by J. K. Wright, "Early Christian Belief in a Flat Earth," in his *Geographical Lore of the Time of the Crusades* (1925; reprint New York: Dover, 1965), pp. 53–54, who nevertheless suspected that some believed the earth to be flat like a disk rather than spherical; and by Charles W. Jones, "The Flat Earth," *Thought*, 1934, 9:296–307, whose copious references to the earth's sphericity in medieval literature of all centuries left no such doubts. See also D. R. Dicks, *Early Greek Astronomy to Aristotle* (London: Thames & Hudson, 1970), pp. 21–22, 72, 177–178, et passim; Germaine Aujac, "L'image du globe terrestre dans la Grèce ancienne," *Revue d'Histoire des Sciences*, 1974, 27:193–210.

[2] Especially valuable is the list of manuscript and facsimiles provided by Marcel Destombes in *Mappemondes*, Vol. I of *Monumenta cartographica vetustioris aevi*, A D 1200–1500, ed. R. Almagia and M. Destombes (Amsterdam: N. Israel, 1964), esp. Ch. II, although the earliest examples were not reproduced.

earliest extant figure said to represent the shape of the earth and the heavens—has been described as displaying a disk floating on a cosmic sea.[3] Evidence of this concept is lacking however from any other Mediterranean people, for example, the Egyptians or Hebrews. Among the Greeks no surviving fragment of pre-Socratic writings requires the notion of a disk-shaped earth, though some modern scholars have accepted ascriptions of the disk or tambourine shape to certain of the pre-Socratics in the later doxographical literature.[4] The dominant Greco-Roman tradition, however, is of a spherical earth. Attributed first to the fifth-century Pythagoreans Parmenides and Empedocles, the concept of the earth as a globe was accepted and promoted by Plato, Eudoxus, and Aristotle; it more or less drove out all other concepts in surviving Greek and Latin literature.[5] In consequence, the surviving Latin manuscripts containing Hellenistic, Roman, and medieval illustrations of the whole earth all depict a sphere.

Figure 1. The *orbis quadratus* ascribed to Krates of Mallos.

Extant representations of the earth projected on a circle are of three types. The first is an ancient *orbis quadratus* similar to one often found in Etruscan and other Mediterranean designs. Its earliest indisputable scientific use was attributed by Strabo to Krates of Mallos, a scholar at Pergamum during the second century B.C.[6] He was said to have projected a globe with four continents divided by great rivers or oceans. Euro-Asia and all the Mediterranean peoples of the *oikoumenê* were located in the upper right hand section, and other peoples were presumed to be found in the other three zones: *sunoikoi* and *antoikoi* in the Eastern Hemisphere; *perioikoi* and *antipodes* in the Western Hemisphere (see Fig. 1). This model of the earth can be conceived as parallel with the astronomical sphere, its divisions being constituted by the equator and any convenient meridian and its hemispheres being symmetrical in any direction, as was explained by Geminos about A.D. 50.[7] It became common in literature of the Roman Empire: a public oration addressed to Constantius Chlorus in A.D. 297 by an Autun schoolmaster compared his majesty to that of the entire four-part universe, comprehending the *orbis quadrifariam duplici discretus oceano*. Julius Honorius relied upon this figure in the fifth century to describe both heavens and

---

[3]London, British Museum, Department of Western Asiatic Antiquities no. 92687; a good photograph of this may be seen in Leo S. Bagrow, *History of Cartography* (1944, 1951; revised ed. R. A. Skelton, Cambridge, Mass.: Harvard University Press, 1964), Plt. VI. The cosmogonic interpretation was proposed by Eckhard Unger, "From the Cosmos Picture to the World Map," *Imago Mundi,* 1937, 2:1–7, with a photograph and illustrative drawings. Actually the text written above this design is less grandiose and merely names exploits by Sargon of Akkad; perhaps it illustrates scenes of his campaigns.

[4]Dicks, *Early Greek Astronomy,* pp. 39–42 on the need to discriminate primary evidence from later reports about the pre-Socratics.

[5]*Ibid.,* pp. 49–55, 72–73.

[6]*Geographia,* I 2. 24.

[7]Fig. 1 is based upon the study by H. J. Mette, *Sphairopodia: Untersuchungen zur Kosmologie des Krates von Pergamon* (Munich: Beck, 1936), esp. pp. 66–78; it is here adapted from Wanda Wolska, *La topographie chrétienne de Cosmas Indicopleustes* (Paris: Presses Universitaires de France, 1962), p. 258. See also Geminos, *Phaenomena* XVI 1–2, ed. Germaine Aujac (Paris: Collection Budé, 1975); Aujac, "L'image du globe terrestre," pp. 205–208; Otto Neugebauer, *A History of Ancient Mathematical Astronomy* (Berlin: Springer-Verlag, 1975), pp. 578–587.

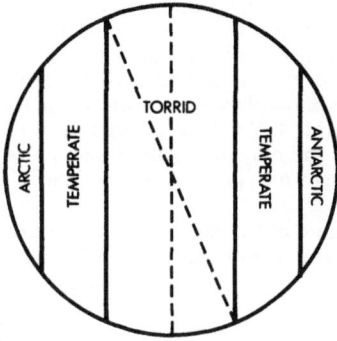 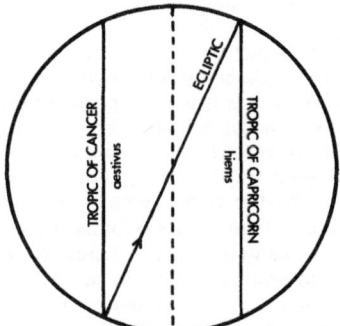

Figure 2. A zonal model of the earth, with five zones.

Figure 3. A zonal model with three zones, adapted for astronomical use.

earth, and it continued to be used until about A.D. 900 as a simplified teaching diagram.[8]

The second model of the earth displays either two or four lines dividing the globe into three or five zones: temperate both north and south of the torrid zone, with arctic and antarctic at the extremes (see Fig. 2). Euro-Asia and the coastal lands of *mare nostrum* fall in the northern temperate zone, and it was assumed that the southern temperate zone was also habitable. But the literature includes debate about whether there really are inhabitants there and in the terrible climate of *perusta* in this model. Any such depiction of the earth also has astronomical connotations, for its two central lines are often labelled *aestivus* for the northerly and *hiems* for the southerly divisions (see Fig. 3), representing the sun's highest point on the meridian at summer solstice and its lowest at winter solstice respectively. The concept was used with reference to the heavens by all astronomers after Eudoxos (ca. 370 B.C.), at first according to the latitude of Cos and Cnidos but then attributed or adjusted to Rhodes under the influence of that great school. Those who assumed this model in reference to the earth made a variety of attempts to identify parallels of *klima* or latitude by using patterns of shadows cast by the sun, longest hours of daylight, and sexagesimal parts of the circumference.[9] This Eudoxian concept was also applied to the earth by Krates of Mallos and apparently by Ptolemy, whence it came to be disseminated among Latin readers through the work of Macrobius and others. Its earliest manuscript dates from the late eighth century.[10]

[8] Árpád Szabó, "Roma quadrata," *Rheinisches Museum*, 1938, *87*:160–169, discussed many references to this figure in Latin literature and established the meaning of the term *quadratus* as four-part (not four-cornered as in the modern rectangular concept), with equal parts meeting in the center where angles or corners occur within the circumference of the figure; he noticed also the common references to hemispheres of the *orbis quadratus* divided by the meridian. Lingering doubts about these matters were laid to rest by his further article, "Roma Quadrata," in *Maia*, 1956:243–274. I gratefully acknowledge the very full bibliography supplied by Dr. Jocelyn Penny Small, Director of the U.S. Center of Documentation, Lexicon Iconographicum Mythologiae Classicae, Rutgers University. For Autun see *XII Panegyrici Latini*, No. 4, ed. Edouard Galletier, Vol. I (Paris: Collection Budé, 1949), pp. 84–85; and Eumenius's appeal (No. 5, pp. 137–138) only a year later for reconstruction of the city's entry walls which had been mostly destroyed in A.D. 269; he visualized great maps with many details depicted. Surviving excerpts from Julius Honorius were edited by A. Riese, *Geographi latini minores* (Heilbronn: Henninger, 1878), pp. 24–55. See further MS Bern, Bürgerbibliothek 45 (ca. A.D. 900) fol. 41, where the *orbis quadratus* occurs along with other figures of the earth in the margin of Lucan's *Pharsalia*.

III

The resemblance of this spherical figure, especially when oriented to the north, to images of the globe now current might suggest that it would encourage increasingly accurate depictions of the earth; but the narrow space of the northern temperate zone provides little scope for anyone wanting to develop a landchart. Characteristic of such limitations is the full-page representation of the globe as a zonal *rota* on folio 29 of MS London British Library Cotton Tiberius B.V.; there is space for only the barest sketch of lands in *Asia maior et minor* and only the northern shore of *Africa* which gives way to the equinoctial zone, an area which in this case is filled with a text concerning the circumference of the globe.

These difficulties are avoided by a third figure of the earth, the *rota terrarum* or *orbis terrae* depicting three great continents: *Asia, Libya, Europa* (see Fig. 4). These terms appear in this sense as early as Aeschylus (525–456 B.C.), and the figure was easily adapted to common use in early *periploi,* which listed the names of ports, rivers, regions, and peoples around the coasts of the Mediterranean, and often those at a very great distance from the coastline. Such lists were not necessarily pictorial; by the early second century however, the Greek Dionysius Periegetes had adapted his *periplous* to a model earth with three continents.[11] Many Latin literary descriptions, such as those of Pliny, Augustine, and Orosius, followed the same pattern and assumed or described this third figure of the earth. It may have been used by Sallust to depict Roman campaigns in the province of *Africa*—a small area which could be expanded on the *rota* to take up most of the space allowed for the whole continent of *Libya*.

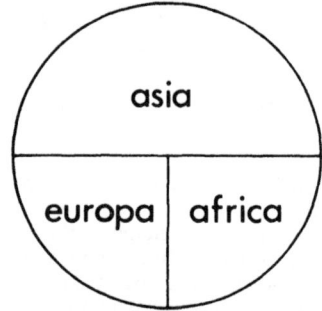

Figure 4. The tripartite model of the earth emphasizing the three known continents.

(Under Latin influences *Libya* eventually yielded to *Africa* as name for the southern continent.) This figure appears to have been used as well by Lucan to describe and depict events in *Hispania* or in *Thessalia*—also cartographically expanding in the space of the continent *Europa*. For these purposes the north-south rivers which divide continents were adjusted to form variant diagrams, now called the Sallust-type or the Lucan-type; unfortunately these special variants occur only late in the tradition. Although Sallust and Lucan must have written about 40 B.C. and A.D., 60–64, the diagrams named after them do not survive in manuscripts earlier than the late

[9]D. R. Dicks, *The Geographical Fragments of Hipparchus* (London: Athlone, 1960), pp. 24–25; Dicks, *Early Greek Astronomy,* pp. 17–21; Aujac, "L'image du globe terrestre," pp. 196–205; Neugebauer, *Ancient Mathematical Astronomy,* pp. 725–748.

[10]Attribution to Krates is found in Ptolemy's *Geographia* VI 6, but it is not certain that this portion derives from Ptolemy himself. See further Destombes, *Mappemondes,* pp. 17, 22, and 30: type C, of which a 13th-century example is shown in Plt. IV; but the earliest extant representation is MS St. Gallen Stiftsbibliothek 237 (1st half of the 11th century), apparently adapted from a Sallust-type *rota* (discussed below).

[11]On Aeschylus see Eric H. Warmington, *The Greek Geographers* (London: E. P. Dutton, 1934), p. xxvii; the continents are named in texts of tragedies and fragments. On Dionysius see Henry F. Tozer, *History of Ancient Geography* (Cambridge University Press, 1935), pp. 281–287, who summarized the contents; he argued that it could not have been written earlier than Vespasian or later than the end of the 2nd century. G. Leue, ("Zeit und Heimath des periegeten Dionysius," *Philologus,* 1884, *42*:175–178), however, had already discovered two acrostics in Dionysius' verses which allowed him to affirm that he was a grammaticus in Alexandria at the time of Hadrian (A.D. 117–138). His "complete guide" (to the earth) was drawn upon for Latin works by Avienus (2nd half of the 4th century) and Priscian (ca. A.D. 500); he and Ptolemy were the Greek sources for tales about monsters in the Indian Ocean who can swallow whole ships.

ninth and early tenth centuries.[12] Some applications of this diagram were even more extensively cartographical. Regions were carefully blocked out around the Mediterranean shoreline in MS Albi 29 (late 8th century); islands of the Mediterranean could all be displayed in MS Vaticanus Latinus 6018 (8th or 9th century), as well as islands of the oceans—the Azores, the Canaries, Ceylon, and many others of great distance—projected with angular extension and foreshortened depth implying curvature of the globe.[13]

Authors and artists also represented both earth and heavens with similar diagrams, each adapted according to need for generality or representative detail. As a geometrical figure also the sphere is represented by a circle, as Cassiodorus explained when presenting the notion of solid numbers (see Fig. 5). It can further be compared to the circle upon which Jahweh sits in the early manuscript illuminations for the text, "It is he that sits upon the circle of the earth . . ." (Isaiah 40:22). Elsewhere Jahweh always stands, and the teaching Christ always sits upon curvature, for no other aspect of the earth is depicted by the circle.[14] As a teaching device the spherical *rota* was simply divided by single lines to represent the waters which separated continents: the horizontal line corresponds with the Tanais (Don) river–Black Sea–Bosphorus on the north side, and on the south with the Nile or sometimes the Red Sea, while the Mediterranean is shown by dropping a perpendicular from the center. Such an image was well known on public monuments, coins, and literary descriptions,[15] but no manuscript diagram survives prior to Isidore of Seville, whose *rota* antedates the cartographical examples by more than a century.

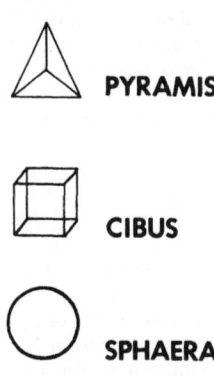

Figure 5. Diagrams of geometrical solids; the sphere is represented by a circle.

Isidore is better known for his *Origines vel Etymologiae,* which supplied general information and interpretations of many aspects of life which would assist the Christian preacher or reader of sacred scriptures, including a *periplus* similar to that of Orosius. Late manuscripts of that book attracted to their margins the tripartite *rota terrarum* and other diagrams. Isidore's earlier work *De natura rerum* was however quite different in purpose and content: it was a schoolbook. The teaching was elementary, but within the context of seventh-century Visigothic Spain it served as a useful introduction to the fields of study; the *rota* first appeared in this earlier work and is extant in the earliest manuscript.

In his excellent edition of *De natura rerum* Jacques Fontaine has identified three manuscript traditions.[16] The first contained the forty-six chapters which Isidore sent in 612 to his protégé Sisebuto, the young king of the Visigoths. Not only was Sisebuto literate and able to answer Isidore in verse, but his letter of reply included an explanation and diagram of a solar eclipse—demonstrating how the elementary textbook could become a stepping stone to more sophisticated knowledge.[17] This

---

[12]See the list of manuscripts in Destombes, *Mappemondes,* pp. 37–39.

[13]Facsimiles were published in *Itineraria et alia geografica,* ed. Fr. Glorie, Corpus Christianorum, Series Latina, Vol. CLXXV (Turnhout, Belgium: Brepols, 1965), p. 470; and Destombes, *Mappemondes,* Plt. XIX.

[14]Cf. also Job 22:14 and Prov. 8:27, in which the same Hebrew word can be rendered circuit, arc, compass, or orb. It seems unlikely that any of the three Biblical authors intended thereby a two-dimensional description of limited space, but rather a figure of enormous range. It is in this latter sense that Greek and Latin commentators and translators took the term, before the Enlightenment's inappropriate demands upon the language.

[15]Destombes, *Mappemondes,* pp. 3–4.

[16]Jacques Fontaine, ed., *Isidore de Séville: Traité de la nature* (Bordeaux: Féret et Fils, 1960), pp. 38–45, 75–82.

version of the work with the verse letter is called the short recension, and it is found in the earliest extant manuscript, El Escorial R.II.18, folios 9–24, written in a single Visigothic hand within the period 636–690. To this Isidore added in 613 a forty-seventh chapter, "De partibus terrae," followed by the *rota* under discussion; this chapter was copied into the El Escorial manuscript on folio 24v in a somewhat different Visigothic uncial script, also within the seventh century, and supplies evidence for Fontaine's putative middle recension. A third version with forty-eight chapters was developed later by other persons and is known through manuscripts from England, Fulda, the region of Murbach, and Verona; it too includes the final chapter "De partibus terrae" with its *rota terrarum* in all completed manuscripts. Fontaine included seven other teaching diagrams from Isidore's short version of 612, but unfortunately he and all other editors published the 613 text of the final chapter "De partibus terrae," lacking its *rota terrarum*. Yet no completed second or third version manuscript exists without it.[18]

The reasons for this omission are not clear. The tripartite or Isidorean *rota* appears to be so simple as to require no discussion. Like the Babylonian clay *imago*, however, it could possibly represent the earth as a disk, and some writers have presumed that it does so—especially in the hands of Isidore. There are several passages from his works which have been construed in accord with that assumption: *De natura rerum* X, XII, XVI, and *Origines* III and IX. If such allegations were true, Isidore would stand as a remarkable exception to the Hellenistic tradition, surviving in both Greek and Latin descriptive literature, which almost uniformly discussed the earth as a globe and the heavens as a sphere.[19] But those passages have been misconstrued. For example Isidore was rebutting a local superstition when he affirmed in *De natura rerum* XVI "De quantitate solis et lunae," that everyone experiences the sun and moon, their sizes, their distances, and their radiation in exactly the same way no matter where they may be on the face of the earth— "Similis sol est et Indis et Brittanis"—and this includes even that unusual enlarged appearance at the moment of rising—*eodem momento* in the East exactly as in the West. But Isidore does not assert that the sun itself is seen to rise simultaneously at eastern and western extremities of the globe— an assertion which would allow an historian to suspect that flatness of the earth's surface was implied, but which would also contradict Isidore's discussions of time-reckoning.[20] Isidore also explains the five climates of earth in parallel with the five bands of the celestial sphere in *De natura rerum* X and XII, as on the model of our Figures 3 and 4. Chapter X, "De quinque circuli mundi," is plainly based upon the Hellenistic model of a spherical universe (= *kosmos* = *mundus*), but it too can be misinterpreted. This is because the language is not at all clear: like Chapters XI and XII it emphasizes the interrelation of all parts of the universe rather than describes the ones in which we are interested; and one of the drawings rearranges the five great

---

[17]*Ibid.*, pp. 329–335 and 151–161.

[18]Manuscripts and editions are described in *ibid.*, pp. 19–38 and 141–145.

[19]The only exception that I know is in late Greek literary polemics: a Syrian merchant (or monk) of the 6th century apparently named Cosmas. As a Nestorian Christian and a Biblical fundamentalist, he distorted texts in order to portray the universe as a doubled tent or tabernacle and to ridicule the dominant Christian orthodoxy of such teachers as John Philoponus. See Wolska, *La topographie chrétienne*, and Marshall Clagett, *Greek Science in Antiquity* (1955; 2nd ed., New York: Collier Books, 1963), pp. 169, 175–176, 179, 207–217. Cosmas' notion cannot be attributed to Christians in any common or general sense of the term, even in the time or region in which he lived.

[20]*De natura rerum* XVI is summarized less clearly in *Origines* III, 47, "De magnitudine solis," but neither version can justify the claims that "Isidore thought the world was flat" or that "he tended to view the earth in the shape of a wheel or flat disk," a tendency also inferred from his doubts that *Antipodae* lived opposite us. See most recently the otherwise commendable article by T. R. Eckenrode, "Venerable Bede as a scientist," *American Benedictine Review*, 1971, *21*:486–507. Eckenrode is certainly correct in affirming that "the scholarly Bede knew more about this universe than the scholarly Isidore of the previous century" (p. 489).

zones (*circuli*) as petals of a flower! However disconcerting it may be as a teaching diagram, that device displays both the northern and the southern temperate zones of the earth as habitable, a touchstone which most modern commentators use as evidence for a concept of sphericity (e.g., Bunbury, Dreyer, Wright, Bagrow).

Again, Isidore joins those among Greek and Latin scholars who doubted that *Antipodae* live opposite them (*Origines* IX 2, "De gentium vocabulis," 133; and XIV 5, "De Libya," 17). *Antichthones* or *antipodes* were people and not poles, as is sometimes assumed in modern discussions of these questions. Their place on the earth was to be opposite the mapmakers, whether Greeks or Latins or Visigoths, not opposite a projected pole. It is worth repeating that there never was a doctrine of the Christian Church condemning the idea that there might be inhabitants of the southern temperate zone or of a presumed fourth continent.[21] But rejection of such speculations as poetic fancies without supporting evidence bears no consequence for concepts of the earth's shape.

On the other hand in *Origines* XIV 2, "De orbe," Isidore describes the shape of the earth directly: "Orbis a rotunditate circuli dictus, quia sicut rota est; unde brevis etiam rotella orbiculus appellatur." Modern readers trying to span the ages might assume that the three terms here juxtaposed—*orbis, rotunditas, circulus*—only refer to a two-dimensional circularity; however the concept of sphericity is not merely implied but is taught overtly in this language. The *Epistula Sisebuti* 38–41 confirms this fact in the plainest way. In order to avoid redundancy in his verses Isidore's student applies the term *globus* to the earth, a *globus* which intervenes with the sun's rays to cast a shadow upon the moon. We thus have no alternative but to agree with Charles W. Jones that "[Isidore's] cosmology, insofar as it has any consistency, is only consistent with a globular earth."[22] "Sicut rota est" then refers to a circular diagram of the *globus terrae*.

II

In the manuscript examples of *rota terrarum* a reverse image sometimes accompanies texts of Sallust, Lucan, and other Latin authors; one is also found in a fourth version of Isidore's schoolbook not previously noticed. Six English manuscripts of *De natura rerum* with the final chapter and *rota* survive, forming an Isidorian tradition which varies significantly from the long version discussed above: the text is altered in many chapters, and the design of the final *rota* is quite different (see Fig. 6). These changes indicate that Isidore's book was actively used in early Saxon and perhaps Celtic schools. But why is the *rota terrarum* in mirror image in some of these manuscripts? Indeed, this reversal goes against schemata in earlier exemplars of *De natura rerum*

---

[21] In *Speculum,* 1976, *51*:752–755, I have discussed medieval evidence and modern assumptions about antipodes in a review of K. Hillkowitz, *Zur Kosmographie des Aethicus,* Vols. I (Cologne, 1934) and II (Frankfurt am Main, 1973), and other related literature.

Among those who ridiculed the idea that the cosmos was spherical, the argument therefrom that the earth too was a globe, and the claim that therefore on the part opposite them there must be animals and men (*Antipodae*) upside down were Plutarch (fl. ca. A.D. 90–125), *De facie in orbe lunae* VII, and the heterodox Christian and rhetorician Lactantius (fl. A.D. 284–317), *Divinae institutiones* III 24. Neither offered any positive description of the shape of the earth, but merely scoffed at absurdities and contradictions of the *philosophi*. Lactantius became important in the fifteenth century both because·Lorenzo da Valla praised the Ciceronean quality of his style and later because Copernicus cited him in the dedicatory letter for *De revolutionibus:* "For it is known that Lactantius—a poor mathematician though in other respects a worthy author—writes very childishly about the shape of the earth when he scoffs at those who affirm it to be a globe." This was repeated by Galileo in his *Letter to the Grand Duchess Christina,* which I quote from the translation by Stillman Drake, *Discoveries and Opinions of Galileo* (New York: Doubleday, 1957), p. 180. My appreciation for the reference to Copernicus and Galileo goes to Professor Bert Hansen, Institute for History and Philosophy of Science, University of Toronto.

[22] Charles W. Jones, *Bedae Opera de temporibus* (Cambridge, Mass.: The Mediaeval Academy of America, 1943), p. 367, and his article "The Flat Earth" cited in note 1 above.

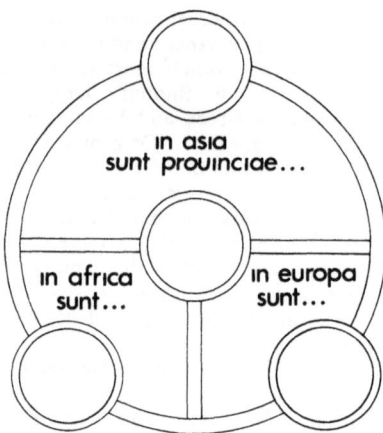

Figure 6. The reversed tripartite rota from the Exeter and Vitellius manuscripts of Isidore's *De natura rerum*.

(see Fig. 4). The earliest English manuscripts are Exeter Cathedral 3507, British Library Cotton Domitian I (both second half of the 10th century), and British Library Vitellius A. XII (end of the 11th century).[23] Their texts have been collated and are nearly identical; but there are variant readings which make it clear that they must have been copied directly from a single exemplar, not from each other. Yet despite the identity of their modified long version texts, Domitian gives a simple Isidorean *rota* (Fig. 4) whereas Exeter and Vitellius give the reverse image— apparently by choice rather than by error. How can this be explained in a scholarly text?[24]

Hellenistic tradition represented astronomical phenomena in alternate ways that may lead us to understand the manner in which the *rota terrarum* was conceived and used. There have always been two perspectives on the *globus caelestis* and its stellar phenomena, which result in two quite different projections of star charts and land charts. If the astronomer looks up at the sky with *Asia* and its *oceanus orientalis* at his head, then *europa est dextera, africa est laeva*. On the other hand if he imagines himself outside the globe and looking east, he would see that *europa est laeva, africa est dextera*.

Both perspectives are found in popular literature. Latin texts of the several versions of Hyginus and Aratus have illustrations of constellations (such as Perseus or Eridanus) which face the viewer or which turn away from and are laterally reversed for him. In the one case the viewer is thought to be within the globe or

---

[23] For the present discussion only these three—the Exeter, Domitian, and Vitellius MSS—will be cited. All six will be analyzed elsewhere.

[24] Historians of geography have sometimes made the peculiar assumption that many unexpected notions during the Middle Ages can be attributed to ignorance or error and need no explanation; for example Destombes, *Mappemondes*, p. 37, commented in 1964 on the reversed *rota terrarum* in some Sallust MSS: "Ces mappemondes sont toutes du type A, la plupart du type A3 avec les noms des trois continents quelquefois interchangés par suite d'une erreur ancienne." Without being aware of any pattern, Gerald R. Crone studied *The Hereford World Map* (London: Royal Geographical Society, 1949), p. 6, and reported that "the draughtsman has carelessly interchanged the names of Africa and Europa." At the cathedral bookshop unsuspecting tourists are offered a replica on which the modern draughtsman has "corrected" the past.

sphere of stars looking out at the figures; in the other he is assumed to be looking on from the outside.[25] The more sophisticated planispheres of astronomers oriented to the north celestial pole could also take either perspective: the early ninth-century Fulda planisphere viewed the skies from without, as had Hipparchus. The equally venerable Salzburg planisphere assumed the perspective of a viewer within the globe looking out, in effect similar to Ptolemy's viewpoint from the south celestial pole.[26]

Orientations for the earth were similarly flexible. When geographers prepared *picturae* of their *sphaerae terrestris,* they often placed Asia and its *oceanus orientalis* at the head of a chart or at the top of a codex leaf, and regions were projected onto the flat surface according to the relationships of the planisphere. But there were alternatives: one could face in any direction while placing each zone or continent in true relationship with the others to the left or right, upper or lower. For example, Aristotle explained in *De caelo* II 2 that the Greeks are to be found on the right of the diagram in the lower hemisphere; the North is lower, and those living in the South are in the upper hemisphere on the left. So far as he was concerned, therefore, the Pythagoreans had everything backwards, and he rejected their apparent practice of depicting *Europa* in the upper part of the sphere. Both may have been assuming the four-part zonal concept (Fig. 2) with meridian dividing hemispheres, several centuries before Krates. Apparently the Pythagoreans depicted the globe so that the stars appeared to move from left to right, whereas Aristotle insists on the viewer facing the heavens and imagining himself turning right (forward) to left with the stars.

Later Hipparchus was critical of Eratosthenes' mirror-image astronomy as well. From the outer Hipparchan perspective on the celestial sphere projected onto the globe of the earth, *Europa* and its *oceanus septentrionalis* would be found on the left, as with the Isidorean *rota*; but from the inner Ptolemaic perspective[27] *Europa* would then be on the right, as with the reversed *rota*. The cartographer who knew a *globus caelestis* or a stellar planisphere of the first type (looking down on it all) could apply his terms for the oceans of stars equally well to the geography beneath them without turning his head, whereas the cartographer who knew a *globus caelestis* or a stellar planisphere of the second type had to look up to the stars and then down to the lands, thus requiring him to distinguish between celestial and terrestrial regions and to keep in mind the distinct perspectives required.

Thus we may recognize two conventions for representing the heavens, each of which had intelligible consequences for the *rota terrarum*: neither the Isidorean

---

[25] Many of these storybook illustrations may be seen in Fritz Saxl, ed., *Verzeichnis astrologischer und mythologischer illustrierter Handschriften des lateinischen Mittelalters,* Vols. I and II (Sitzungsberichte der Heidelberger Akademie der Wissenschaften, Philosophische-Historische Klasse, Vols. VI, 1915, Nos. 6–7, and XVI, 1925–1926, No. 2); Fritz Saxl, H. Meier, and H. Bober, eds., *Handschriften in englischen Bibliotheken* (London: Warburg Institute, 1953); and Patrick McGurk, *Catalogue of Astrological and Mythological Illuminated Manuscripts in Italian Libraries other than Rome* (London: Warburg Institute, 1966). McGurk appears to be the first to notice a pattern of perspectives.

[26] The Fulda planisphere is MS Basel Univ.-Bibliothek A.N. IV. 18 (Fulda, early 9th century), fol. 1v. The Salzburg planisphere is CLM 210, fol. 113v (810–818?), known in Regensburg during the ninth century. Further planispheres which take each perspective have been listed by Patrick McGurk, "Germanici Caesaris Aratea cum scholiis, a New Illustrated Witness from Wales," *The National Library of Wales Journal,* 1973, *18*:197–216, especially pp. 200–201.

[27] As above, I refer to Hipparchus and Ptolemy only with respect to viewpoints in observing the heavens. It is worth recalling that the so-called "Maps of Ptolemy," so often cited by historians of Renaissance or Enlightenment science, were drawn probably in the 14th century on the basis of a Byzantine "Geographia" composed in the 10th or 11th century. See Lev. S. Bagrov [Leo S. Bagrow], "The Origin of Ptolemy's Geographia," *Geografiska Annaler,* 1945, 318–387, especially pp. 368–372 and 387; Bagrow, *History of Cartography,* pp. 34–37; and Otto Neugebauer, *The Exact Sciences in Antiquity,* 2nd ed. (Providence, R.I.: Brown University Press, 1957), p. 227; Neugebauer, *Ancient Mathematical Astronomy,* pp. 885–886. A rebuttal of Bagrow was undertaken by Erich Polaschek, "Ptolemy's Geographia in a New Light," *Imago Mundi,* 1959, *14*:17–37; but recent studies by David Thomason at the Warburg Institute indicate that the "Maps of Ptolemy" probably were composed in the 14th century.

III

image nor the reverse image of the globe indicates a lack of understanding, a failure of observation, or a repetition of error. Both may be best understood in terms of a priority of the spherical cosmos in the *picturae* of literati, cosmographers and cartographers. The results of this enquiry may be summarized thus: In the tradition of Hellenistic science there were several diagrams for presenting the shape and features of the earth's globe. All continued to be cited in Roman and medieval literature, and all show the earth to be spherical. But by far the earliest of these diagrams extant in dated manuscripts is the *rota terrarum* with three continents. It properly belongs together with the final chapter which Isidore of Seville added in 613 to his own work *De natura rerum*. A copy of the long version was then used, adapted, and elaborated in schools of eighth- and ninth-century England, whence it passed also to Fulda, Verona, and elsewhere. Among the teachers in England were scholars who knew some astronomy and thought of heavens and earth as two spheres—from which the *rota* model first derived. One of them saw his *rota* as oriented to an inner or Ptolemaic perspective on the heavens; thus looking up, he found continents of earth beneath continents of stars. Turning to his work table, he maintained those relationships. Although he did not intend to confuse modern readers, the resultant figure became a riddle for generations of moderns. They have labeled as error what in fact was astronomy.

IV

SCIENTIFIC INSTRUCTION IN EARLY INSULAR SCHOOLS

The science to be discussed in this paper has to do with calculations
and observations of both earth and skies. The evidence has not heretofore impressed modern philologists and historians who have searched for
literary activities and their foundation in *grammatica* or who have
scrutinized the patterns of prayer and psalms that one finds throughout
European monasticism in the development of *cantica*. Yet the Celtic
culture of striking images, fanciful word-plays, and astonishing holiness cannot be discussed very thoroughly without at least acknowledgement of a certain problem, like a thorn in the flesh: the date of
Easter.[1] In Ireland, England, Spain, and Gaul during the entire seventh
century, that problem demanded all available resources: resounding arguments ensued; traditions were born, defended, or undermined. Some
scholars, like Bede, considered that the question had been resolved by
A. D. 664; but later writers, like Abbo of Fleury, knew that it had
not.[2] From the fragmentary materials surviving from arithmetical
calculations, geometrical models of heaven and earth, and arguments about
solar and lunar cycles it is possible to recognize in *computistica* not
only occasional concerns but also concerted and significant scientific
labours in the schools of the Scots of Ireland, Britons of Wales, Angles
of Northumbria, Saxons of Southumbria and Wessex. If there were fanciful images they begin in *grammatica*; if there were prayers they were
reinforced in *cantica*; but practical study of natural objects and the
reckoning of relationships and motions of things were gathered together
in *computistica*.[3] For science we look to the *computus*.

In the year A. D. 630 by our reckoning, a synod was held in Ireland
at Campus Lenis (Magh Léna), and it was decided that beginning with the

Reprinted from *Insular Latin Studies. Papers on Latin Texts and Manuscripts of the British Isles:
550–1066*, ed. Michael W. Herren, pp. 83–111, by permission of the publisher. © 1981 by the
Pontifical Institute of Mediaeval Studies, Toronto.

next year all should keep Easter with the universal church. It is not at all certain by what system this was to be achieved: whether it would be a simple table of dates; a sequence of memory verses; or a cycle of 8 or 19 or 76 or 84 or 95 or 112 years, with the *saltus* placed in the twelfth or the fourteenth or the sixteenth or the nineteenth year.

We know none of this. But we do know that there was serious objection to this decision and that someone brought about division among those who had agreed to celebrate Easter with the rest of the church. This objection seems to have led to a second synod a year later at Campus Albus (near Slieve Margy, Queen's County). According to the *Vita S. Munnu Siue Fintani* (d.636), the new *ordo Paschae* was defended there by Laisrén, abbot of Leighlenne, under whom there were 1500 monks. But the old computus was defended by the holy Fintan, abbot of Tech Munnu (Taghmon, Co. Wexford), who kept the assembly waiting for a considerable time and finally made his entrance just before evening. Then he suggested how the Easter controversy could be settled:

> There are three options, Laisrén:
>
> Two books could be put in the fire, the old and the new,
> so that we may see which of them survives the fire;
>
> or two monks, one of mine and one of yours, could be
> closed up in a hut and the hut could be set afire, and
> we should see which of them comes through the fire unhurt;
>
> or we could go to the tomb of a just monk who has died
> and bring him back to life, and he will tell us by which
> *ordo* we should celebrate Easter this year.[4]

Laisrén discreetly refused the ordeal in any form on the grounds that Munnu's sanctity was so great that God would grant whatever he asked.

In the face of overwhelming holiness, Laisrén's decision was not an act of weakness, for the southern abbots were coming over to the new *ordo Paschalis*. Within only one century the northerners, British, and even the holy foundation of Columba on the island of Iona had abandoned the older computus "whose author or place or time we do not know," *cuius auctorem locum tempus incertum habemus* (as Cummian said), in favour of an Alexandrian series of Easters, which appeared to be a true reckoning universally observed.

Rather than rehearsing the Paschal controversies of the seventh century, I want to discover how it came about that sanctity yielded to science among the Irish and British as well as among the Saxons and Angles. Cummian will lead us:

Dominis sanctis et in Christo venerandis, Segieno
abbati Columbae sancti et caeterorum sanctorum successorum,
Beccanoque solitario, charo carne et spiritu fratri,
cum suis sapientibus; Cummianus...(*Epistola Cummiani*)[5]

There are many Ségénes, Beccáns, and Cummians named in Irish annals and
documents containing reports from the seventh century. This Ségéne was
abbot of Iona (A. D. 623-652), and this Beccán was presumably a relative
of the author of the letter. Beyond this letter our author is scarcely
known; this Cummian must have been living in a region from which
prominent abbeys would respond to his call for a synod at Mag Léna, a
plain near Durrow.[6] They came from

Emly (Co. Tipperary) where Ailbe had been bishop;
Clonmacnois (Co. Offaly), founded by Ciaran;
Birr (Co. Offaly), the place of Brendan;
Mungret (Co. Limerick) where Nessan was known;
Clonfert-Mulloe (Co. Laois) begun by Lugaid, that is, Muloa;[7]

Cummian reports and comments on the controversy of new and old
reckonings in which Laisrén avoided the ordeal proposed by that *paries
dealbatus,* by which he presumably means Fintan (d. A. D. 635/6), successor
to Munnu of Taghmon.[8] Against the ruler of the monks of Munnu, Cummian
quoted St. Paul's response after the high priest Ananias had commanded
him to be struck across the mouth: "God shall strike you, you white-
washed wall..." (Acts 23.3), for Fintan pretended to observe the traditions of his elders and created discord from the unity which had been
achieved at the previous synod of Mag Lena. "The Lord, I hope, will
strike him down in whatever way he sees fit!" added Cummian, after
Paul.[9] Against the discord of the old holy man, Cummian argues for
Christian unity in celebrating the central feast of the year; and this
unity was depicted in ancient terms as well as in terms of contemporary
experience.

1. It makes no sense for Britons and Irish to observe
   an Easter in conflict with the four apostolic sees:
   Rome, Jerusalem, Antioch, Alexandria;[10]

2. It occurs that the whole world (Hebrews, Greeks, Scythians,
   and Egyptians) observes Easter at the same time, as found
   by a delegation sent to Rome in 631 whose Irish tables
   gave them a date an entire month later than the others.[11]

Are they all wicked, Cummian asked the holy Ségéne and the wise Beccán,
or is it we who are on the outer edge of things? Are not the Britons
and Irish in these matters but a pimple on the face of the earth?[12]

Ségéne, like Fintan, was unmoved by appeals to Christian unity either
in apostolic or universal terms. But Cummian was able to present also
a series of detailed arguments based upon resources which, though lacking

IV

an appeal to piety and apostolicity, nevertheless served those ends.
He had studied the Easter cycles; for a whole year he had withheld judgment, sought out the explanations of different cycles, and analyzed
what each race thought about the course of sun and moon. Could he really
do this at the outer edge of the whole wide world? Here are the books
which he names:

> 1. Primum illum quem sanctus Patricius papa noster tulit et facit;
>   in quo luna a XIV. usque in XXI. regulariter, et aequinoctium
>   a XII. kalend. Aprilis observatur.[13]

This was the usage of the ecclesiastical *civitas* of Milan, apparently
following an Alexandrian table of the fourth century. Its limits for
observing the Paschal moon were luna XIV–XXI, and luna XIV was the vernal
equinox of XII Kal. Aprilis (=21 March) rather than XI Kal. Aprilis
(=22 March) – accepting an adjustment to the Julian calendar which
stemmed either from Alexandria or perhaps from the Council of Nicaea.
The Easter festival would thus occur on the first Sunday after the first
full moon after the vernal equinox. This was the system which Augustine must have learned from Ambrose. As the imperial government
increasingly used Milan as a residence and seat of administrative
government, the church there came to surpass other *civitates* in influence
for northern Italy and Provence from the fourth to the eleventh century.
The reckoning survived at some places in Gaul for several centuries,
particularly at Auxerre, and it could have been learned by Patrick if
he visited Gaul or Provence.

Cummian, however, also quoted from the acts of the Synod of Arles
(assembled in A. D. 314 by Constantine to settle the disputed election
of Caecilian as bishop of Carthage) concerning the need to celebrate
Easter in common with the whole church:

> Item Arelatensi synodo sexcentorum episcoporum confirmante
> primo in loco de observatione Paschae, ut uno die et uno tempore
> per totum orbem terrarum a nobis conservetur: ut universalis
> Ecclesia uno ore, juxta apostolum, honorificet dominum unum.[14]

The canons of the Synod of Arles survive in at least two versions, and
there are also short and long versions of a letter from Bishop Marinus
of Arles to newly elected Bishop Silvester of Rome, who had not been
present. Conflation of these versions results in an initial canon
which states:

> Primo loco de observatione Paschae Domini, ut uno die
> et uno tempore per omnem orbem a nobis observatur
> [et iuxta consuetudinem literas ad omnes tu dirigas]...[15]

It would appear, therefore, that the Synod of Arles had given some
thought to the diversity of Easter dates then practiced and had
affirmed the need for unity and consistency. The evident interpolation
to the above canon, as it appears in a manuscript of Marinus' letter,
affirms that the bishop of Rome would fulfil this need through annual
letters, but that expectation was not found in Cummian's source and was
indeed misplaced. Doubtless some of them did prepare and circulate
such letters in successive years within their area of influence,[16] but
that area was most uncertain and their influence was insecure for good
reason. At the time of the Synod of Arles, the Roman ecclesiastical
civitas was using the older *Supputatio Romana* which had limits of luna
XIV-XX, with Easter Sunday following equinox VIII Kal. Aprilis (=25
March). After A. D. 342 this was abandoned in favour of an 84-year
cycle with lunar limits XVI-XXII. A further attempt to apply limits for
the Sunday of Easter itself, as XI Kal. Aprilis - XI Kal. Maii (=22
March - 21 April), was introduced, and other adjustments suggest that
the Romans may never have understood how to compute a date. Dependence
upon regular correspondence with the bishop of Alexandria in addition to
the tables was frustrating, as Leo (bishop of Rome, A. D. 444-461)
pointed out in several letters. Thus the many changes of local usage by
Roman bishops would not fulfil the intention of the Synod of Arles at
all, and it is not surprising that Milanese usage and Roman variations
came into conflict during the fifth century in Roman Africa, as will be
discussed below.

Five British clergymen were reported to have been present at the
Synod of Arles in A. D. 314: Eborius of York, Restitutus of London,
Adelfius *de civitate Coloniae Londinensium* (perhaps Caelon on Usk),
Sacerdos Presbyter, and Arminius Diaconus.[16a] What system for reckoning
Easter this British delegation may have learned at Arles and taken
home is, therefore, most uncertain, and all attempts to reconstruct
tables on the basis of their attendance there have been contradictory.[17]
However, the system attributed to Patrick[17] by Cummian[17a] was probably
never used in Rome but was used in the Milanese *civitas* and its area of
influence.

  2. Secundo Anatolium, quem vos extollitis quidem ad veram Paschae
     rationem numquam pervenire eos qui cyclum LXXXIV. annorum observant.

The Pseudo-Anatolian *Canon Paschalis* also advocated the vernal
equinox at the early Easter limit but took it to be VIII Kal. Aprilis
(=25 March) of the earlier Julian calendar, even altering an authentic

paragraph from Anatolius of Laodicae for this purpose.[18] It held limits
for the Paschal moon to be luna XIV-XX, and it strongly rejected the
system associated with Patrick. The author may have been a Welsh Briton
in the tradition of those British bishops at the Council of Arles in
A. D. 314, and his reckoning must have been passed to the Irish long
before A. D. 590, for these were the criteria which Columbanus knew at
Bangor and which he promoted among the hostile bishops of Gaul.[19] Also
apparently accepted at Iona, this Irish canon was an intelligent attack
on a 19-year lunar table which was not being used properly, which seems
to mean that the table was being applied with the Milanese or "Patrici-
an" criteria of Luna XIV-XXI with equinox XII Kal. Aprilis. The canon
further rejected an *amplior circulus,* deriving from certain African
teachers, which not only advocated those limits but also required
Easter Sunday to fall within XI Kal. Aprilis and XI Kal. Maii (=22
March - 21 April). This range was too broad when the equinox was
observed on VIII Kal. Aprilis according to Pseudo-Anatolius. Thus the
Acta Synodi Caesareae, which Cummian cited in the next item as "Theophilus," ought to be rejected, *detestandos ac succidendos esse.*

Both of the usages under attack were adjustments of lunar tables
which had been generated to meet local needs elsewhere. The older
Supputatio Romana, with limits of luna XIV-XX and Easter Sunday following
equinox VIII Kal. Aprilis, had been used in the Roman *civitas* only
during A. D. 312-342. A later Roman attempt to set limits for Easter
Sunday itself of XI Kal. Aprilis - XI Kal. Maii was cited by Hilarianus
in A. D. 397 and was probably known to the Irish Pseudo-Anatolius
indirectly through its adaptation to a special case in fifth-century
Roman Africa. There the *Acta Synodi Caesareae* was produced in an attempt
to reconcile divergent criteria, and this too was subject to a reaction
in the *Computus Carthagininsis* of A. D. 457. Either one could have
struck our Irish author as the sophistry of certain Africans.

The *Canon Paschalis,* however, emphasized solar data, and this
could have gained the respect shown it by later scholars, like Bede,
even though they advised against following it. Although Pseudo-Anatolius
rejected the system of Victorius of Aquitaine, he insisted that movements of the solar year could be accounted for only by use of an 84-
year cycle.

3. Tertio Theophilum

This is probably a tract headed *Epistola Theophili,* which was a
version of the *Acta Synodi* often quoted in Cummian's letter. It

repeatedly named a Theophilus, bishop of Caesarea, a metropolitan see
said to have displaced Jerusalem in authority; and it affirmed that he
presided over the synod by direction of Victor, *papa Romaneque urbis
episcopus*. Nevertheless it does not appear to have derived either from
Caesarea or from the Theophilus who had disputed Paschal questions with
Victor, bishop of Rome (189-98). Nor was it from another Theophilus,
bishop of Alexandria (385-412), who had indeed developed a Paschal table
but not the one prefaced by this document. The first *Acta Synodi* may
have originated in Roman Africa, from where it travelled and was
modified in Spain, Celtic Britain, and Ireland. The situations from
which each version arose have been accounted by   C. W. Jones:[20]
Augustine apparently had brought the lunar limits XIV-XXI from Milan to
Carthage and Thagaste in A. D. 389/90 and thence to Hippo Regius where
he became bishop, A. D. 395-430. These were the limits in use at
Alexandria and Magna Graecia, including the bay of Naples and most of
southern Italy; they were followed in Milan, northern Italy and parts
of southern Gaul, but not in Rome. Churches of the Roman *civitas* and
its sphere of influence were normally dependent upon annual correspondence between the bishop of Rome and the bishop of Alexandria. From
time to time the Romans undertook to generate their own tables, though
without much luck. The Romans also developed the custom of limiting
the Sunday on which Easter would be celebrated to the period XI Kal.
Aprilis - XI Kal. Maii (=22 March - 21 April) in order to avoid uproar-
ious pagan festivities celebrating the founding of the city. Many
cities in the Roman province of Africa accepted Roman guidance and, thus,
the two kinds of Easter limits came into conflict there. The first
version of the *Acta Synodi* appears to report speeches and discussion
from a consultation and answers to a specifically localized problem in
fifth century Africa; it may be authentic, and it certainly offers
reconciliation. There is evidence that this settlement was rejected by
a reassertion of "Roman" criteria in the *Computus Carthaginiensis* of
A. D. 457.

The *Acta Synodi* travelled to Spain under the rubric *Epistola
Philippi de Pascha* in some manuscripts, but there it was attributed to
a Caesarean synod presided over by Theophilus. Sixth-century modifica-
tions include reference to certain churches in Gaul which celebrated
Easter every year on the Julian equinox, VIII Kal. Aprilis (=25 March),
according to an old Quartodeciman custom which was much berated by
Cummian as well as by Pseudo-Anatolius. The same phenomenon was also

reported by the Spanish *Prologus* attributed to Cyril and by the *Tractatus de Ratione Paschae* attributed to Cyprian.

Variants of this Spanish version of the *Acta* were later complicated in the manuscripts by the introduction of sentences that were intended to emphasize conformity with Dionysian reckoning, which was also becoming known in Irish and Saxon schools during the seventh century. Such a variant was used by Bede in Northumbria by A. D. 725 in his *De Temporum Ratione Liber* (47: 87-94) in order to refute the *sententia vulgatum* that the crucifixion and resurrection had occurred on VIII Kal. and VI Kal. Aprilis (25 and 27 March), the same problem that plagued Spanish and Irish computists. There is no proper edition of any of these variant versions.

4. Quarto Dionysium.

Dionysius Exiguus (ca. A. D. 470-550) was a Scythian scholar living in Rome when he was asked to prepare new Paschal tables, anticipating the expiration of a Latin version of Alexandrian tables with the year 531. He appears to have issued his 95-year continuation with the same incipit and with eight columns of data with headings similar to those of the Alexandrian usage.[21] However, he also introduced a new column for *anni Domini,* and he made the historical assumption that Jesus had been born on the first year of a 19-year cycle, corresponding to the consulship of C. Caesar and L. Paullus, which we call the year 1 B. C. This invited much criticism during successive centuries.[22] By the year A. D. 525, his tables had been given to Petronius of Africa as well as to officials of the Roman curia; the latter group affirmed to Johannes (bishop of Rome, A. D. 523-526) that the date of Easter could be determined and announced on this basis.[23] Despite the fact that Dionysius had shown, for the first time, how to reconcile Alexandrian Paschal limits of luna XV-XXI with the Roman custom that restricts Easter Sunday to the period XI Kal. Aprilis - XI Kal. Maii, there is no evidence that his tables were used in the Roman *civitas* by the time of Cummian, or of Bede, or even by the tenth century.[24] Cummian did not cite data from Dionysius' *Tabula Paschalis* or his nine authentic *Argumenta Paschalia*.[25] However, the terms used to describe his tenth source will demonstrate that he had at hand the *Praefatio* to those tables and both dedicatory letters.

5. Quinto Cyrillum.

The *Epistola de Pascha,* which Cyril (bishop of Alexandria, A. D. 436-444) sent to the Council of Carthage concerning the Easter of A. D. 420, survives in thirteen computistical manuscripts, often along with materials related to Irish usage. This authentic letter was also available in the *Canones Ecclesiastici* 38, prepared by Dionysius Exiguus (PL 67: 145). However, the Latin version has been put to later use.[26] It seems to have been adapted at the behest of Boniface (bishop of Rome, A. D. 607) for the Roman mission in England, which may have been trying to use an 84-year cycle from either Rome or Gaul, or possibly even a Victorian cycle, but which had neglected the extra embolismic month – precisely the question to which Cyril had addressed himself. Boniface repeated Cyril with a slight adaptation to the year 607, repeated some of the queries he had received, then discussed various computistical matters in terms which provide indications of theories that are later than those of Cyril or Dionysius. It further appears that the elaborated version of this Cyril-Bonifatius letter was sent by Laurentius (bishop of Canterbury) to Celtic centres, both British and Irish. It was from the portions added to Cyril's early letter in Rome or in Canterbury that Cummian quoted.[27]

The *Epistola de Pascha* (or *Epistola Cyrilli*) is found together with Dionysian tables in extant computistical collections. It may be that they accompanied it from Rome to Canterbury, or from Canterbury to Celtic centres; but there is an absence of evidence that Dionysian usage had been accepted thus far either in Rome or in Canterbury.[28]

6. Sexto Morinum.

Disputatio Murini Episcopi Alexandrini de Ratione Paschali.... survives in eight manuscripts. The first part introduced Victorian tables in the sixth century but is scarcely readable. The second part is intelligible and can be dated to either A. D. 604 or 632 by reference to the Victorian *saltus lunae,* but it favours Alexandrian and Dionysian usage against the Pseudo-Anatolian arguments. Thus attribution of the *Disputatio* to a bishop of Alexandria should be ignored, and at least the second part is probably of Irish origin from the early seventh century.[29]

All extant manuscripts are in Carolingian scripts which misread insular *r, s, n,* and *p;* thus the forms *Murinus, Morinus,* or *Morianus* could correspond to the two reported names of Irish computists from this

period: *Monino* and *Mosinu.* Monino Mocan was learned in the science of
computus, according to Columbanus who had studied at Bangor before
A. D. 590, and considered himself competent to challenge bishops of
Gaul or Rome concerning Easter reckoning.[30]   Mosinu maccu Min is said
to have been a scribe and abbot of Bangor who "was the first of the
Irish who learned by rote the computus from a certain learned Greek";
he is also named as Sillan, abbot of Bangor who died A. D. 610.[41]
Palaeographically, the names *Monino* and *Mosinu* could be one and the
same name. The computus which he memorized could also have been the
memory verse which came to be attributed to Pachomius, allowing one to
keep track of the lunar epacts in accordance with Alexandrian usage.
This would have been an innovation at Bangor after Columbanus' depart-
ure.

7. Septimo Augustinum (Agustinum *MS*).

Cummian named Augustine in three other places (along with Paul,
Jerome, and Gregory) as an advocate against heresy and for unity, and
he seems to be rather well-versed in Augustinian thought. Reference
to Augustine as a source for computus offers several possibilities,
however. The African bishop's best contribution to calendar problems
was his long *Epistola ad Januarium,* which might have been know from
Eugippius' *Excerpta ex Opera Augustini.* This work abbreviates the
Epistola, with passages on the phases of the moon added from *De Genesi
ad Litteram.*[32] Augustine had brought Alexandrian Paschal usage of Milan
into Roman Africa, where it conflicted with local Roman custom as
explained above. The reconciliation sought in the *Acta Synodi Caesareae*
could have maintained his name in relation to collections of Spanish
origin, which were further developed in Ireland as in the *sententiae
Sancti Agustini et Isidori in Laude Compoti.*[33] While the surviving
four parts of that accumulation have very little Augustine and much
Isidore - the reverse of what one finds in Cummian's letter - its
capitulatio enumerates fifty-six chapters of a computus no longer found
in the fifteen extant manuscripts. Cummian's spelling of the name,
Agustinus, could certainly be Spanish[34] and was retained by Irish
scribes in various works.

8. Octavo Victorium.

This Victorius was a teacher of arithmetic in Aquitaine during the
fifth century and wrote a *Calculus* which served to instruct students in
arithmetic without arithmology. In A. D. 455, the Alexandrian date of

Easter fell on VII Kal. Maii (=24 April) during Roman pagan celebrations
and beyond the limits there preferred. Archdeacon Hilarus requested
that Victorius study this problem, and he answered two years later with
a letter accompanied by a *Cursus paschalis* based upon the 19-year lunar
cycle of Alexandria which was known in Rome *de Graeco translata*.
Victorius did not concern himself at all with avoiding the problems
which had led to Roman discomfiture: an Easter later than 21 April and
the need for celebrating the feast together with the other apostolic
sees on the same date. Rather he pointed out discrepancies in basic
assumptions about rules for calculating the first month (Nisan) and the
limits for the Paschal moon, which were lunae XV-XXI for Alexandria but
had become by the fifth century lunae XVI-XXII for Rome. In consequence,
all tables for 84, 95, or 112 years were not in accordance with the
others, were not cyclic for the other period, and should be set aside
in favour of reckoning full moons (luna XIV) on the basis of the *annus
Passionis*. This he found from the Eusebius-Jerome *Chronicle* as
equivalent to the *annus mundi* 5229 (=A. D. 28). From this point he calculated Easters backwards to creation and forwards to his current A. P.
430 (=A. D. 457). At A. P. 532 (=A. D. 559) it appeared that his data
began repeating themselves.[35]

Victorius' use of the *annus Passionis,* however, required him to
apply the 19-year cycle out of phase with that of Alexandria, a problem
accentuated by the *saltus lunae*. Although he provided alternate dates
for *Latini* when reckonings for *Graeci* passed Roman limits, he attributed
the *Graeci* Paschal limits to luna XXII, which they never observed.
Furthermore, he did not take into account the change of vernal equinox
from VIII to XII Kal. Aprilis, which added confusion to the application
of his tables.

The Victorian cycle was used only once in Rome (probably by coincidence), but it became popular in Gaul. According to Columbanus, it had
been tested and rejected in his part of Ireland before A. D. 590, but
it appears to have been favoured still in the mid-seventh century by
others. Cummian spoke well of Victorius, apparently in appreciation
of his application of the Alexandrian 19-year cycle and of his reject-
tion of the other kinds of Easter tables.[36]

9. Nono Pacomium monachum, Aegypti coenobiorum fundatorem; cui ab
    angelo ratio Paschae dictata est.

Pachomius was a cenobite of Upper Egypt who lived ca. A. D. 290-347.
His activities began about 320 at Tabennisi in the Thebaid, where he

IV

94

linked cells of hermits into great colonies. The reference by Cummian
is to nineteen verses, which may have been composed by a Visigoth in
Spain (ca. A. D. 600) whence they soon came to be known in Ireland and
Gaul: Nonae Aprilis norunt quinos/...[37] They provide in memorable form
the Julian date for the Paschal moon and the ferial regular for each
year of the Alexandrian 19-year cycle in Latin usage. The ferial
regular for each year is a number which indicates the difference between
the weekday on IX Kal. Aprilis (=24 March) and the weekday on which the
Paschal full moon occurs. To this one adds the concurrents and thus
learns the weekday of the Easter limit, beyond which Easter does not
cocur in that year. The verses seem to have been known at Bangor about
A. D. 605, and the *Epistola Cyrilli,* cited above, received an addition
which could have had an insular source prior to the Easter problem of
607, and which reported angelic voices to Pachomius, pertaining *ut non
errorem incurrerent in solemnitatis paschalis ratione...,* on the basis of
Gennadius' fifth-century account.[38] With these verses, the dates and
epacts for nineteen years of an Alexandrian cycle could be remembered
anywhere, anytime.

   10. Decimo trecentorum decem et octo episcoporum decennovenalem
        cyclum (qui Graece Enneacedeciterida dicitur) in quo kalendae
        Januarii lunaeque ejusdem diei et initia primi mensis
        ipsiusque XIV. lunae recto jure ac si quodam clarissimo tramite,
        ignorantiae relictis tenebris, studiosis quibusque cunctis
        temporibus sunt adnotate, quibus paschalis sollennitas
        probabiliter inveniri potest.

No Paschal table survives from the Council of Nicaea, though
epistolary references to it suggest that there had been agreement on
certain rules which are simple and memorable: Easter will be celebrated
on the first Sunday after the first full moon after the vernal equinox,
and that equinox was assumed to occur on XII Kal. Aprilis (=21 March).
This became a tradition which attached itself to certain Easter tables,
among them the Dionysian ones. In the preface to his tables, Dionysius
had been satisifed to cite the *sancti patres* as the source for the
Alexandrian tables, upon which his own were based; but in both his
Epistola ad Petronium and his *Epistola ad Bonifatium et Bonum* he identified them with the 318 *venerabiles pontifices* or *sancti* at Nicaea
accepting a tradition which had been developing for a century and a
half.[39] Cummian cited *Nicena sinodus CCCXXIII episcoporum,* and he also
quoted one phrase, *decennovenalem qui Grece enneacedeciterida (dicitur),*
which could have been found either in the preface to Dionysius' *Tabula
Paschalis* or in his *Epistola ad Petronium.* The second phrase, *ignorantiae*

relictis tenebris, studiosis, is from the *Epistola ad Bonifatium et*
Bonum.[40] However, neither these writings nor the tables deriving from
Dionysius offer data for the first and fourteenth moons of the first
month (Nisan) or for the *kalendae Januarii lunaeque ejusdem diei.*
Dionysius reckoned the age of the moon or the number of epacts on the
Kalends of September, *sedes epactarum,* in keeping with Alexandrian
usage; and he ignored the Kalends of January. Other tables generated
under Roman influence, such as the *Laterculus* of Augustalis which ran
for a hundred years (213-312) on the basis of an 84-year cycle,
reckoned the age of the moon on the Kalends of January. Thus, it was
probably not a Dionysian table to which Cummian referred.

The ten computistical items named by Cummian have very intricate
histories of their own and are now found in a variety of versions that
are really quite amazing. Each requires a critical edition which would
show the efforts of many highly trained men to conceive of relations
between sun and moon and earthly history such that the Christian festival could be celebrated everywhere and by everyone on the same date,
year after year. But each attempt failed. Each attempt, nevertheless,
survived and was adapted to new circumstances in new places. The calculations that these computations require are multiple but not very
sophisticated in themselves, whereas their applications and their
relationships are complex. Cummian assessed these complexities and was
faced with many difficulties. The 84-year cycle of "our cycle," he
found, was out of phase with the rest of the world. Moreover, the cycle
set forth different weekdays on the Kalends of January (both for
common and bissextile years), a different first month (Nisan) and thus
a different Paschal moon (luna XIV), and finally a different equinox
(that is, Julian 25 March rather than Nicaean 21 March). Even the work
of Anatolius, upon whom the Irish *ordo Paschae* depended, could be cited
as admitting that no true cycle could be achieved through his rules
of computation. On the other hand, Alexandrian Easters could be known
either from the table of Victorius or from the Pachomian verses
promoted by Mosinu, abbot of Bangor. There survive tables of Irish
origin which show that different systems of Victorius and Dionysius
were being compared by scholars on various occasions.[41] But finally
the cycle of the 318 bishops, said Cummian, is correctly calculated in
a very clear table so that, leaving behind the darkness of ignorance
and taking note of all the assembled data, anyone can come to a certain
knowledge of the date of Easter.[42] Could one give a rational account of
them?

What was needed was education in cycles of sun and moon and in basic arithmetic. This is what Cummian argued would lead to the truth, and it is what he called upon Ségéne to provide when his letter ended with the exhortation not to err, but especially not to err knowingly.[43] It was knowledge to be found by study both of the scriptures and of the cycles. Bede's great respect for Irish scholarship depends upon this science which flourished within a context of piety.

Old science is dull reading and mercifully tends to disappear. There survives, however, some fragmentary evidence in many manuscripts that drills in arithmetic were done, even though mistakes are occasionally found. Especially numerous are sections of computistical cycles and computistical *argumenta,* which either explain how to apply the principles or how to get the answer directly without any principles. Broader studies of natural phenomena made their appearance in this way, attracted to pages of computistical texts and other schoolbooks.

One of the most important schoolbooks of our period was written in Seville in A. D. 612 by Isidore. I am not referring to his encyclopedic *Origines*, first drafted in 612 x 620 and developed in several versions to supply all sorts of information to those who read or preached the scripture. Rather, I mean his schoolbook, *De Natura Rerum,* which was intended to guide the basic scientific education of anyone and seems to have done so for almost everyone for several centuries.[44] The extent to which it was known in insular schools is not at all clear, however. Within the Celtic range from Britons to southern Irish, one finds the use of one or another word or phrase which could have derived from the *De Natura Rerum;* but Fontaine could only assume that it had reached Ireland on the basis of verbal images in the added chapter 44. However, it was transcribed in Clm 396 by an early ninth-century scribe who certainly was trained in the Hisperic practices, as revealed in his colophon. And the commentary on *Donatus Maior* in MS St. Paul-in-Karnten Stiftsbibliothek 25.2.16 (s.VIII in.) ff. 21-42 contains an epilogue *Ad Cuimnanum* with further passages reported to have Isidore's schoolbook as a source, copied by an Anglo-Saxon hand.[45]

 * What then of Saxon schools? In the monastery of Malmesbury Aldhelm was educated by the Irish Máeldubh, according to William of Malmesbury. Although Aldhelm's somewhat abstruse prose often invites comment, his was not a purely literary education. He wrote of three *disciplinae philosophorum: Physica, Ethica,* and *Logica,* of which the seven *fisicae artes* were listed by him in three separate works and cited in a fourth:

Arithmetica, Geometrica, Musica, Astronomia, Astrologia, Mechanica, Medicina.[46] He learned a little arithmetic but admitted that *diffici-llima rerum argumenta et calculi supputatione, quas partes numeri appellant, lectionis instantia repperi.*[46a] This probably referred to those number groups which one has to keep in mind when multiplying or dividing. Roman *calculi* were the pebbles or roundels moved across a board for calculations, in other words, a form of *abacus*. The reckoning learned at Malmesbury may not have satisifed his later master at St. Augustine's Canterbury in the monastic school of Hadrian (ca. 670-672), and like most students, Aldhelm regarded "all my past labour of study as of little value," since he could keep those *partes numeri* in mind only with a struggle and when "sustained by heavenly grace."[47] But the fuller range of *arithmetica* was included with the *computus,* a subject with which Aldhelm was very much concerned. He seems to have known the letter sent by Bishop Vitalinus of Rome (657-672)[48] to King Oswiu of Northumbria before the Council of Whitby, as is indicated in his letter to Geraint, king of Devon and Cornwall and to the bishops of Devon and Cornwall. He wished to convince them that the 84-year cycle with lunae XIV-XX should be abandoned in favour of the rules of the 318 fathers. Those fathers at the Council of Nicaea sanctioned a 19-year cycle in a perpetual *laterculus* which ran *per ogdoadam et endicadam* with Paschal limits luna XV-XXI, according to Aldhelm. On the other hand the 19-year computation of Anatolius and the 84-year rule of Sulpicius Severus "observe the paschal solemnity on luna XIV along with the Jews" which is an Oriental heresy; they should be set aside "since neither follows the bishops of the Roman Church in their perfect method of computation." Before continuing his explanation of the heretical "quartodecimans," Aldhelm added that the Roman bishops "also declared that the paschal computus of Victorius, which observes a cycle of 532 years, should not be followed in future.[49] This appeal to Geraint and his bishops is set within a general appeal to unity with the church fathers and to their charity, in contrast to the bishops of Dyfed. His appeal to the see of Peter is made with respect only to the question of Easter and, in that regard, echoes Wilfrid's claim as repeated by Bede's *Historia Eccleiastica* 3, 25. As Michael Herren has pointed out,[50] the letter carries out the appeal of Theodore to the Synod of Hertford (September, 672) and its first canon "that we all keep Easter Day at the same time, namely on the Sunday after the fourteenth day of the moon of the first month" (Bede, *HE* 4.2), and it

could have been written at the behest of any widely attended council
with the same concerns thereafter, during Aldhelm's abbacy of Malmesbury (673/4? - 705/6). The Britons in Wales, subject to the kingdom of
Wessex, accepted the new *ordo Paschae* in the year that Aldhelm became
bishop of Sherborne (705/6).

The English settlement of Easter problems at Whitby in 664 favoured
an Alexandrian reckoning in Dionysian terms, which was justified partly
in the language of Petrine mythology that must have appealed to Irish
Romani as well as to bishops of Canterbury. But it was accepted in
Iona only in A. D. 716. British participation in these matters has
left almost no record until the tenth century.[51] Some of those subject
to Wessex accepted it in 884, but others, only in 909, under the
authority of the Saxon see of Crediton.

The terms for the seven *fisicae artes* used by Aldhelm could have
been found in Isidore's *Differentiarum Libri Duo,* sections 149-152.
Aldhelm's references to *geometria, astronomia,* and *astrologia* were
augmented from Isidore's *Origines* 3.10.3 (geometry concerns measure of
lengths and the terms of measurement) and *De Natura Rerum* 18 and 24-26.
Astrologia pertains to rising, falling, visibility, and circular
courses of stars through the heavens as they circle the earth, while
astronomia or the non-Isidorian term *lex astrorum* accounts for the
movements of planets and their periods of orbit. Although the terms
themselves had been used by Roman *mathematici* (that is, astrologers and
magicians), in the hands of both Isidore and Aldhelm there is a complete
absence of number mysticism and astral influences.[51a] On the other
hand, Aldhelm seems to have used the knowledge gained at Canterbury of
"arithmetic and all the mysteries of the stars in the heavens." In
this letter from Canterbury to a West Saxon bishop, he emphasized
"intense disputation of computation"; the profound subject of the
"zodiac, the circle of twelve signs that rotates at the peak of heaven";
and "the complex reckoning of the horoscope" which "requires laborious
investigation of the expert."[52] *Horoscopus* in this context means an
instrument for reckoning astronomical phenomena, such as the sundial
moonfinder, or *horologium nocturnum* - a timetable of stellar night
movements.[53] Furthermore, there is extant in several manuscripts a
Cyclus Aldhelmi de Cursu Lunae per Signa XII Secundum Grecos, based
upon a synodical moon of 29½ days. From this cycle was generated the
lunar A-P series, which survives in four manuscripts without the table.
This is one of several efforts to determine the relation of the moon to

the zodiac: that band of heavenly space through which sun and moon pass
and which was arbitrarily described in twelve equal sections by names
of constellations (some of which are extremely difficult actually to
see). The *Cyclus Aldhelmi* was not successful and dropped out of use
very early because it was based upon the synodical rather than the
sidereal moon (27 1/3 days). Bede made this correction, while computists at Fleury and Auxerre attempted further adaptations, though none
of these tables made a lasting contribution to knowledge. This series
of efforts at observation and calculation of the moon's course could
not attain sufficient precision for lack of an adequate instrument, a
need which began to be met only in the last quarter of the tenth century with the first Latin astrolabe.[54]

     The works of Bede could supply much more evidence, of course, for
arithmetica and *astronomia,* and his works are being assessed very fruit-
fully in terms of scientific assumptions and accomplishments which have
been heretofore cited more often than understood.[55] Beyond Jarrow and
Wearmouth, however, there was a study of these essential scientific
subjects in many schools, which has been little suspected and which may
normally be traced by attention to evidence of *computistica*. To what
has, thus far, been presented may be added yet another scientific
concern which may be illustrated by a description of a fourth and
especially English version of Isidore's *De Natura Rerum*. It is concern
with the shape of the whole earth, its continents, and its discovery
with reference to the shape of the heavens.

     Cummian led us through the groundwork of Paschal questions in
Ireland, and he will also lead us to Isidore among the West Saxons as
well as to the earth as a globe.[56] There is only one surviving manuscript of Cummian's letter, MS London BL Cotton Vitellius A.XII, f.
79-83. In the binding it was given after the Robert Cotton fire of
1731 seven different books were put together, and the first of these
is itself a small collection which was transcribed in great haste
towards the end of the eleventh century, f. 4-72 (new foliation). It
contains Hrabanus' *De Computo,* a set of computistical memory verses,
Isidore's *De Natura Rerum,* with the whole enclosed by a short work by
Abbo of Fleury. At the very beginning is the *Penitential* of Egbert,
and at the end is a *Martyrology* from the West Country. This is another
instance of science within a context of piety; where it was all written,
I do not know. However, its central contents duplicate those of MS
Exeter 3507, written a century earlier in large and beautiful Anglo-Saxon

cursive, from about A. D. 960 to 986.   There are a half dozen English
manuscripts with excerpts of Hrabanus' *computus,* and there are also six
English manuscripts from the tenth to the fourteenth centuries
containing Isidore's *De Natura Rerum.*   There is, however, no overlap
between the two groups,[57] save the Exeter and Vitellius copies.

It can be shown that both Aldhelm and Bede used the short version
of Isidore.[58]   From the text and studies presented by Jacques Fontaine,[59]
we know that the short forty-six-chapter version was sent by the author
to young King Sisebut of the Visigoths in A. D. 612, and that the king
responded in a verse letter that shows how useful the work would be by
providing a systematic explanation of a solar eclipse, illustrated by
an excellent diagram.   But Isidore added another chapter in A. D. 613
concerning the sphere of the earth in three continents, and added
drawings of the *rota terrarum.*   Someone else added a chapter on the seas
and winds and elaborated parts of other chapters for the long version.

It was the last described long version which I expected to find
in Exeter and Vitellius, but this text has been elaborated even more in
dozens of places.   Moreover, this fourth version has been glossed with
lines taken from the Epistle of Sisebut but attributed to Isidore; this
implies that Aldhelm's short version of *De Natura Rerum* with the anonymous letter had been used to gloss the English fourth version.   Finally,
the name of Isidore had been set aside in favour of attribution to
Gildas, a change which occurred also in Hrabanus' *computus.*   In fact,
these works were receiving a great deal of attention and development
during the eighth, ninth, and tenth centuries – doubtless in schools
where Irish, Saxon, and Briton interacted with each other as student,
master, monk: schools such as Oswald's Gloucester, Aldhelm's Malmesbury,
or Glastonbury, where interest in Gildas flourished.   In these establish-
ments argument could and did occur about the correct computus, cycles
of sun and moon, and positions of planets through the zodiacal band.

Since A. D. 613, Isidore's *De Natura Rerum* concluded with chapter
48, *De partibus terrae,* and a *rota terrarum,* one of seven illustrated
chapters.   The *rota terrarum* is a well-known diagram and, generally,
not taken seriously by cartographers.   But its use by Isidore in the
early seventh century left us the earliest exemplars which are still
extant.   The diagram was oriented not to the north (as modern cartographers prefer) but to the east, normally.   It was the basis for the
earliest real maps in the late eighth century, maps which display
known (but very distant) islands in the seas, according to a distorting

curvature.[60] For *partes septentrionales*, this gave a visual context
for Cummian's image of *mentagrae orbis terrarum*.[61]

In the two earliest manuscripts of the fourth version, the English
version of *De Natura Rerum*, Domitian I (s.X^2), has the standard tripartite *rota*, but the Exeter 3507 (s.X ex) has a more elaborate figure
which includes three lists of place names. These names are grouped
under continent headings *Asia, Africa, Europa* - but with an unexpected
reversal: *Africa* is on the lower left, *Europa* on the lower right.
Historians of cartography have been satisfied to say that this reversal
was merely an ancient error when it occurred in manuscripts of Sallust
or Lucan, and especially of Isidore.[62] Among historians of science it
is a commonplace that medieval scribes are notorius for silly errors,
and that nothing was too gauche for Isidore. My own collations of
computus texts have turned up errors, but sometimes they may reveal a
busy mind. How could intelligent scholars accept and use a reversed
rota?

A possible explanation may be sought in astronomy, both past and
present. There have always been two perspectives on the *globe of the
heavens* and its stellar phenomena which would result in two quite
different projections of star charts and land charts.[63] If the scholar
looks up at the sky with *Asia* and its *Oceanus orientalis* at his head,
then *Europa est dextera, Africa est laeva*. On the other hand, if he
imagines himself outside the globe and looking east, he would see
Europa est laeva, Africa est dextera. Both perspectives are found in
the popular literature of Aratus and Hyginus, as well as in sophisticated planispheres of Fulda and of Salzburg, which I have elsewhere
called Hipparchan and Ptolemaic.[64] Geographers prepared *picturae* of
their terrestrial spheres, normally with *Asia* and its *Oceanus orientalis*
at the head of a chart, and the other parts were projected on the flat
surface according to either perspective. From the outer perspective
Europa est laeva, as with MS Domitian; from the inner perspective
Europa est dextera, as with MS Exeter, requiring that one keep in mind
a distinction between regions of the heavens and regions of the earth.

Thus, we may recognize two conventions for representing the heavens,
each of which had intelligible consequences. Neither the Isidorian
image nor the reversed image of the globe indicates a lack of understanding, a failure of observation, or a repetition of error. Both may
be best understood in terms of a priority of the spherical cosmos and
spherical earth.

IV

In conclusion, therefore, we have abundant evidence of piety but only scraps of science. The holy Fintan is remembered by a *Vita* which mentions Laisrén only because he was awestruck by sanctity. But the warning of Cummian to Ségéne of Iona was forceful. His argument for a new *ordo Paschae* was heeded slowly but surely by others, mostly anonymous, who learned their *calculi supputatio* on reckoning boards with Aldhelm, observed the equinox with Bede, and counted epacts and tested various *computi* until they found that the Dionysian system really did work. This was *computistica* which encompassed the cycles of sun and moon and planets for their own sake and certainly went well beyond the needs of Easter dates, in the service of Christian unity and apostolicity.

NOTES

[1] The scholarly conflict about the Alexandrian, Milanese, Roman, Victorian, and Celtic systems for determining the date for Easter - all of which flourished in the seventh century - have generated much literature in the twentieth century as well. The best historical introduction is by C. W. Jones, *Bedae Opera de Temporibus* (Cambridge, Mass., 1943), pp. 3-122, to be supplemented below.

[2] The Synod of Whitby was presented as a closed case by Bede *HE* 3.25. The monks of Iona celebrated Easter correctly (according to Bede) on the 24 April, A. D. 729, as a result of Egbert's preaching and holy life among them; Egbert who had studied in Ireland thus completed the work of the Scots Aidan, who had brought holiness from Iona to Northumbria (*HE* 5.22), and Adomnan who spread the Northumbrian teaching among most northern Irish except his own Iona.

Abbo studied in the monastery of St. Benoît-sur-Loire at Fleury, long an important centre of computistical studies. His letters to Gerald and Asper in A. D. 1003-1004 challenged Bede's acceptance of Dionysius Exiguus' tables for A. D. 532-611 and their historical assumptions, and justified a 22-year shift in the year of the Incarnation. One of his two letters on this topic was published by Pierre Varin (1849) and reprinted by A. Cordoliani, "Abbon de Fleury, Hériger de Lobbes et Gerland de Besançon, sur l'ère de l'incarnation de Denys le Petit," *Revue de l'Histoire Ecclésiastique* 44 (1949): 463-87; there are three Geralds to be distinguished. A second letter survives in five manuscripts but has not been edited.

[3] Grammar, song, and reckoning are broad fields which included those studies which were essential for monastic schools in their vocational purposes, just as the seven arts of Hellenistic instruction had been considered useful for the very different purposes of the Roman elite. Carolingian romanizers attempted to reinforce the learning of their schools by citing the categories of Roman pedagogy from Boethius, Cassiodorus, or Isidore; but the working patterns of their schools are better expressed through the practices of *grammatica, cantica, computistica*. See W. Stevens, Introduction to *Rabani De Computo* CCCM 44 (Turnhout, 1979): 165-68. This emphasis has been explored in many essays by C. W. Jones.

[4] "Tres opciones dantur tibi, Lasreane: id est, duo libri in ignem mittentur, liber veteris ordinis et noui, ut videamus, quis eorum de igne liberabitur; vel duo monachi, unus meus, alter tuus, in unam domum recludantur; et domus comburatur, et videbimus, quis ex eis evadat intactus igne. Aut eamus ad sepulcrum mortui iusti monachi, et resuscitemus eum, et indicet nobis, quo ordine debemus hoc anno Pascha celebrare." *Vita Sancti Munnu* 27, ed. C. Plummer (Oxford, 1910; rep. 1968), 2: 237.

Concerning the manuscript see P. Grosjean "Notes sur quelques sources des Antiquitates de Jacques Ussher," *AB* 77 (1959): 154-87; and W. M. Stevens, "Introduction: Manuscripts and Editions," *Rabani Mogontiacensis Episcopi De Computo* CCCM 44 (Turnhout, 1979): 194.

This event and Cummian's letter have been discussed many times: cf. B. MacCarthy, "Introduction," *The Annals of Ulster* (Dublin, 1901), 4: cxxxviii-cxlii; J. Schmid, *Die Osterfestberechnung auf den britischen Inseln vom Anfang des vierten bis zum Ende des achten Jahrhunderts* (Regensburg, 1904), pp. 37-41; D. J. O'Connell, "Easter Cycles in the Early Irish Church," *Journal of the Royal Society of Antiquaries of Ireland* 66, 1 (1936): 80-81; and C. W. Jones, *Bedae Opera de Temporibus* (Cambridge, Mass., 1943), pp. 89-98.

* [5] *Epistola Cummiani* survives in a unique transcript from the late eleventh or early twelfth century: MS London BL Cotton Vitellius A.XII, ff. 79-83. It was edited by James Ussher, *Veterum Epistolarum Hibernicarum Sylloge* (Dublin, 1632), pp. 24-35; rep. *Works* (1864), 4: 432-43; the much emended version in PL 76: 969-978 was also based on Ussher but is unreliable. See also the textual corrections by M. Esposito, "Latin Learning and Literature in Medieval Ireland," *Hermathena* 45 (1930): 240-45. A much needed critical edition and translation has been prepared by Maura Walsh O'Cróinín and should be published.

[6] No specific evidence has been presented to support various suggestions to identify the author of *Epistola Cummiani* with either Clonfert or Durrow, or with other persons of this name. There was, however, a Cummaine Fota (Cumianus or Comianus or Cumineus Longus, the "tall") who died in A. D. 662 at the age of 72 as abbot of Clonfert. He wrote a hymn, *Celebra Iuda* (and a penitential (*expl.*: "Finitus est hic liber scriptus a Comminiano"), ed. L. Bieler and D. A. Binchy, *The Irish Penitentials* (Scriptores Latini Hiberniae 5 [Dublin, 1963]: 108-35; see also pp. 5-7 and 17-19.) It is not necessary to assume that Cummaine Fota's reputation for learning was based exclusively on *Sacra Scriptura,* yet nothing reveals on his part an interest in the earlier debates about Easter or in the disciplines necessary for resolving those problems.

[7] See sketch maps 1 and 4 provided by Kathleen Hughes, *Early Christian Ireland: Introduction to the Sources* (Ithaca, 1972), pp. 111, 139.

[8] MacCarthy, *Annals of Ulster* 4: cxlii; J. K. Kenney, *Sources for the Early History of Ireland: I Ecclesiastical* (New York, 1929; rep. 1966) p. 224, n. 187; J. Ryan, *Irish Monasticism* (Dublin, 1931), pp. 198, 200, 208-9, 215.

[9] "Sed non post multum surrexit quidam paries dealbatus, traditionem seniorum servare se simulans; qui utraque non fecit unum, sed divisit, et irritum ex parte fecit quod promissum est: quem Dominus (ut spero) percutiet quoquo modo voluerit," ed. Ussher, *Works,* 4: 442.

[10] "Quid autem pravius sentiri potest de Ecclesia matre, quam si dicamus Roma errat, Hierosolyma errat, Alexandria errat, Antiochia errat, totus mundus errat: soli tantum Scoti et Britones rectum sapiunt." ed. Ussher, *Works* 4: 438-39: "What more derogatory can be thought of mother church than to say: Rome errs, Jerusalem errs, Alexandria errs, Antioch errs, the whole world errs; only Scots and Britons think rightly?"

[11]"Misimus quos novimus sapientes et humiles esse, velut natos ad
matrem, et properum iter in voluntate Dei habentes, et ad Romam urbem
aliqui ex eis venientes, tertio anno ad nos usque pervenerunt...et in
uno hospitio cum Graeco et Hebraeo, Scytha et Aegyptiaco in Ecclesia
sancti Petri simul in Pascha (in quo mense integro disjuncti sumus)
fuerunt: et ante sancta testati sunt nobis, dicentes: 'Per totum
orbem terrarum hoc Pascha, ut scimus, celebratur.'" ed. Ussher, Works
4: 442-43: "We sent men we knew to be wise and humble, as sons to a
mother; and having completed the journey successfully by God's will,
some of those who reached Rome returned to us in the third year...In
one hospice they were with a Greek and a Hebrew, a Scythian and an
Egyptian, and they were united together in the Church of St. Peter on
Easter (at which time we Scots are different by a whole month). And
before holy objects they swore this, saying: 'Throughout the whole
world this Easter is celebrated, as we know.'"

[12]"Vos considerate quae sunt conventicula quae dixi: utrum Hebraei
et Graeci et Latini et Aegyptii simul in observatione praecipuarum
solennitatum uniti; an Britonum Scotorumque particula, qui sunt pene
extremi, et, ut ita dicam, mentagrae orbis terrarum; hoc mihi judicate."
ed. Ussher, Works, 4: 436: "Consider who are the conventicles! Judge
for me whether they are the Hebrews, Greeks, Latins, and Egyptians -
all united in observance of the chief holy days, or the particle of
Britons and Scots who are almost at the ends and (I might put in this
way) are protrusions upon the world."
Mentagrae was used literally for pimples in Pliny's Historia Naturalis 26.2, but is otherwise rare; see M. Esposito, "Latin learning,"
pp. 240-45.

[13]Ed. Ussher, Works, 4: 440.

[14]Ibid., 435.

[15]The conflation of these traditions in the edition of the
Benedictines of Saint-Maur, Collectio Conciliorum Galliae (1789), was
preferred by K. J. Hefele, Conciliengeschichte 1 (1855): 201-19; French
trans. by H. Leclercq, Histoire des Conciles 1 (Paris, 1907): 275-98.
Although Hefele printed an additional six canons (beyond the usual
twenty-two) found in MS Lucca B. Capitolare 124 (s. IX) by J. D. Mansi,
Sacrorum Conciliorum Nova et Amplissima Collectio (Florence, 1759) 2:
469-74, he rejected the shorter version of Marinus' letter found there
and apparently in all surviving manuscripts with the exception of MS
Paris BN lat. 1711. Annotations by Leclercq indicate reasons to question
the authenticity of the long version of the letter as well as the
heading and portions of the first canon. The improbable reference to
DC episcopi appears only in a few manuscripts, and the alternate CC
episcopi occurs as a variant in Augustine's Contra Epistolam Parmeniani
1. 5 (PL 43: 40-41), rather than XXXIII in the usual heading and a
somewhat larger number of actual names of bishops, priests, and deacons
listed. Thus the search for Cummian's source could be narrowed somewhat.

[16]Paschal letters from metropolitan bishops: see C. W. Jones, ed.,
Bedae Opera, pp. 10-11, 17, 24-25, ff.

[16a]A. W. Haddan and W. Stubbs, eds., Councils and Ecclesiastical
Documents Relating to Great Britain and Ireland. 3 vols. (Oxford, 1869-
78), 1: 7; also Mansi, 2: 466-67.

[17]For British bishops at Arles see B. Krusch, "Die Einführung des griechischen Paschalritus im Abendlande," *Neues Archiv* 9 (1884): 167.

[17a]For the use of Patrick as an authority in the Easter controversy, see Jones, ed., *Bedae Opera,* pp. 86–87; 89; 91, n. 3; 93; 95, n. 1; 101; 109.

[18]Pseudo-Anatolius' *Canon Paschalis* or *Liber de Ratione Paschali,* ed. B. Krusch, *Studien zur christlich-mittelalterlichen Chronologie: Der 84 jährige Ostercyclus und seine Quellen* (Leipzig, 1880), pp. 316–27. Nine manuscripts of the text, whole or fragmentary, have now been identified. Authentic passages from Anatolius were given by Eusebius, *Ecclesiastical History* 7.32, and the Latin translation of A. D. 395 by *Rufinus* 8.28, cited by Bede, *De Temporum Ratione* 4 (ed. Jones, p. 211), 31 (p. 237), 42 (p. 256).

[19]Jones, ed., *Bedae Opera,* pp. 78–87.

[20]Ibid., pp. 87–89 et passim. A new attempt to sort out the several versions will be outlined in W. M. Stevens, *Catalogue of Computistical Tracts,* soon to be completed. Notice that there was a Caesarea in Mauretania to which Augustine travelled in A. D. 418 to deal with some dispute; see Gerald Bonner, "Augustine's visit to Caesarea in 418," in *Studies in Church History* 1, ed. C. W. Dugmore and C. Duggan (London, 1964) pp. 104–13.

[21]The works of Dionysius are cited from the edition by J. W. Jan, *Historia Cycli Dionysiani cum Argumentis Paschalibus* (1718; 2nd ed., Halle, 1769), rep. PL 67: 19–23, 453–520. For the *Tabula Paschalis* and *Praefatio* Jan used MS Berlin Staatsbibl. Phillipps 1830 [cat. 129] (s. IX^2) f. 2^v and 4–10^v; for the letters he used MS Oxford BL Digby 63 (x. IX^2) f.63–67 and apparently MS London BL Cotton Caligula A.XV (s.VIII^2) f. 84–86^v. Th. Mommsen collated the Digby manuscript once more with Jan's edition, and this was published by B. Krusch, *Die Entstehung unserer heutigen Zeitrechnung,* Abhandlungen der Preussischen Akademie, Jahrgang 1937, Philol.-hist. Klasse, Nr. 8 (Berlin, 1938). However, I have now identified fourteen additional manuscripts of the *Epistola ad Petronium* and twenty-three of the *Epistola ad Bonifatium,* many of these earlier than MS Digby 63. Critical editions are needed.

[22]The problem was formal, in that it avoided the Roman habit of reckoning the beginning point twice, that is once from each direction, but it was also practical in that "a year 0 is intolerable in everyday thinking," (K. Harrison, *The Framework of Anglo-Saxon History to A. D. 900* [Cambridge, 1976] p. 37.) Bede's *De Temporum Ratione* 47 was able to rebut the criticism of this solution only with difficulty, and many later scholars were stimulated to offer alternate years for the *annus Incarnationis,* e.g., Florus of Lyons, Abbo of Fleury, Marianus Scotus, Robert of Hereford, et al.

[23]*Exemplum Suggestionis Bonifatii,* ed. B. Krusch; "Ein Bericht der päpstlichen Kanzlei," in *Papsttum und Kaisertum, Festschrift P. Kehr,* ed. A. Brackman (Munich, 1926), pp. 56–58.

[24]F. Rühl (*Chronologie des Mittelalters und der Neuzeit* [Berlin, 1897], p. 171) thought that he had found Dionysian indications in Roman documents by A. D. 584, thus implying that his *Tabula Paschalis* was being used. Indictions, however, were an Alexandrian usage which had been present in Paschal tables previous to Dionysius and could be continued without his changes.

[25]Only the first nine of the *Argumenta Paschalia* are authentic, of those attributed to Dionysius in the editions of Jan and Krusch. Many of those which are authentic do not have the proper terms and reckonings for Dionysius' *annus praesens*. Many new manuscripts have been identified since 1937, but all texts have been adapted to later applications. A critical edition is needed.

[26]This explanation follows the proposals of C. W. Jones, *Bedae Opera*, pp. 93-97; cf. A. Cordoliani, "Computistes insulaires et les écrits pseudo-Alexandrins," *Bibliothèque de l'Ecole des Chartes* (Paris, 1945/1946), vol. 106 [= 107 (publ. 1948)]: 24-28. The best edition is that of B. Krusch, *Studien* (Leipzig, 1880), pp. 344-49; but thirteen manuscripts have now been identified which would allow for an improved edition and a better account of this rather murky situation. There were two Bonifaces (February to November 607, and 607/8 to 615); and the most prominent question in this letter is the Easter date VIIII Kal. Maii (=23 April), as emphasized by K. Harrison, "Luni-Solar Cycles: their Accuracy and Some Types of Usage," in *Saints, Scholars & Heroes: Festschrift in Honour of C. W. Jones*, ed. M. H. King and W. M. Stevens, 2 vols. (Collegeville, Minn., 1979), 2: 73-74.

[27]"'Scrutaminique, ut Cyrillus ait, 'quod ordinavit synodus Nicena lunas quartasdecimas omnium annorum per decemnovennalem cyclum...'" Ussher, *Works*, 4: 440. Cf. *Epistola Cyrilli de Pascha*, ed. Krusch, *Studien* p. 347.

[28]Two others items which also bear the name of Cyril were composed either in Africa, *Praefatio Cyrilli* (A. D. 482?), or in Spain, *Prologus Cyrilli* (A. D. 577-590?), to oppose Victorius of Aquitaine and to promote Alexandrian reckoning. The *Praefatio Cyrilli* cites Theophilus of Alexandria and both were originally accompanied by tables; but there is no indication that either was known to Cummian. See Krusch, *Studien*, pp. 89-98; Jones, *Bedae Opera*, pp. 38-54. A Cordolani, "Textos de Computo Español del siglo IV: El Prologus Cyrilli," *Hispania Sacra* 9 (1956): 127-39.

[29]The best available edition is not very good: L. Muratori, *Anecdota Latina* (Milano-Padua, 1713), 3: 195-96, rep. PL 139: 1357-1358, from MS Milano Bib. Ambrosiana H.150 inf. (x. IX in) f. 80-81[V]. This was reprinted in parallel with a second version from MS Tours B. mun. 334 (x. IX) f.16[V]-18 by A. Cordoliani, "Computistes insulaires..." p. 30-34. See also C. W. Jónes, "The 'Lost' Sirmond Manuscript of Bede's 'Computus,'" *EHR* 52 (1937): 216, and *Bedae Opera* (1943), p. 97.

[30]Columbanus, *Epistola ad Gregorium*, ed. G. S. M. Walker, *S. Columbani Opera* (Dublin, 1957) pp. 2-13.

[31]For details see Kenney, p. 218. The reference to Mosinu maccu Min occurs in an addition to the Gospel of St. Matthew in the Irish MS Würzburg Univ.-Bibl, M.p.th. F.61 (s. VIII[2]) f. 29. There is no support for the notion that this Irishman learned his tables from Dionysius Exiguus, proposed by Paul Grosjean, "Recherches sur les debuts de la controverse pascal chez les Celtes," *AB* 64 (1946): 215.

[32]Eugippius' version was edited from five manuscripts by P. Knoell, *Eugippii Excerpta ex Operibus S. Augustini* (Vienna, 1885), CSEL 9, 1: 425–44, no. 118: *Ex libro secundo ad eundem Ianuarium. Quid sibi velit in celebratione paschae observatio sabbati et lunae; et de quadragesima vel de consonantia pentecostses cum die legis datae.*

[33]The Herwagen edition of 1563, rep. PL 90: 647–57, is the only version in print and is inadequate.

[34]On this orthgraphy see J. Fontaine, *Isidore de Séville, Traité de la nature* (Bordeaux, 1960), pp. 94–95.

[35]Victorius' prologue and cycles were edited by B. Krusch, *Studien,* pp. 17–26 and 27–52; rep. PL *Supplementum* 3 (1963): 381–6 and 387–426. See also the discussion by Krusch, pp. 4–15; and C. W. Jones, *Bedae Opera,* pp. 61–67.

[36]The *Laterculus Victorii* was preferred by bishops in Gaul at the Synod of Orleans in A. D. 541, although it would be an exaggeration to assert that it was uniformly used in the region. It was recognized as a *cyclus recapitulans* by the Spanish monk Leo in A. D. 627 and cited in the *Vita Iohannis Abbatis Reomaensis* of seventh century Bobbio, as well as in the Irish work *De Mirabilibus Sacrae Scripturae* 2.4. See Krusch, *Studien,* p. 299; Jones, "The Victorian and Dionysiac Paschal Tables in the West," *Speculum* 9 (1935): 408–13; Harrison *Framework,* pp. 33–34, 64–65. There is no specific evidence that Roman bishops recommended the Victorian usage at any time, so far as I am aware.

[37]The verses occur in the margins of numerous computistical tracts and calendars and were printed from MS Milano B. Ambros. H. 150 inf. (s. IX in) by L. A. Muratori, *Anecdota Latini,* vol. 3; rep. PL 129: 1283; and many other editions, most recently C. W. Jones, "Carolingian Aesthetics: Why Modular Verse?" *Viator* 6 (1975): 388.

[38]Jones, "A legend of St. Pachomius," *Speculum* 17 (1943): 198–210, and *Bedae Opera* pp. 32–33; A. Cordoliani, "Les computistes insulaires," pp. 24–28. Bede's *De Temporum Ratione* 43. 67–75 quoted the passage from *Ep. Cyrilli* but did not repeat the verse.

[39]C. W. Jones, ed. *Bedae Opera,* pp. 71–72, notes that the pious attribution of Alexandrian tables to the Nicene fathers rather than to the more general *sancti patres* was perhaps an incorrect inference by many, but not a fraud by Dionysius.

[40]Ussher, *Works,* 4: 440. Cf. Dionysius, PL 67: 496.

[41]Victorian and Alexandrian tables were collated in MS Milano B. Ambrosiana H. 150 inf (ca. A. D. 805) f. 125; St. Gallen 248 (s. IX) p. 77; Köln Dombibl. 83[II] (ca. A. D. 800) f. 59[v]; Rouen B. mun. 1.49 [Cat. 524] (s. IX) f.82[v]; and others. The terms "Dionysius" and

"Victorius" came to refer generally to Alexandrian or old Roman methods. In MS Köln 83^{II} f. 211^v a table for epacts on the first day of each month in Alexandrian terms has been adjusted to the Victorian cycle.

[42]Ussher, *Works* 4: 449: "...in quo kalendas [*sic*] Januarii lunaeque ejusdem diei et initia primi mensis ipsiusque XIV. lunae recto jure ac si quodam clarissimo tramite, ignorantiae relictis tenebris, studiosis quibusque cunctis temporibus sunt adnotatae, quibus paschalis sollennitas probabiliter inveniri potest."

[43]Ussher, *Works* 4: 443: "Nefas est enim errata tua non agnoscere, et prolata certiora non approbare. Haereticorum est proprie, sententiam suam non corrigere malle perversam quam mutare defensam."

[44]J. Fontaine, *Isidore de Séville, Traité de la nature* (Bordeaux, 1960).

[45]Fontaine, ed., pp. 42-45 (additions) and 75-78 (manuscripts and citations). For Isidorian influences in Hisperic literature see M. Herren, *Hisperica Famina* I (Toronto, 1974): 11-22 and 134 (especially use of word *tollus*); the colophon of Clm 396 will be edited by Herren in vol. 2. Concerning the epilogue called *Anonymus ad Cuimnanum,* see B. Bischoff, "Eine verschollene Einteilung der Wissenschaften," *Mittelalterliche Studien* I (Stuttgart, 1966): 282-83.

[46a]In the Letter to Leuthere: see the following note.

[47]K. W. Menninger, *Number Words and Number Symbols* (Eng. trans., Cambridge, Mass., 1969), pp. 305-06, 315-31; E. Alföldi-Rosenbaum, "The Finger Calculus in Antiquity and in the Middle Ages, "*Frühmittelalterliche Studien* 5 (1971): 1-9. Aldhelm's Letters and other works were edited by R. Ehwald, MGH *Auctores Antiquissimi* 15 (Berlin, 1919); see now the English translation by Michael Herren in *Aldhelm, The Prose Works,* trans. M. Lapidge and M. Herren (Cambridge, 1979), pp. 152-53.

[48]A portion of that letter was quoted by Bede, *Historia Ecclesiastica* 3.29. Another part of the letter from Vitalinus is in MS Oxford BL Digby 63 (s. IX^2) f. 59^v; cf. Aldhelm's reference to Victorius in the Letter to Geraint (ref. in next note).

[49]Ehwald ed. pp. 483-84; Herren trans. pp. 158-59.

[50]Herren, trans., pp. 140-43.

[51]Two fragments in Welsh explain an astronomical table which shows the position of the moon in the heavens every day of the year (cf. Bede *DTR* 19): MS Cambridge Univ. Library Addit. 4543 (A. D. 930-1039?), translated by Ifor Williams, "The Computus Fragment," *BBCS* 3 (1927): 245-72.

[51a]The pagan mathematical philosophy was strongly opposed by Ambrose, Augustine, Boethius, Cassiodorus, Isidore, and almost all Christian teachers, who nevertheless continued to use number and numerical relationships as a heuristic device for elucidating levels of meaning for a literary text. Christian schools rejected Pythagorean number mysticism, however, in favour of what we recognize as basic arithmetic, as found for example in the *Calculus* of Victorius (ca. 450) and the *Arithmetica* of Boethius (ca. 500) - both very popular in monastic and cathedral schools. See further W. M. Stevens, "Compotistica et

110

Astronomica in the Fulda School,"in *Saints, Scholars and Heroes* (cited above), 2: 27-28 and notes 2-5.

[52]*Epistola* no. 1 (perhaps addressed to Leutherius of Winchester, A. D. 670-676), Ehwald, ed. pp. 475-78; Herren, trans. pp. 152-53.

[53]For Bedan observations to determine the equinox by means of *horologica inspectione,* see K. Harrison, "Easter Cycles and the Equinox in the British Isles," *ASE* 7 (1978): 1-8.

[54]C. W. Jones, *Bedae Pseudepigrapha* (Ithaca, 1939), pp. 68-70; Bede's *DTR* 19, ed. Jones, *Bedae Opera,* pp. 219 and 354; cf.also the Welsh fragment cited in n. 51 above. For the Latin astrolabe, see Marcel Destombes, "Un astrolabe carolingien et l'origine de nos chiffres arabes," *Archives Internationales d'Histoire des Sciences,* nos. 58-59 (1962) pp. 3-45. Rejection of a date before x.XII by Guy Beaujouan, "Enseignment du quadrivium," *Settimane di Studio* 19 (Spoleto, 1972): 658-62, was based upon a lack of literary citation and upon divergent speculations by Destombes and Beaujouan about the early forms of numerals. Ths Latin words on all parts of the instrument however appear to me to be in the style of tenth-century Visigothic script. I wish to express great appreciation to Professor Destombes for inviting me to handle this astrolabe in his apartment, 24 April, 1975.

[55]See especially the essays by K. Harrison cited above; and also T. R. Eckenrode, "Venerable Bede as a Scientist," *American Benedictine Review* 21 (1971): 486-507, and other essays, together with a dissertation which should be published: "Original Aspects in Venerable Bede's Tidal Theories" (Saint Louis University, 1970).

[56]Germaine Aujac, "L'image du globe terreste dans la Grèce ancienne," *Revue d'Histoire des Sciences* 27 (1974): 193-210; W. M. Stevens, "The Figure of the Earth in Isidore's 'De Natura Rerum,'" *Isis* 71 (1980): 268-77.

[57]The six English manuscripts are Exeter Cathedral 3507 (A-S minuscule post A. D. 960) f. 67-97^v; London BL Cotton Domitian I (English Caroline minuscule s.X^2) f. 3-37; Vitellius A.XII (Norman minuscule s.XI ex) f. 46-64; Oxford, BL Auct. F.3.14 (s.XII) f. 1-19^v; Auct. F.2.20 (s.XII) f. 1-16^v; St. John's College CLXXVIII (s. XIV) f. 9-37^v. A study of these is in preparation.

[58]Two verses of the *Epistula Sisebuto* were quoted in letters of Aldhelm without identification. One of them is repeated in the Exeter and Vitellius MSS as a gloss and attributed to Isidore. Bede's works draw upon Isidore's *De Natura Rerum* but not from any of the chapters which form the median or long version. Note especially Bede's *DNR* 51, "Divisio terrae," which has contents parallel with Isidore's chapter 48 but proceeds to describe the diagram in the opposite direction and does not quote or paraphrase him. Thus it is likely that Aldhelm and Bede used the short and not the long version.

[59]Fontaine ed., pp. 38-45 (n. 34 above).

[60]MS Albi Bibl. Rochegude 29 (x. VIII^2) f. 57^v of Languedoc provenance, ed. Fr. Glorie. *Itineraria et Alia Geographica* CCSL 175 (Tournhout, 1965): 467-68, facsimile. MS Vat. Lat. 6018 (s.VIII/IX) f. 63^v-64 ed. Glorie, ibid.; and *Monumenta Cartographica Vetustioris*

Aevi, A. D. 1200-1500, ed. R. Almagia and M. Destombes, vol. I: *Mappemondes* (Amsterdam, 1964), pl. XIX.

[61]In addition, we note that the statement from the Synod of Arles that Easter should be observed on the same day and at the same time *per omnem orbem* has been given by Cummian as *per totum orbem terrarum* (quoted above).

[62]Destombes, *Mappemondes* (1964), p. 37 (Sallust); G. R. Crone, "The Hereford World Map" (London, 1949), p. 6: "The draughtsman has carelessly interchanged the names of Africa and Europa."

[63]P. McGurk, "Germanici Caesaris Aratea cum Scholiis, a New Illustrated Witness from Wales," *National Library of Wales Journal 18* (1973): 197-216, esp. 200-1. Of course anyone may observe modern astronomers using the reversal of East and West in current star-charts on the assumption that one looks up at the stars.

[64]See note 56, above.

V

Sidereal Time in Anglo-Saxon England

From the manuscripts and the fragmentary materials that survive, it is possible to recognize the active scientific labors of Scots of Ireland, Britons of Wales and Cornwall, Saxons of Wessex, and Angles of Northumbria. These evidences of intellectual activities show that they were not merely occasional concerns or literary allusions. They are arithmetical calculations, geometric models of heaven and earth, complicated *argumenta* about solar and lunar cycles. The study of natural phenomena and the reckoning of their relationships were necessary for the organized life and the social relations of Christians in the British Isles, and those studies have left many manuscripts from the Anglo-Saxon period, especially medical, computistical, and astronomical tracts and tables. In this discussion the emphasis will be upon *computus* and astronomy. [1]

Some of these scientific materials survive because of Benedict Biscop's five or six trips to the markets of Gaul and of Italy and his fortunate return each time with another load of books for libraries of Wearmouth and Jarrow.[2] Other works survive because quite a few Franks and Scots visited each others' schools and traveled to Angle and Saxon schoolmasters. Aldhelm and Bede each knew students who had traveled for study to Ireland,[3] and books certainly moved in both directions along the same routes and may have been exchanged by the same persons. Almost all of the computistical sources used by Bede have been identified, and they are extremely diverse and of no single origin.[4] These and many other computistical materials known and used from the seventh through the eleventh centuries in Anglo-Saxon schools will allow a broad assessment of scientific labors in all parts of England, from which the particular study of lunar and solar cycles will permit us to identify sidereal time as a new theoretical development in astronomy prior to introduction of the astrolabe.

Computus up to the Age of Bede

In England the reckoning of dates did not begin with Bede. He was taught the Scriptures by Trumberht and probably learned *computus* from Ceolfrith, whose letter about 706 to Nechtan IV, king of the Picts,[5] explained the new Easter reckoning that had been accepted in Northumbria—

gradually by some, grudgingly by others. There were at least four computistical systems represented in 664 at the famous Synod before King
Oswiu, his son Alchfrith, king of Deira, and their host the noble abbess
Hild of Streanæshalch (Whitby).[6] Spokesmen for the different Easter tables and *computi* were the Irish Colman, the Frank Acgilbert, and the Angle Wilfrid. When the teachings of each of the three learned men are put
into their particular contexts, a rather complicated picture of scientific
knowledge emerges for their several schools.

Colman was a monk of Iona before becoming bishop of Northumbria
(661?-64) with his seat on the Lindisfarne. At Whitby he presented the
case for a *Canon paschalis* that dated the series of Easter Sundays by noting the full moon nearest vernal equinox within the limits of that
month's *luna* XIV-XX (the days of the lunar month were calculated from
the first appearance of the crescent of the new moon); and the equinox
was taken to be March 21, though earlier it had been March 25 therein, as
Bede detected. The *Canon* (referred to below as the "pseudo-Anatolian
system") elaborated a table of eighty-four years and was thought to be in
accord with teachings of Anatolius, supposed by Colman to have been a
disciple of the apostle John. The actual Anatolius had been a teacher in
Alexandria and became bishop of Laodicea (269-80).[7] Colman's practice
had been brought to Northumbria from Iona by Aidan and was known
also in parts of northern Ireland, Scotland, and Wales. Other Irish centers
nevertheless were using the Alexandrian or the relatively similar Victurian paschal tables.[8]

Another notable personage in the presence of King Oswiu, Alchfrith,
and Hild at Whitby was Acgilbert, who wielded considerable influence.
Acgilbert had been raised in a noble Frankish household in Gaul near
Soissons before coming about 635-40 to study in southern Ireland, possibly at Ráth Maélsigi. We lose track of him for ten years, but during some
part of the period from 650 to 664 he served as a bishop in Wessex (wandering without a seat?) and later as a bishop in Paris from 667/68 until his
death about 680. Though his life and pastoral concerns were completed in
Paris, this bishop helped his cousin Adon to found the abbey of Jouarre
near Fleury and Orléans about 635, and his sister Teodechildis was abbess
of Jouarre, where Acgilbert was buried. Furthermore, his cousin Adon or
Audoenus (St. Ouen) was bishop of Rouen, and his nephew Leuthere succeeded him as bishop of Wessex.[9] Many tracts and tables were available to
a Frankish bishop with such wide-ranging experience on the Continent,
southern Ireland, and southern England. But what Easter table did this
family and this bishop use?

It should be assumed that Acgilbert's Easter practices were in accord
with his circumstances, unless there is supporting evidence to the contrary. But what were these circumstances? During the sixth and seventh
centuries some Frankish bishops in council had accepted the explanations
and tables created about 457 by the calculator Victurius of Aquitaine,[10]

though by the eighth century several were trying out the Dionysian tables.[11] What the Frankish student could have studied at Ráth Maélsigi was the great diversity of computistical documents found in the Sirmond group of manuscripts, which contained many texts supporting the Alexandrian cycle and included selections from both Victurian and Dionysian applications of that nineteen-year cycle and their assumptions for coordinating lunar and solar cycles. But at the present there seems to be no evidence for Acgilbert's practice in any of the places he served. This is unfortunate, for it was he whom Oswiu addressed first and who deferred to Wilfrid at Whitby.

Wilfrid had been ordained priest by Acgilbert, and then after the Synod of Whitby he was sponsored by King Alchfrith of Deira, who sent him to Acgilbert for consecration as bishop of York, though he was never able to exercise that office. The views expressed by Wilfrid at Whitby on the dating of Easter had doubtless been rehearsed with Acgilbert before addressing the Synod. His argument was a good account of the basic Alexandrian terms of reference for a nineteen-year cycle with *luna* XIV-XXI as limits for the Easter moon, with the Resurrection celebrated on the Sunday following. Both Bede and the *Vita sancti Wilfridi* have also described how cleverly he trapped Colman into the unanswerable question of whether Peter holds the keys to the kingdom of heaven. Oswiu drew the conclusion, rather flippantly, that therefore the Roman bishop must surely know the right date for Easter, smiling as he did so, according to the author of the *Vita*.[12] But does any historian know what system was then used in Rome? About sixty years after Whitby, Bede could argue that the Dionysian system was best, with its eight columns of data for ninety-five years, lunar limits of XV-XXI for the Easter moon, and he could imply that it was this which Wilfrid had meant by his appeal to Rome. But that implication is hard to credit, when the evidence is assessed for the Roman practice at that time.

During the previous century Roman bishops were defending the use of the Victurian tables in Irish schools.[13] When Columbanus brought the Bangor *computus* to Gaul for use in his new foundations of Annegray, Luxeuil, and Bobbio, he found that the Frankish bishops expected him to follow the Victurian *computus*. He could object to Gregory the Great forcefully in 599/600 that his northern Irish school had tested the Victurian cycles and rejected them, but he was not going to receive any support from Rome.[14] A century later and over thirty-seven years after Whitby, Bede's friend Huaetbercte was probably one of those monks of Wearmouth and Jarrow who traveled to Rome and observed at Christmas 701 in St. Peter's Basilica that the great candle bore an inscription proclaiming the day not in terms of the Dionysian *aera Incarnationis* but in terms of an *aera Passionis*.[15] From the second to fifth centuries there were quartodeciman and other early forms of reckoning from the year of the Passion being used in Christian communities, but at the beginning of the eighth century such a practice in Rome probably required the system of Victurius, which observed *luna* XVI-XXII for the Easter moon during a ninety-five-year cycle.

Thus Acgilbert's withdrawal from the Whitby debate on grounds of inability to express himself in local language and Wilfrid's appeal to Rome may not be as informative as historians have hoped. These two must have agreed to oppose the pseudo-Anatolian system of the Irish Colman and to promote Alexandrian paschal usages, but that is not sufficient to inform us whether each one preferred (1) the Victurian tables prevalent in Gaul, (2) the Victurian form still considered viable in some schools of Ireland though rejected in Bangor, or (3) the Dionysian form then being studied in some schools of Gaul and southern Ireland and preferred in England by students of Bede, though possibly not yet either in Canterbury or Whitby.

Major differences between these four systems of computing—the pseudo-Anatolian, the Alexandrian nineteen-year cycle, and the Victurian and Dionysian systems based upon the Alexandrian cycle—were well known in Anglo-Saxon England, and they have been clarified by modern scholarship. Each was difficult and required long instruction and practice to understand. The choice between them cannot have been made on literary grounds. It is necessary to recognize many schools at work behind the personalities named at Whitby, each school or group with its own network of communications and influence. As Ceolfrith said to Nechtan, there was a crowd of calculators who "were able to continue the Easter cycles and keep the sequences of sun, moon, month, and week in the same order as before."[16]

That "crowd of calculators" for each paschal system had to be able to keep the bissextile or extra solar day every fourth year, a task that is actually not so easy to do as it sounds. They must know how and when to omit the one day for *saltus lunae* at the correct time during any nineteen-year cycle (and the *saltus* varies for each system) in order to keep a calendar in close accord with the actual course of sun and moon. Therefore they must know well how the sun moves relative to the stars and how the circle of the sun itself moves to extremes, the solstices, as that circle tilts about 23 1/2 degrees to the north and 23 1/2 degrees to the south of the celestial equator. They ought to be able to explain how the solar cycle twice annually passes a midpoint, the equinoxes, on each of which it may be observed anywhere on the face of the earth that the hours of light and the hours of darkness are the same. An understanding of Hellenistic astronomy as taught in Anglo-Saxon schools would allow anyone to travel safely for thousands of kilometers across continents of land or desert spaces of sea and to return home safely by other routes and without confusion.[17] At Jarrow a great monastic scholar could carefully observe, count, and measure the lunar cycle on the zodiacal scale and draw the relations between it and the mysterious motions of the sea to create a scientific theory of tides[18] that remains to this day the basis for British Admiralty Tide Tables, used by all harbor pilots in the world without acknowledging their debt to the work of the Venerable Bede.[19] These several paschal systems were each set within a broad range of Hellenistic science, and some of them brought forth new scientific knowledge of considerable significance.

The Need for Sidereal Reckoning

Yet another bit of ancient scientific research in astronomy from the schools of Anglo-Saxon England is the measurement of time, both synodical time and sidereal time. In the Hellenistic model of the heavens, lunar and solar cycles were usually correlated in terms of the number of solar days between recurring appearances of the moon above the earth's horizon. As the first sliver of the new moon may be observed sometimes after 29, sometimes after 30 days, and usually in the regular alternation of 29- and 30-day periods, the mean lunar month was taken as 29 1/2 solar days in synodical time. Twelve such lunar months, however, results in only 354 days and does not complete a solar year. For calendar purposes, therefore, the position of the moon in synodical time was accounted for annually as 365 less 354 or 11 days, called *epactae*; after the second year, 11 more days would accumulate, making up 22 epacts, and so forth. Further adjustments are also made in keeping track of epacts in the Alexandrian nineteen-year cycle. This synodical time was assumed in all four of the calendar systems discussed above.

Synodical time, however, did not allow computists to reckon lunar positions with the accuracy many of them desired. Therefore, sidereal time was introduced in order to coordinate the lunar cycle not with the terrestrial horizon but with the position of the stars. This position was expressed in terms of the zodiacal scale, whether or not the moon was visible at the time. On this scale, the lunar month in sidereal time was discovered to be 27 solar days and 8 hours. Computistical tracts will often discuss the Hellenistic model of the heavens and its applications in greater detail and complexity than this, and they give many sets of *argumenta* that explain the problems of synodical and sidereal time, along with procedures for solving them. Therefore, *computus* manuscripts often gather further tracts devoted exclusively to basic astronomy for study in Anglo-Saxon schools.

In three of his works, Aldhelm (640?–709/10) listed the major categories of knowledge, the *disciplinae philosophorum*, to be *physica*, *ethica*, and *logica*. The first of those categories included seven *fisicae artes*, which he gave as *arithmetica*, *geometrica*, *musica*, *astronomia*, *astrologia*, *mechanica*, and *medicina*. All seven terms are found and defined in the several works of Isidore of Seville that were known to Aldhelm. *Geometria* concerns measure of lengths and the terms of measurement with reference to the earth's surface — it sounds something like civil engineering; *astrologia* pertains to the rising, falling, and visibility of the stars and their courses through the heavens as they circle the earth, while *astronomia* accounts for the movements of planets and their periods of orbit. Although these terms had been common to Roman *mathematici* and *magi* (that is, astrologers and magicians), they do not suggest any number mysticism or astral influences when mentioned by either Isidore or Aldhelm, who stay with descriptive accounts of stellar phenomena.[20]

When he went to study at the Canterbury abbey of Saints Peter and Paul for the first time about 670-72, the mature Aldhelm must have received additional instruction "in the art of meter, astronomy and *computus*."[21] Although he may already have been elected abbot of Malmesbury in 670, the arithmetic that he had learned and practiced there may not have given immediate satisfaction to his master Hadrian, newly arrived at Canterbury. One of Aldhelm's complaints was about those *partes numeri* to be carried over in addition, subtraction, multiplication, or division, which could be kept in mind only with difficulty and when "sustained by heavenly grace." But he must have learned well, for the *computus* was a subject he pursued further. As a student writing to his West Saxon bishop, he emphasized the "intense disputation of computation" at the Canterbury school, the profound subject of the "Zodiac, the circle of twelve signs that rotates at the peak of heaven," and "the complex reckoning of the horoscope," which he said "requires of the expert laborious investigation."[22] *Horoscopus* in this context means an instrument for reckoning *astronomica phenomena*, perhaps a sundial or a table of data recording stellar motions.

Did Aldhelm pursue this sort of study of arithmetic and of astronomical phenomena later in his life? He seems to have known the letter sent by Bishop Vitalianus of Rome (657-72) to King Oswiu of Northumbria before the Council of Whitby.[23] He heard the appeal of Theodore of Canterbury "that we all keep Easter Day at the same time, namely on the Sunday after the fourteenth day of the moon of the first month," and he appears to have carried out the directions of the Synod of Hertford about this question by writing to Geraint, king of Devon and Cornwall, and to his bishops, who were using an eighty-four-year cycle with different lunar limits for setting the times in order (probably *luna* XIV-XX) and who were willing to observe Easter Sunday on the fourteenth day of the moon.[24] From this evidence one should not expect from Aldhelm a profound knowledge of Hellenistic astronomy.[25] But after studying with Hadrian he must have exercised his school astronomy to some extent before becoming bishop of Sherborne in 705, for there is extant in several ninth-century manuscripts a *Cyclus Aldhelmi de cursu lunae per signa XII secundum grecos*.[26] The earliest transcription of the *Cyclus Aldhelmi* that survives was written between 836 and 848 in MS Karlsruhe Landesbibliothek Aug. 167, perhaps at Peronna Scottorum.[27] The purpose of this table of data is to determine the relation of the moon to the stars throughout the nineteen-year Easter cycle by use of the zodiac, and this relation is based upon a synodical lunar period of 29 1/2 days. There are nineteen vertical columns and thirty horizontal rows. The rows are designated by fifteen letters of the alphabet alternated with blanks; year 2 begins with a blank space on the top row and continues with letter L; epacts are given at the foot of each column for use at commencement of each successive year. This cycle thus generated the lunar A-P series, which survives in many calendars without the table.[28] Several other efforts were made in Continental schools to achieve the same goal, and none were successful if they were based upon the synodical period of 29 days and 12 hours, rather than upon the sidereal period of 27 days and 8 hours.[29] Thus the *Cyclus Aldhelmi* and its synodical time adjustments dropped out of use very early in most Anglo-Saxon schools. Bede, however, recognized the problem of synodical time and proposed to use sidereal time.

Astronomia in the Age of Bede

The works of Bede could supply much evidence for the practice of *astronomia*. In *De natura rerum liber* (hereafter cited as *DNR*) written about A.D. 701, he explained the axis of the cosmos and its poles as theoretical constructs; that is, the true North Pole is near to the pole star but must be located geometrically.[30] He described the five-zone model of the heavens and its projection upon the earth with equator, two tropics, and two arctic circles.[31] All constellations and planets of course were seen to pass daily from east to west.[32] Thus he could describe motions of sun and moon in detail and gave a classic description of both lunar and solar eclipses.[33]

For anyone seeking the date of Easter, the west to east motions of planets other than sun and moon against the stellar background would be of no importance at all. Nevertheless, Bede informed his students and his readers of the long periods required by each planet to complete its west to east circuit, along with its much shorter periods of visibility.[34] He used the scale of the zodiacal band of *ccclx partes* (360 degrees) to track these longitudinal motions, but he also used its breadth of *xii partes* for an account of latitudinal *apsidae* (apogees and perigees) for the planetary orbits on that oblique band of stellar space.[35] Writing in 701, he seems to have assumed with Pliny the Elder that the stars are brighter or dimmer because of their various distances from the earth[36] and that planets move about the earth on eccentric circles.[37] But from Isidore he added a description of retrograde motion and stations for some planets,[38] implying an epicyclic theory of planetary motions that may not be consistent with eccentric circles without some mechanism for reconciliation, but he offered no explanation. Bede also noted the maximum elongations of the orbits of Venus and Mercury relative to the sun,[39] but he did not mention the model of planetary cycles by which those two planets could be thought to circle the sun, while the three of them circled the earth together as a subsystem—a model that became very popular in ninth-century Frankish schools and that came to Wales from Limoges, perhaps in the tenth century.[40]

With some of Bede's astronomy thus in mind, we may turn again to the coordination of lunar cycles with solar cycles by use of the zodiacal band. In *DNR* XXI, "Argumentum de cursu lunae per signa," Bede had accepted from Pliny a formula for the moon's circuit through the zodiac thirteen times during twelve lunar months. By the use of Roman fractions, he found that the moon would travel through one sign of the zodiac in two days, six and two-thirds hours. He considered that he could simplify the operation of finding the number of signs of the zodiac through which the moon had progressed by using a ratio of 4:9. Thus he would multiply the number of lunar days that had passed from the starting point by four and then divide by nine, in order to find the number of signs and *partes* through which the moon had progressed from the stellar starting point on the zodiacal scale.[41] But even at the time of writing in 701, he recognized that this solution was not adequate for the purpose.

Coordination of lunar and solar cycles was becoming known in the schools through excerpts of Pliny and also through attention to the writings of Ambrose and Isidore. There were young men at Malmesbury and at Sherborne during Aldhelm's lifetime who were trained in the disciplines of *arithmetica*, *astronomia*, and *astrologia*, such as Pechthelm and possibly Pleguina, both of whom later came north,[42] and they could well have brought Aldhelm's new table with them to Hexham and Jarrow. How would Bede respond to it?

Early in the period 721-25 when his *De temporum ratione* was being composed, Bede had become more careful about his reckonings and insistent about the necessity of avoiding miscalculations concerning the course of the sun and the moon through the zodiac and its twelve signs (chapter XVI, lines 23-25). The turning of the spherical ball of the heavens can be followed by taking the circuit of twelve distinct and large signs, as divided each into 30 separate and small parts (lines 25-27), since the sun gives light for 30 days in each sign. When this is observed for a full 24 hours, however, the sun's course will have actually run 10 1/2 hours more per sign than you might have expected, but you do not want to include those several 10 1/2s until the sun has passed through all twelve signs and a total of 126 hours or 5 1/4 days have accumulated. Then you add the 30 parts per sign and the extra 10 half-hours per sign twelve times, and you get 365 1/4 days for the sun's course (lines 27-33). This is an approximation that will be used to coordinate 365 1/4 days with 360 parts of the zodiac in one year.

Bede then proceeded to describe something about each of the twelve signs (lines 33-64). But he also returned to discuss their movement in quantifying terms and introduced the significance of sidereal months (lines 65-79). He noted that the sun's course through the total zodiac requires 365 days and a quarter (or 6 hours), while the moon takes only 27 days and one-third (or 8 hours). More precisely, the sun passes through one sign in 30 days and 10 1/2 hours, while the moon needs only 2 days and 6 2/3 hours per sign (lines 68-69).

There are those who assert that the moon takes 30 days to complete the same course that the sun completes in 365 days, but they err if they do that, according to Bede (lines 74-75). This general notion could have been found in Ambrose or Isidore, for example,[43] but he could have also been referring to the *Cyclus Aldhelmi*, which attempted to apply that relationship. He also warns in this case against using the data for synodical lunar months of 29 or 30 days, which only coordinate the appearance of sun and moon without reference to the stars, for it is the more exact sidereal data that are gained by reference to the stars and applied on the zodiacal scale (lines 76-79). He reaffirmed the lunar sidereal cycle of 27 1/3 days, which is obviously nearer to the fraction 1/13 than to 1/12 of a solar cycle, and actually the solar cycle would correspond with about 13 1/3 lunar sidereal cycles (lines 71-74). If this scientific effort to account for the sun's movement relative to the stars still seems unclear, that is because it is indeed confusing. It was an explanatory effort that failed. Happily Bede did not give up his efforts as a teacher and scholar of applied astronomy at this point.

After some further approximate guidelines for tracking the sun and moon across the face of the heavens, Bede finally offered in chapter XIX the *pagina regularum* and its explanation[44] as an improvement upon the *Cyclus Aldhelmi*. It guides the user by the letter series A-O, alternating fourteen letter spaces with thirteen blank spaces, coordinated with twelve zodiacal signs on the left and twelve solar months on the right. This table or cycle was offered with Bede's rather plain words that it had been created *ad capacitatem ingenioli*, and indeed it was surely an improvement for those like myself and many of my readers who may feel unsure of their ability to follow those previous calculations. Bede's *pagina regularum* became a very popular table and seems to have been taken into many other *computi* written during the next four centuries. But as Bede himself pointed out, this too is inadequate; and he invited others to try their hands at it and bring forth a better rule.[45]

Manuscript Evidence for Anglo-Saxon Computistical Activity

Unfortunately, there is in England no longer a single copy of any of Bede's scientific works written before the end of the eleventh century or the beginning of the twelfth.[46] Norse invasions and the wars of Alfred and many others were not good for schools and libraries. But probably more destructive than enemy attack were the many fires that resulted from ordinary human work, accident, or carelessness. Yet many coals are still glowing amongst the ashes, and much science survives in Anglo-Saxon manuscripts. In 1981 Helmut Gneuss included, in his "Preliminary List of Manuscripts," not only manuscripts containing Anglo-Saxon texts and glosses but also Latin manuscripts, fragments, and charters, written or imported, with practically no exclusions.[47] Because of the cooperation of a great number of scholars, this work was able to rely less upon older catalogs and more upon the direct studies of individual scholars concerning scripts, contents, and codicological detail that reveal dating and provenance than had Neil R. Ker's 1941 *Medieval Libraries* (revised 1964, 1987) and his 1957 *Catalogue of Manuscripts*.[48] As a result, the number of books that are known to have been available for study in Anglo-Saxon England has grown spectacularly. Supposing (wrongly) that Ker's 1957 list of manuscripts were representative, one could have counted only 37 books containing Anglo-Saxon texts or glosses that may have been written or used in Canterbury. From Gneuss's list in 1981, however, one may locate 190 texts known during that period at Canterbury: about 90 from the library of Christchurch and about 100 from the monastic library of Saints Peter and Paul (later called Saint Augustine's). With only slightly better indication of contents, it may easily be seen that the theological tracts expected in that library have been joined by 23 books in the scientific categories of seven *fisicae artes* named by Aldhelm. While some of these surviving manuscripts are fragmentary, the texts contained in many of them are multiple in number, diverse in content, and often very rich in scientific interest.

In the schools of other parts of England many and diverse works of *computus* and astronomy could reveal further studies that developed better understanding of natural phenomena in our period. Only a few examples are necessary to illustrate the breadth and depth of scientific studies in Anglo-Saxon England. The first 39 folios of MS Oxford BL Digby 63 contain computistical materials apparently written during 844-55 but transcribed about 867 (*annus praesens*, fol. 20v) by the Frankish scribe Raegenbold, who brought them with him to England. There he transcribed other computistical tracts from both south and north of the Humber estuary during successive periods of his life until his final entry at the end of the ninth century.

The first eight folios of this booklet supply Dionysian Easter tables with eight columns of data from the third year of a nineteen-year cycle in 229 through thirty-three more cycles to the year 872. This was followed by a *Computatio Grecorum*, a series of computistical *argumenta* known as *Lectiones*, a set of *Versus computistica* perhaps from Verona and Saint Gallen, and a *Tractatus de solesticia*. None of these contents were known in Britain in another manuscript before the middle of the tenth century, so far as I know, and it should be assumed that Raegenbold brought folios 1-39v with him from the continent.[49] But the last folio or two of his book may have been blank when he finished. A table is found on folios 38v-39v that is headed *Transitus lune per signa* and is based upon the sidereal lunar month of 27 days and 8 hours. Until I am able to inspect this manuscript again, I could not attempt to say whether Raegenbold also brought the sidereal table with him from the continent or found and added it in England.

Folios 40-48, however, contain a calendar whose origins have attracted considerable interest; historians may wonder whether it derives from Northumbria and came into the southern area of Wessex during the reign of King Alfred.[50] I do not wish to intervene in that discussion but should note at least that the calendar page for January concludes with "Nox horam XVI die hora VIII" (*sic*), which must indicate a southern latitude, but the page for December concludes with "Nox hora XVIII die hora VI," which could correspond with a northern latitude such as Aberdeen. One might suppose that the writer began in the south and migrated north, but he also might have adapted some of the latitudes of a calendar to the lunar table on folios 46-48 without traveling at all. Thus it cannot prima facie be taken to indicate origin of the codex or even of this part.

The third booklet in MS Digby 63 comprises folios 49-71 and contains some of the African and Spanish computistical materials that seem to have passed through Ireland (Ráth Maélsigi?) to the library of Jarrow and that were used by Bede.[51] They include *Acta synodi*, version III; letters of Pascasinus, Proterius, and Cyril; two letters of Dionysius Exiguus; *Argumentum de nativitate: Querenda est nobis nativitas lune XIIII* ... ; the African letter of Felix and the related *Prologus Sancti Felicis Chyllitani*, also known as "Successor Dionisi"; along with other items of interest. But there is also a part of the letter from Vitalianus of Rome to Oswiu of Northumbria, omitted by Bede and apparently unknown to him. These tracts are not in the same order as they appear with other *computi* in the so-called Sirmond manuscript, Oxford BL Bodley 309 (s. XI 2 Vendôme), folios 82v-90 and 94v-95v, which has been called the computistical library of Bede, nor were they copied in the same order as they appear in the related MS London British Library Cotton Caligula A.XV. Charles W. Jones therefore could speculate that MS Digby 63 represented a separate line of transmission from Ireland to Canterbury. Unfortunately the evidence cannot settle the question of whether the writer had gone north or the calendar and *computi* had come to him in the south of England from Ireland.

In 892 Raegenbold was still working on these materials and decided to quit. Thus he wrote on folio 71:

Finit praestante. Finit liber de conputacio.
Raegenbold[us].Sacerdos. [... ... ...]
Scripsit.Istum.Libellum. et q[u]icumque legit.
Semper pro illum orat. // //! // : —

Nevertheless, the same writer added yet a fourth booklet on folios 72-83. He copied the nine authentic Dionysian *argumenta*, as well as three more from other writers; a *Calculatio quomodo . . .* ; the *Disputatio Morini*; an entry associated with Cummian; the beginning of a *Tractatus de celebratione paschae* (completed, however, by a second writer on folios 83-87); and other items that were also scattered through either MS Bodley 309 or MS Caligula A.XV. By 1000 this set of four booklets in 83 plus 4 folios was being used at Winchester, where several words were erased after Raegenbold's name and *de Pentonia* was added, as was the name of Baernini, who asked the reader to pray for him and wrote a few more words on folio 71 that are now illegible.[52] It was quite a respectable collection of computistical literature and was much studied.

Another important manuscript for the *fisicae artes* in Anglo-Saxon England was the Exeter Cathedral MS 3507, written about 960-86 in a beautiful English square minuscule.[53] This codex contains two significant works, Isidore's *De natura rerum* and Hraban's *De computo*, along with a collection of computistical verses[54] and geographical texts of interest for scientific instruction in the schools. A complete transcription of these materials was made toward the end of the eleventh century in MS London BL Cotton Vitellius A.XII, folios 10v-64, not taken directly from the MS Exeter 3507 but from a common exemplar no longer extant. Analysis of the text of Isidore's schoolbook *De natura rerum* in the Exeter and Vitellius manuscripts has shown that this is not the "recension longue" expected from the theories of three versions and their diffusion offered by Jacques Fontaine in his excellent critical edition of Isidore's work.[55] Rather it is a more elaborate version than has been found anywhere else. Especially notable for Anglo-Saxon studies is the earliest copy of this English version of Isidore's schoolbook in an Insular Carolingian script from mid–tenth century, MS London British Library Cotton Domitian I; its writer shared the same exemplar soon to be used for MS Exeter and then later for MS Vitellius. Further comparison with the manuscripts used by Fontaine also had interesting results, which may only be summarized here. It can be demonstrated that Bede used the earliest and shortest version prepared by Isidore in Seville in 612, whereas Aldhelm used a version in the next stage of development that had been augmented in 613 by Isidore with information from Pliny's *Historia naturalis* and that also contained astronomical verses from which Aldhelm selected a few lines. The verses have been identified in a letter from Isidore's student to whom the work had been addressed, the young Visigothic king Sisebuto.[56] The versified epistle explained both lunar and solar eclipses and included an excellent diagram of these phenomena of a type still used by astronomical societies and their publications.[57] Although Fontaine speculated that this "recension longue" could have been composed in Ireland and passed through southern England or moved directly to Northumbria, there is no evidence to connect it with either Ireland or Northumbria at all. No manuscripts survive for Fontaine's putative "recension moyen," which Aldhelm could have quoted. What he used, however, would have been present in the "recension longue" manuscripts that were written in Anglo-Saxon minuscule or in Fulda Insular minuscule during the eighth century.[58] None of these show any relationship with Northumbria or Bede. Thus we have identified four different versions of an important schoolbook for basic instruction in the understanding of natural phenomena, versions that were being used by Angles and Saxons in several parts of England during the eighth, ninth, tenth, and eleventh centuries.

The *computus* of Hraban, written at Fulda during 819 and 820, was fairly well known in Carolingian schools. Although only two copies survive in England, in the Exeter 3507 and Vitellius A.XII manuscripts, two more copies are of English origin: the MS Avranches Bibliothèque Municipale 114 (post-1128), folios 98-132, perhaps from Mont-Saint-Michel, and MS Firenze Biblioteca Medicea Riccardiana 885 (s. XIV in), folios 312-46, with the Prologus and partial capitula repeated on folio 349; both are glossed like the earlier Exeter and Vitellius copies.[59] There are also excerpts from Hraban's *computus* in marginal glosses to a calendar in MS Oxford Saint John's College Library XVII (1102-10), folios 16-22, copied at Ramsey Abbey for Thorney Abbey, and in the Cerne Abbey MS Cambridge Trinity College O.2.45 (post-1248), pages 19-20, 182-86.

However, Hraban had not explained the difference between synodical and sidereal lunar months. He did select many of the problems and the calculations set out by Bede's *De temporibus ratione* XVII and XVIII, but he passed over all references to sidereal time and omitted the *pagina regularum* and its practical applications. This aspect of Bedan science was not furthered by the books in MS Exeter 3507 and apparently not by the school and library of Leofric at Exeter, though it was taught in other schools of tenth- and eleventh-century England.

Tenth to Eleventh Centuries

Communications between the several parts of Great Britain and the several parts of Francia were lively during the tenth and eleventh centuries, and several travelers from the Continent came north by invitation of Oswald of York. One of them was Abbo (ca. 945–1004), who had known both Oswald and Germanus before 985 when they were unwilling *peregrini* in the abbey of Saint Benoît-sur-Loire at Fleury near Orléans. There Abbo taught with particular attention to mathematical and scientific subjects. At Ramsey Abbey in Lincolnshire near East Anglia and elsewhere in England from 986 to 988, he wrote *Quaestiones grammaticales* and four *Sententiae*. Both works are found in English manuscripts from our period.[60] For example, the exemplar from which MS Exeter 3507 had been copied was later wrapped by several more bifolia before being copied again in southern England over a century later into MS Vitellius A.XII (s. XI ex), folios 10v-64; the wrapper was transcribed into MS Vitellius, folios 8-10v and 64-64v. Those bifolia contained the *Sententiae* of Abbo: "De differentia circuli et sperae," 'De cursu septem planetarum per zodiacum circulum" (with figure), "Signa in quibus singuli planetarum morentur scies," and "De duplici signorum ortu vel occasu."[61]

It may also have been Abbo who introduced to English masters and students another schoolbook, the *De computo lunae* or *Ars calculatoria*. Probably written or at least completed about 883 by Heiric or Helperic (ca. 841-903) in the Abbey of Saint Germain at Auxerre and presented about 900 to Asper, *dominus* of Grandval, this *Ars calculatoria* was much used in Anglo-Saxon schools from the tenth century onward. Two English manuscripts used *annus praesens* 975, and the year 978 was given in ten more; the latter date probably stems from Abbo's own copy from Fleury.[62] In his chapter XVI: "Cur nunc XXX, nunc XXIX pronuncietur, vel quantum in unoquoque signo, vel in toto zodiaco moretur," that author included his own discussion of the sidereal lunar month of twenty-seven days and eight hours as the moon moved through the zodiac. His intention was to explain how there could be thirteen lunar months during approximately the course of twelve solar months, rather than to provide Bede's quantifying explanation of the moon's observed annual progress relative to the starry background.[63] He did not provide or discuss the *pagina regularum* or its A-O series. Thus a student was introduced to the concepts of sidereal time by Helperic's *computus* but was not carried forward to understand their usefulness.

Having been introduced to Anglo-Saxon England in 986, the *Ars calculatoria* was actively studied in its several regions and continued to be used during two subsequent centuries. *Anni praesentes* of 900, 903, 975, 977, 978, 980, 990, 994, and so on to 1155 are variants that indicate that scribes were not merely copying a text but were putting its contents to current applications. There are several versions of the text, and the manuscripts provide quite diverse series of final chapters and explicits. No careful analysis of contents has yet distinguished these versions properly.[64]

Abbo taught arithmetic from the *Calculus* of Victurius of Aquitaine and wrote a commentary on its prologue,[65] and he used the writings of Dionysius Exiguus and Bede heavily in works that were copied into MS Berlin Staatsbibliothek Phillipps 1833 [Cat. 138] (s. X ex, Fleury), folios 1-58. He was one of the earliest scholars who proposed to revise the *aera Incarnationis*, and he identified several reasons for doing so. When he projected the ninety-five-year Dionysian Easter tables from year 531 back to their beginning, Abbo did not pass over from +1 to −1 in the Roman style, as had Dionysius, but used the nul point to enumerate the year of the Incarnation. Adjustments to the Bedan calendar that this required for three nineteen-year cycles, together with two of Abbo's letters explaining them, survive in MS Vat. Regin. lat. 1281 (s. XII), folios 1-3, from the school at Hereford.

He did not accept the Dionysian A.D. XXXIII as the putative year of Christ's death. He had noticed that the death in the year DXXVIIII of Dionysius's contemporary, Benedict of Nursia, was reported to be XII Kal. Aprilis, feria VII (= 21 March, Saturday); unfortunately that weekday and that day of the month would not correspond with DXXVIIII but required year DVIIII in the Dionysian cycle, a difference of twenty years. If this difference were applied to the Passion, its year should have been given as A.D. XIII rather than XXXIII in that cycle. Therefore Abbo implied that Bede must have erred by accepting Dionysian chronology without correction.

On the other hand, Abbo took Bede's approach to sidereal time quite seriously. In his *Computus vulgaris qui dicitur ephemerida*, Abbo devoted ten paragraphs to the sidereal lunar month of twenty-seven days and eight hours, the *pagina regularum* with its letter series A-O for twenty-seven lunar days, and especially problems of continuing from O to A for only eight hours, or one-third of the twenty-eighth day, in order to track the moon's course relative to the stars.[66] Sidereal time provided astronomical justification for Abbo's corrections of epacts, concurrents, and lunar regulars in the Bedan calendar.

One of Abbo's English students at Ramsey Abbey, Byrhtferth, undertook to prepare an *Enchiridion* or *Handboc*, which survives in MS Oxford Bodleian Library Ashmole 328 (s. XI med), pages 1-247, in four books.[67] This may have been a useful schoolbook because it was written in the form of commentaries upon works of Bede, Hraban, and Heiric/ Helperic — often quoting or epitomizing a passage in Latin and then translating or summarizing in English. There are also passages similar to parts of the *computi* by Abbo and Ælfric. This work contains not only computistical materials but also various bits of natural history, grammar, rhetoric, and exhortations against worldly evils, but the surviving portions of the text have omitted any reference to the sidereal lunar months and year, even the leading query by Helperic. Thus, despite its positive values for school teaching, it is a disappointing work for modern research, as there are really no advances in the level of scientific learning, and the sources were more sophisticated than the teacher who was making use of them. The *Enchiridion* together with its sources, however, represents the development of a substantial library of mathematical, astronomical, and specifically computistical texts in the school at Ramsey Abbey after Abbo left in 988 and through the entire eleventh century. Those many tracts and tables gathered for study at Ramsey were then copied into MS Oxford St. John's College XVII about 1102 to 1110 and must have been intended to stimulate further such studies in the school of Thorney Abbey.[68]

Another *computus* that alternates Latin with English materials is Ælfric's *De anni temporibus*. Ælfric (ca. 955–1020/25) apparently came from the North, entered the monastic school at Winchester under Æthelwold (abbot ca. 971-1005), and then lived at Cerne Abbey, where this short treatise was written between 992 and 995. He discussed divisions of the year along with a bit of astronomy and other natural phenomena by quoting passages from the two *computi* of Bede, augmenting this sometimes from the two school-books *De natura rerum* of Bede and Isidore. He also discussed corrections to be made in other Anglo-Saxon *computi*, supplied his own computistical *argumenta* concerning the beginning of the year and the bissextile (a day added in leap year), and firmly distinguished between effects of the moon on persons and nature and some superstitions about these effects. But he was not concerned with adjustments of the calendar to a new year for the Incarnation. Nor did he teach about the sidereal lunar month or its significance for lunar and other planetary positions on the zodiac that reveal sidereal time. Eight manuscripts and surviving fragments show that Ælfric's work was used in several different schools during the tenth and eleventh centuries.[69]

Abbo was not the only scholar who directly challenged the Bedan *computus* concerning the date of the Incarnation. Gerlandus compotista (ca. 1015-1102) wrote his *Computus Bedam imitantis* in England before the mid–eleventh century and taught there for twenty years, from 1038 to 1057; then he moved to Besançon, where he was teaching before 1066. After acknowledging the authority of Dionysius, Bede, and Helperic, he set forth causes and techniques for revising the Bedan *aera Incarnationis* by seven years. Gerland's work went through several versions within the author's lifetime, including revisions on the Continent that used the *annus praesens* 1081 or reference to an eclipse, possibly of 1086. Gerland's first version listed twenty-seven chapters, but he added others, including "Item de cursu solis et lunae," which elaborates his teaching in astronomy. From many detailed calculations of the sun and the moon moving through the zodiacal band and with data refined in terms not only of *horae* and *puncti* but even of *momenta*, *ostenta*, and *athomi*, this chapter demonstrated that the moon passed through the zodiac in twenty-seven days and eight hours, explaining its periodic displacement.[70]

Gerland was cited as an authority by 1102, and parts of this *magister's* *computus* and his tables of data were being used and commented upon by scholars pursuing their studies in the *corpus astronomicum* at Oxford and Paris during the twelfth and thirteenth centuries. The *computus* has not yet been published in any version, but it certainly stirred controversy in the schools.[71] His proposals to revise the calendar were incorporated into many other *computi*, some of which undertook to apply, while others wanted to refute, his system.

Abbo and Gerland raised their question about the *aera Incarnationis* on the basis of chronology but supported their proposals to correct the Bedan calendar with arguments from the data of sidereal time, tracking the movement of sun and moon against the stars and using for that purpose the ecliptic scale of the zodiac. This had a further consequence. Whether or not Dionysius had set the year of the Incarnation rightly, their astronomy demonstrated that the Bedan calendar had eventually departed from its stellar point of reference.

Another migrating scholar computist was Robert de Losinga, who came to England in the wake of or possibly as chaplain for William the Conqueror and who became bishop of Hereford (1079-95). In the school of Hereford he could have used the works of Heiric/Helperic and Gerland, but he also brought from Thuringia and Lotharingia an interest in Marianus Scottus. Marian was a sometime hermit and *peregrinus* who wished to revise the reckonings of Dionysius and Bede in order to relocate the birth of Jesus properly, this time by a change of twenty-two years. About 1086 Robert quoted a large part of Marian's computistical arguments in twenty-two of the twenty-four chapters in his *De annis domini*.[72] Marian and Robert, however, did not make use of the difference between synodical and sidereal lunar months and did not support their argument for calendar revision by sidereal time references. Nearby, however, a new instrument and new tables of data would open a different era of astronomical science for these Anglo-Saxon schools.

By use of an astrolabe, Walcher, prior of Malvern Abbey (fl. 1092-1112), could determine the dates of eclipses and could be certain that an eclipse that he had seen as a traveler in Italy on 30 October 1091 had been noted later on that same day at home in his English abbey. Lunar tables *De naturali cursu cuiusque lunationis*, which he calculated in A.D. 1092 for a seventy-six-year period (A.D. 1037-1112), provided a new form of cycle that would change computistical studies significantly. For some time he worked with Peter Anphus or Alphonsus, a Jewish astronomer and physician who had become a Christian and migrated from Spain to England. One of Peter's works, *De dracone*, gave the motions of sun and moon through the zodiac and discussed lunar nodes, not in Roman fractions as Walcher had used, but in degrees, minutes, and seconds.[73] This type of analysis would be found also in the work of Roger of Hereford (fl. A.D. 1176-1195), who used astronomical tables of Marseilles,[74] and seen again in the scientific works of Robert Grosseteste with his development from the basic terms of *Compotus* I (A.D. 1195-1200) to the Ptolemaic assumptions of his more sophisticated *Computus correctorius* (about 1220).[75]

Conclusion

From Ceolfrith and Aldhelm in the seventh century to Gerland and Robert in the eleventh, computistical and astronomical studies were actively pursued in many schools of Anglo-Saxon England. The masters and students were not only concerned with predicting the dates of future Easter Sundays but were also interested in identifying stellar and planetary regularities and in applying them to the study of natural phenomena in the context of *computus*. Manuscript records of their studies survive in every period from many parts of Britain.

Normally the masters in these schools taught their students to account for the mean lunar cycle from the first appearance of the new moon until its reappearance and often to account for the periods of appearance for the other planets, and their positions were often remarked on the Zodiacal scale. This system is called synodical time, especially with reference to the longitudinal course of the moon. With a mean lunar cycle of twenty-nine solar days and twelve hours, the system of synodical time was assumed by all those computists who met at Whitby in A.D. 664. In order to track the course of any planet by synodical time, the solar months and the zodiacal signs were coordinated in the tabular form of the A-P series by Aldhelm during the last quarter of the seventh century.

Bede was not satisfied with synodical time, however, when it was applied to the continuous motions of the planets, and he offered an alternative way to account for them, which we call sidereal time. In this system lunar and other planetary positions were also read on the zodiacal band, but with reference to the stars the moon was observed to have a complete cycle of twenty-seven solar days and eight hours. Bede presented his system also in the tabular form of a *pagina regularum*, coordinating months and signs with the A-O series for improved observations of the planets.

Considerable knowledge of the moon's cycle and its correlation with the solar cycle is necessary for the *computus*, but not knowledge about the cycles of other planets. In fact, most questions about lunar motions are adequately handled in terms of synodical time. The *Cyclus Aldhelmi* was tried by various computists in Anglo-Saxon and Carolingian schools during the eighth and ninth centuries, and many other schemes of synodical time were known, including Bede's. In the long run, however, they were displaced by sidereal time, and Bede's table and relatively difficult explanations were often transcribed and used.

Because basic procedures of the Bedan *computus* can be followed in terms of synodical time, it is not surprising that many of the masters offered brief explanations of sidereal time without including the *pagina regularum*, or that some may have copied the table without explaining it, and that others may even have omitted all reference to sidereal time from their teaching.

Abbo's two years in English schools toward the end of the tenth century, however, were different. Not only did he teach the system of sidereal time in the *computus*, but he also offered new examples of how useful it could be in astronomy and justified his questions about the Dionysian and Bedan *aera Incarnationis*. Likewise Gerland used it as the basis for radical revision of the Bedan calendar in the eleventh century.

Bede had not been satisfied with his own explanations and applications of sidereal time, and he had invited other scholars to improve upon them. Not even Abbo or Gerland had been able to do that. More exact observations with the astrolabe, however, did allow both Englishmen and Continental immigrants to England to make more refined observations. They also began to develop longer data series, along the lines of those used by Syrian, Greek, Hebrew, and Arab astronomers in Toledo and Marseilles.

One important result of this more accurate astronomy was the use of a seventy-six-year period for the lunar cycle and a new *computus* being taught in the arts curriculum of new universities during the twelfth and thirteenth centuries, a *computus* to which Robert Grosseteste made a primary contribution.

As with all research enterprises, some of the labors of Anglo-Saxon schools succeeded, some did not, and the best scholars could have both successes and failures. The manuscripts offer evidence that scientific research abounded in the schools of England, as the chronological uncertainties of historical events interacted conceptually with computistical and astronomical studies. One aspect of their research, concerning which the manuscripts will have more to reveal, was the discovery and application of sidereal time.

Notes

1. Some indication of scientific labors by Scots, Britons, Saxons, and Angles has been offered by the author in "Scientific Instruction in Early Insular Schools," in *Insular Latin Studies*, ed. Michael Herren (Toronto, 1981), pp. 83-111. Also in preparation is "A Catalogue of Computistical Tracts, A. D. 200–1200," which will include at least forty-five manuscripts written in Anglo-Saxon England containing sixty-two independent *computi*, in addition to many different *versus computistica* and diverse series of *argumenta paschalia*.

2. Biscop Baducing (ca. 630-89) became a monk and took the name Benedictus about 652. He founded the North Sea monasteries of Wearmouth (674) and Jarrow upon Tyne (681), which Bede later entered as a child. Concerning Biscop's trips to Lyons, Arles, Lerins, and Rome, see W. Levison, *England and the Continent in the Eighth Century* (Oxford, 1946), pp. 38-42, and chapter 6, "Learning and Scholarship," especially pp. 132-34 and 143; P. Hunter Blair, *The World of Bede* (New York, 1970), pp. 155-87, especially pp. 156-60, 168-69, 176-77, and 186-87; J. Campbell, "The First Century of Christianity in England," *Ampleforth Journal* 76 (1971): 12-29; H. Mayr-Harting, *The Coming of Christianity to Anglo-Saxon England* (London, 1972); P. Wormald, "Bede and Benedict Biscop," in *Famulus Christi*, ed. G. Bonner (London, 1976), pp. 141-69, especially pp. 149-52; and J. M. Wallace-Hadrill, "Rome and the Early English Church: Some Problems of Transmission," *Settimane di studio del Centro italiano di studi sull'alto medioevo* 7 (Spoleto, 1960): 519-48, reprinted in his *Early English History* (Oxford, 1975), pp. 115-37.

3. See the collection of materials sent by Aldhelm to Aldfrith, king of Northumbria (685–705), and addressed *ad Acircium*, as well as the letters to Wihtfrith and Heahfrith, ed. R. Ehwald, in *Monumenta Germaniae Historica, Auctores Antiquissimi* 15 (1919) (hereafter cited as *MGH, AA*), and the English translations and annotations of Michael Herren in *Aldhelm, the Prose Works*, trans. M. Lapidge and M. Herren (Ipswich, 1979); and Bede, *Historia ecclesiastica* III.27, IV. 4, 26, and V.15, ed. Charles Plummer (Oxford, 1896) (hereafter cited as *HE*).

4. Many of the sources for Bede's computistical studies were copied into the MS Oxford, Bodleian Library, Bodley 309, probably of Vendôme provenance in the second half of the eleventh century. They were identified by Charles W. Jones, "The 'Lost' Sirmond Manuscript of Bede's Computus," *English Historical Review* 52 (1937): 204-19, and described more fully in section 6 of the introduction to his *Bedae Opera de temporibus* (Cambridge, Mass., 1943), pp. 105-13.

5. Ceolfrith was a teacher of Bede and then abbot of Wearmouth and Jarrow, 688-715. His letter to Nechtan provided a clear explanation for the significance of *luna* XIV and of the *termini paschales luna* XV-XXI in order to celebrate Easter during the third week of the first

month of the Jewish year, and it rejects other limits as unbiblical. It was transcribed by Bede, *HE* V.21, and is found separately in MS Montpellier B. Fac. Méd. 157 (s. IX), fols. 55v-60.

6. Christine E. Fell, "Hild, Abbess of Streonæshalch," in *Hagiography and Medieval Literature*, ed. H. Bakker-Nielsen et al. (Odense, 1981), pp. 76-99; and P. Hunter Blair, "Whitby as a Centre of Learning in the Seventh Century," in *Learning and Literature in Anglo-Saxon England: Studies Presented to Peter Clemoes*, ed. M. Lapidge and H. Gneuss (Cambridge, Eng., 1985), pp. 3-32, especially pp. 3-14.

7. Bede, *HE* III.25-26 on Colman's teaching. There was indeed a bishop Anatolius who set forth a *computus*, some of which survives in Eusebius, *Historia ecclesiastica*, ed. E. Schwartz in *Eusebius Werke* (Leipzig, 1903-4), VII.32, 14-19; this was discussed by C. H. Turner, "The Paschal Canon of Anatolius of Laodicea," *English Historical Review* 10 (1895): 699-710. But the *Canon paschalis* used by Colman had been attributed falsely to Anatolius, either by the Brittoni in Wales or the Scotti in Ireland, perhaps in order to fend off the teaching introduced earlier by Patrick; see C. H. Turner, "The Date and Origin of Ps-Anatolius De Ratione Paschali," *Journal of Philology* 28 (1901): 137-51; and Jones, *Bedae Opera de temporibus*, pp. 82-87. Three manuscript fragments of the *Canon paschalis ps-Anatolii* have been identified from eighth-century Echternach and early ninth-century Flavigny, and the full text is found in MS Köln Dombibliothek LXXXIII/2 (Cologne, ca. 805), fols. 188-91v, Geneva University Lat. 50 (second quarter s. IX Massai), fols. 127-28, and four later manuscripts. There is no adequate edition, but see B. Krusch, *Studien zur christlich-mittelalterlichen Chronologie: Der 84jährige Ostercyclus und seine Quellen* (Leipzig, 1880), pp. 316-27; and August Strobel, *Texte zur Geschichte des fruhchristlichen Osterkalendars* (Münster in Westfalen, 1984), pp. 1-42.

8. MS Köln Dombibliothek LXXXIII/2 (Cologne, ca. 805) is a collection of texts from Irish and Anglo-Saxon schools; a table of comparisons of the Victurian and Dionysian data for Alexandrian cycles was added to a *Computus Grecorum* in the same manuscript on fol. 59v, and fol. 211v has a table of lunar epacts for the first day of each month in Alexandrian terms but adjusted to the Victurian cycle. The Victurian cycle was worked out by Victurius of Aquitaine in 457. See further C. W. Jones, "The Victorian and Dionysiac Paschal Tables in the West," *Speculum* 9 (1934): 408-21; and more recently M. Walsh and D. Ó Cróinín, introduction to *Cummian's Letter De controversia paschali and the De ratione conputandi* (Toronto, 1988), pp. 42-45 and 87 n. 216.

9. Acgilbert's family abbey of Jouarre may have been within the range of Irish influence through Luxeuil, according to J. Guerout, "Les origines et le premier siècle," in *L'abbaye royale Notre Dame de Jouarre* (Paris, 1961), pp. 41-45, cited by Wormald, "Bede and Benedict Biscop," pp. 145-46. However, Acgilbert had studied for some time in southern Ireland, possibly at Ráth Maélsigi, according to D. Ó Cróinín, "The Irish Provenance of Bede's Computus," *Peritia* 2 (1983): 229-47, especially pp. 224 and 245. See also Hunter Blair, "Whitby as a Centre of Learning," pp. 30-32.

10. Bishops of Gaul, meeting at Orléans in 541, preferred the *Laterculus Victurii*; according to the earliest surviving letter of Columbanus, some bishops in that region were using Victurian tables about 599-600 and complained mightily that he would not do so. According to his second letter they wished to question him about his practice in a synod at Chalons-sur-Saône in the summer of 603 or possibly winter of 603-4. A late copy of each letter is found in MS St. Gallen Stiftsbibliothek 1346 (s. XVII in), pp. 109-19; and the second also survives in MS Biblioteca Apostolica Vaticana Vat. Lat. 9864 (s. XVIII), fol. 28; they were edited and translated by G. S. M. Walker, *Sancti Columbani Opera* (Dublin, 1957), pp. 2-15.

11. Victurian and Alexandrian Easter tables were set in parallel columns in the MS Milano Biblioteca Ambrosiana H. 150 inf. (s. IX in), fol. 125; most texts in this manuscript appear to derive from Irish and Anglo-Saxon schools, but these tables may be Frankish, and there are later Frankish manuscripts with similar comparisons. See also note 8 above. Historians have usually designated paschal tables as Dionysian when the columns of data were organized in Alexandrian style, but there are significant variations of tables within that

style, and those that were specifically designed by Dionysius Exiguus have not yet been distinguished.

12. Bede, *HE* III.25; *Vita sancti Wilfridi* 10, ed. W. Levison, *M. G. H. Scriptores rerum Merovingicarum* 6 (1913): 202-4; see also D. P. Kirby, "Northumbria in the Time of Wilfrid," and D. H. Farmer, "Saint Wilfrid," in *Saint Wilfrid at Hexham*, ed. D. P. Kirby (Newcastle upon Tyne, 1974), pp. 8-12, 18-20, and 43.

13. Ó Cróinín, "The Irish Provenance of Bede's Computus," pp. 229-47.

14. Note the response conveyed to Columbanus by Candidus, Gregory's diocesan officer in Gallia: " . . . temporis antiquitate roborata mutari non posse," *Columbani Epistola ad Gregorium*, ed. Walker, p. 12; this letter was written within the years 597-600.

15. Bede, *De temporum ratione* XLVII.61-71, ed. C. W. Jones, *Bedae Opera de temporibus* (Cambridge, Mass., 1943) (hereafter cited as *DTR*). Bede dedicated his work to Huaetbercte, who had become abbot of Jarrow in 716.

16. Bede, *HE* V.21. About the Dionysian cycle Ceolfrith said, "Quibus termino adpropinquantibus, tanta hodie calculatorum exuberat copia, ut etiam in nostris per Brittaniam ecclesiis plures sint, qui mandatis memoriae veteribus illis Aegyptiorum argumentis facillime possint in quolibet spatia temporum paschales protendere circulos, etiam si ad quingentos usque et xxx duos voluerint annos; quibus expletis, omnia quae ad solis et lunae, mensis et septimanae consequentiam spectant, eodem quo prius ordine recurrunt," according to Bede, *HE* V.21, p. 341.

17. Two practical systems for calculating latitudes were provided by Bede, *De natura rerum* XLVII (hereafter cited as *DNR*), and enlarged upon in his *De Temporibus* VII, *DTR* XXX-XXXIII, and *HE* I.1. The two systems have been explained in modern terms by Wesley M. Stevens, *Bede's Scientific Achievement*, the Jarrow Lecture for 1985 (Jarrow upon Tyne, 1986), pp. 7-10 and 43-44.

18. Stevens, *Bede's Scientific Achievement*, pp. 10-18.

19. See the *Admiralty Tide Tables*, vol. 1 (1985), *European Waters Including Mediterranean Sea*, published by the Hydrographer of the Navy (London, 1984), pp. vi-xi, "Introduction," and pp. xii-xxx, "Instructions for the Use of Tables." This volume was presented to me in Dublin on 10 February 1986 by Willie Donahoe, harbor pilot, Dublin Bay.

20. Isidore, *Differentiæ* 149-52, *De natura rerum* XVIII. 24-26, and *Etymologiae* III.10.3. These could have been the sources for Aldhelm, *Epistola* 1, written in 675, possibly to Leuthere (bishop of Winchester, 670-76?), ed. Ehwald in *MGH, AA*, p. 475-78, trans. Herren in *Aldhelm, the Prose Works*, pp. 152-53. Concerning the absence of astral influences in Isidore, see J. Fontaine, "Isidore de Séville et l'astrologie," *Revue des Etudes Latines* 31 (1954): 271-300. Both Isidore and Aldhelm have been discussed by Stevens, "Scientific Instruction in Early Insular Schools," pp. 96-98.

21. Bede, *HE* IV.2: "Ita ut etiam metricae artis, astronomiae et arithmeticae ecclesiasticae disciplinam inter sacrorum apicum volumina suis auditoribus contraderent." In general see V. R. Stallbaumer, "The Canterbury School of Theodore and Hadrian," *American Benedictine Review* 22 (1981): 46-63; the use of glossaries in the Canterbury schools has been explained from the manuscripts by Michael Lapidge, "The School of Theodore and Hadrian," *Anglo-Saxon England* 15 (1986): 45-72.

22. Aldhelm, *Epistola* 1, ed. Ehwald, in *MGH, AA*, pp. 475-78, especially pp. 476-77, trans. Herren in *Aldhelm*, pp. 152-53.

23. Aldhelm, *Epistola ad Geruntium* (690?), ed. Ehwald in *MGH, AA*, pp. 483-86, especially pp. 483-84, trans. Herren in *Aldhelm*, pp. 158-59. Other parts of that letter survive in Bede, *HE* III.29, and MS Oxford Bodleian Library Digby 63 (s. IX/2), fol. 59v, ed. B. Krusch in *Studien zur christlich-mittelalterlichen Chronologie: Die Entstehung unserer heutigen Zeitrechnung* (Berlin, 1938), p. 86. See also Jones, *Bedae Opera de temporibus*, pp. 18 and 102-4; and Herren, *Aldhelm*, pp. 140-43.

24. Aldhelm, *Epistola ad Geruntium*, ed. Ehwald in *MGH, AA*, pp. 483-84, trans. Herren in *Aldhelm*, pp. 158-59.

V

25. It should not be supposed that his knowledge was limited to poetic imagery such as he used in the *Carmen rhythmicum*, ed. Ehwald in *MGH, AA*, pp. 524-28, trans. M. Lapidge in *Aldhelm: The Poetic Works* (Cambridge, Eng., 1985), pp. 177-79; Lapidge has pointed out (p. 262) that similar imagery could have been found by Aldhelm in Isidore, *Origines* III.71, 6-9 and 13-15, and also XIII.11.

26. *Cyclus Aldhelmi* is found in six manuscripts of the ninth, tenth, eleventh, and twelfth centuries and was edited by C. W. Jones, *Bedae Pseudepigrapha* (Ithaca, N.Y., 1939), p. 70; see also his comments in *Bedae Opera de temporibus*, pp. 353-54.

27. Michael Lapidge, "The Present State of Anglo-Latin Studies," in *Insular Latin Studies*, ed. Herren, pp. 45-82, notes that the *Cyclus Aldhelmi* is found "in a number of manuscripts either of English origin or from continental centres with English connections" and states that "although this table was not known to Ehwald, I see no reason not to regard it as Aldhelm's" (p. 49).

28. An early user of the *Cyclus Aldhelmi* also created a lunar series of data organized by the fifteen letters A through P. The Series A-P helps one coordinate the synodical month of 29 1/2 lunar days with 30 degrees per sign and 360 degrees of the zodiacal band. It has been noticed in MS Paris, Bibliothèque National Lat. 5543 (847 Fleury), fols. 95-100v; Vat. Regin. lat. 309 (s. IX/2 St. Denis), fols. 128-40; Chartres, Bibliothèque de la Ville 19 (26) (s. X), fols. 3v-8v (destroyed in 1944); Biblioteca Apostolica Vaticana, Vat. Lat. 644 (s. X St. Gallen?), fols. 24v-28v; Milano, Biblioteca Ambrosiana D. 48 inf. (s. X/2), fols. 101-8; Leiden, Bibliotheek der Rijksuniversiteit Scaliger 38 (s. XI), fols. 7v-13; and ed. *Patrologia Latina (PL)* 90 (1850), 759-84, col. 2, but with a line of data omitted. See Jones, *Bedae Pseudepigrapha*, pp. 108-10.

29. Lunar letter series are often found with twelve-month calendars. See H. Leclercq, s.v. "calendars," in *Dictionnaire d'archéologie chrétienne et de liturgie*, ed. F. Cabrol and H. Leclercq (Paris, 1907 seq.), 8:624-67, and Jones, *Bedae Pseudepigrapha*, pp. 108-10.

30. Bede, *DNR* VI, from Pliny, *Historia Naturalis (HN)* II.71, 177, and 76, 183-85.

31. Bede, *DNR* IX and XLVII.

32. Bede, *DNR* III, and V-VI.

33. Bede, *DNR* XXII and *DTR* XXVII.2-22. Both explanations were drawn from Pliny, *HN* II.47-50 and 56-57, rather than from Isidore, *De natura rerum* XX-XXI or *Origines* III.58-59. Added by Bede, *DNR* XXIII, from Pliny, *HN* II.72-73, 180-81, was the example of observation of the longitudinal passage of an eclipse that had been recorded at two distant places on the eleventh night before Alexander's victory over Darius at Arbela (Gaugamela), 20/21 September 331 B.C.

34. Bede, *DNR* XII.2-8, from Pliny, *HN* II.4, 12 and 6, 32-33 (planetary circuits); *DNR* XIII.1-11 on the time required for each circuit, from Pliny, *HN* II.6, 32-41; and *DNR* XIII.11-14 on the periods of visibility, from Pliny, *HN* II.15, 78 and 12, 59-61. These excerpts from Pliny were available in several manuscripts and were edited by Karl Rück, *Auszüge aus der Naturgeschichte des C. Plinius Secundus* (Munich, 1888), pp. 34-35.

35. Bede, *DNR* XVI.5-7, used Pliny, *HN* II.13, 66-67 to explain the range of wandering for the seven planets within the zodiacal band, or beyond it in the case of Venus. Bede, *DTR* VII.31-32 and XXVI.15-19, states that the band has a breadth of *xii partes* (twelve degrees), which were alluded to in *DNR* XVI only as *superior et inferior*. The two Bedan texts written in 701 and then 721-25 repeat from Pliny the faulty notion that the sun varies its course from the median by two degrees: "Sol deinde medio fertur inter duas partes flexuoso draconum meatu inaequalis" — perhaps a misunderstanding of the Pliny excerpt. However, it is uncertain whether observation could correct this mistake with instruments available prior to the seventeenth century.

36. Bede, *DNR* III, from Pliny, *HN* II.4, 10-11; Bede, *DNR* V.7-9, from Augustine, *De Genesi ad litteram libri* II.10, 23, and Pliny, *HN* II.13, 63; and Bede, *DNR* VI.2-8, from Pliny, *HN* II.71, 177 and II.76, 183-85.

37. Bede, *DNR* VI and XIV.2-20, the latter from Pliny, *HN* II.13, 63-64 and 68; as he often did elsewhere, Bede then referred the reader to his source: "De quibus si plenius scire velis, lege Plinium Secundum ex quo et ista nos excerpsimus." Similar excerpts were published by Rück, *Naturgeschichte des C. Plinius Secundus*, pp. 39-40. The colors of the planets helped in understanding these phenomena of various and variable distances, according to Bede, *DNR* XV, from Pliny, *HN* II.16, 79.

38. Bede, *DNR* XIV.20-21, from Isidore, *De natura rerum* XXII.3, about retrograde motion and stations of planets.

39. Bede, *DNR* XIII.13-14, from Pliny, *HN* II.15, 78.

40. Concerning the origins and early medieval interest in the planetary configuration of Venus and Mercury with the sun, as a group that travels together around the earth, see Bruce S. Eastwood, "Plinian Astronomical Diagrams in the Early Middle Ages," in *Mathematics and Its Applications to Science and Natural Philosophy in the Middle Ages*, ed. E. Grant and J. Murdoch (Cambridge, Eng., 1987), pp. 141-72; Eastwood, "Plinian Astronomy in the Middle Ages and Renaissance," in *Science in the Early Roman Empire: Pliny the Elder, His Sources and Influence*, ed. R. French and F. Greenaway (London, 1986), pp. 197-251; Eastwood, "The Chaster Path of Venus (orbis veneris castior) in the Astronomy of Martianus Capella," *Archives Internationales d'Histoire des Sciences* 32 (1982): 145-58; idem, "Mss Madrid 9605, Munich 6364, and the Evolution of Two Plinian Astronomical Diagrams in the Tenth Century," *Dynamis, Acta Hispanica ad Medicinae Scientiarumque Historiam Illustrandam* 3 (1983): 265-80; Patrick McGurk, "Germanici Caesaris Aratea cum scholiis, A New Illustrated Witness from Wales," *National Library of Wales Journal* 18 (1973): 197-216; and Eastwood, "Notes on the Planetary Configuration in Aberystwyth N. L. W. ms. 735 C, f. 4ᵛ," *National Library of Wales Journal* 22 (1981): 129-40.

41. Bede, *DNR* XXI.2-7: "Luna zodiacum tredecies in duodecim suis conficit mensibus, duobus scilicet diebus et sex horis et besse, id est octo uncii unius horae, per singula signa decurrens. Si ergo vis scire in quo signo luna versetur, sume lunam quam volueris computare, utpote duodecim, multiplica per quattuor, fiunt XLVIII. Partire per novem (novies quini quadragies quinquis)."

42. *HE* V.13. 18, 23: Pechthelm had been with Aldhelm as deacon and monk; he was well known to Bede and was probably his primary informant for West Saxon affairs; he may have been active in Galloway and Dumfriesshire before he became bishop of Whithorn (ca. 730-34). Bede, *Epistola ad Pleguinam*: Pleguina came north of the Humber about 707.

43. Ambrose, *Hexaemeron* IV.5, 24; Isidore, *De natura rerum* XIX.1 or *Origines* V.36, 3.

44. Bede, *DTR* XIX and accompanying table, pp. 218-20, and brief notes on p. 354.

45. The *pagina regularum* will be further considered on another occasion.

46. Stevens, *Bede's Scientific Achievement*, pp. 33-38, Appendix 1: "Bede's Scientific Works," is organized by title, incipit and explicit, extant manuscripts, and editions of his scientific tracts and letters, with a brief guide to some of the questions of text and authorship of each. Several fragments of Bede, *DTR*, in Northumbrian uncial and minuscule scripts survive at Bückeburg, Münster in Westfalen, Darmstadt, and Vienna, but none in Great Britain; see ibid., pp. 39-42, Appendix 2: "Bedae De temporum ratione liber, Manuscripts Dating from the Eighth and Ninth Centuries." Several historians have cited the complete text of *DTR* in MS Salisbury Cathedral 158 (s. IX 2), fols. 20-83; the earlier part of that manuscript may have been written in England, but those folios containing *DTR* display Carolingian script from the Continent (pp. 35 and 41).

47. Helmut Gneuss, "A Preliminary List of Manuscripts Written or Owned in England up to 1100," *Anglo-Saxon England* 9 (1981): 1-60. At least seventy-nine of those manuscripts are reported to contain scientific works, and many of them are said to contain several such works.

48. Neil R. Ker, *A Catalogue of Manuscripts Containing Anglo-Saxon* (Oxford, 1957); idem, *Medieval Libraries of Great Britain: A List of Surviving Books*, Royal Historical So-

ciety Guides and Handbooks 3 (London, 1941; 2nd ed., 1964; supp. ed., Andrew G. Watson, 1987).

49. E. W. B. Nicholson made several entries in the "official copy" of the Digby Catalogue kept in the Bodleian Library, but I should be loath to follow them: what he saw as underlining of a date on fol. 8, for example, is actually the reverse of a trough from dry point lining done on the verso, and the tiny black points of ink under the numerals for year 844 on fol. 36v cannot provide "evidence that the MS was copied from an earlier one written in 844, the first year of Adelhard abbot of St. Bertin's," as he thought. There are similar black points beneath numerals in most nineteen-year cycles, especially from 855 to 876, and they do not necessarily suggest an *annus praesens* for either scribe or user. On fol. 36v and fol. 37 is a list of indictions from 844 to 892, corresponding with the end of the nineteen-year cycles, and I assume that Raegenbold thought his work done by 892 or 893. Nevertheless he added another indiction for each year until 899. In the meantime he or someone else had begun to recopy the Dionysian Argumentum I from fol. 72v onto fol. 8, applied it to the year 867 but with incorrect data, erased, and then abandoned the item. These several observations leave us with the period from 844 to 892/93 within which Raegenbold worked, and they suggest that he may have lived until 899.

50. There is a special feast of John of Beverly and another for Cuthbert in the calendar, which suggest Northumbrian origins. Thus Kenneth Sisam, *Studies in the History of Old English Literature* (Oxford, 1953), p. 71, thought that this collection was Northumbrian and contained "a calendar that reached Southern England about Alfred's time." But this is uncertain.

51. Jones, "The 'Lost' Sirmond Manuscript of Bede's Computus," pp. 204-19; Jones, introduction to *Bedae Opera de temporibus*, pp. 105-13; Stevens, introduction to *Rabani mogontiacensis episcopi de computo*, in *Corpus Christianorum continuatio mediaevalis*, vol. XLIV (Turnhout, Belgium, 1979) pp. 170-71, and 177-84; and Ó Cróinín, "The Irish Provenance of Bede's Computus," pp. 229-47.

52. Raegenbold's odd practice of writing *qi* for *qui*, *qando* for *quando*, and so on, is just one of his peculiarities of orthography and grammar, which have not found parallels in other ninth-century manuscripts. Despite the possibilities outlined by David Dumville, "Motes and Beams," *Peritia* 2 (1983): 249-50, the script does not allow one to expect an English background for this writer.

53. David Dumville has shed much light on the development of this script and may soon clarify the origin of this manuscript. His "English Square Minuscule Script: The Background and Earliest Phases," *Anglo-Saxon England* 16 (1987): 147-79, is a brilliant discussion of "the complex evidence attesting scribal efforts at reform and progress in the period c. 890-c. 920 . . . " and "a beginning made in charting the early history of the new Square minuscule" (p. 179). He finds that "the creation of a new canonical English script form had been achieved by c. 930" (p. 178). MS Exeter 3507 should be included soon in his studies.

54. Some of the same verses were also found in MS Oxford BL Bodley 579, fols. 53v-54v: "Linea Christe . . . ," "Prima dies Phoebi . . . ," "Bes sena mensium . . . ," and "Primus Romanas . . . " – the last from Ausonius but with an extra line; the verses "Ianus et october binis . . . " and "September semper quinis . . . ," however, are not found in the Exeter manuscript, although they were copied into the later MS London BL Sloane 263 (s. XI), fols. 21v-22, and other English manuscripts. MS Bodley 579 is commonly called the Leofric Missal because fols. 1-37 contain a service book and it was listed with Bishop Leofric's gifts to Exeter Cathedral . That part of the codex was written about s. IX/X in northeastern France (perhaps in the region of Arras/Cambrai) and may have been brought from that region to England by Leofric himself. But fols. 38-58 compose a separate codex that was also written in a different Continental minuscule probably of late tenth century. Fol. 49 displays a large *dextera domini* with dates written on each finger of the hand, and this opens a section that applies some computistical knowledge to liturgical practices. The very common nineteen *Termini Paschales* ("Nonae Aprelis . . . ") are found on fol. 52. There are a calendar and Eas-

ter tables for 969 to 1006 on fol. 53, but fols. 54v-58 have additional prose passages with *argumenta* that serve specifically liturgical rather than computistical interests. This part of MS Bodley 579 therefore cannot properly be called an "Anglo-Saxon *computus,*" and its presence or use at Exeter has not yet been demonstrated. MS Bodley 579 was published as *The Leofric Missal* (Oxford, 1883) by F. E. Warren, who hoped that the calendar and Easter tables had been prepared for "Glastonbury before 978." His wish has often been repeated as if it were fact by other scholars without citing evidence to support the proposal.

55. J. Fontaine, *Isidore de Séville, Traité de la nature,* (Bordeaux, 1960) pp. 164-327, Latin text and French translation. See also the sections of his introduction, pp. 38-45: "Les 'accidents massifs' et les trois recensions"; pp. 69-83: "La diffusion du De Natura rerum en Europe de Sisebut à Charlemagne"; and pp. 75-78: "Le problème de la 'méditation irlandaise.' "

56. Fontaine, *Isidore de Séville,* pp. 328-35.

57. The earliest manuscript of Isidore, *De natura rerum* (612, first version), and of the verse *Epistola Sisebuti* (613, response) is MS El Escorial, Real Biblioteca de San Lorenzo, R.II.18, fols. 9-24, in a single Visigothic script within the period 636-90. The drawing on fol. 24 represents the positions of sun, moon, and earth in positions of either solar or lunar eclipses; later users of this manuscript also noted on fol. 66 the eclipses of 778 and 779. Unfortunately, Fontaine did not reproduce the manuscript diagram but created his own explanatory schema on p. 337.

58. The manuscripts are Basel Universitätsbibliothek F.III.15f (Anglo-Saxon minuscule s. VIII), fols. 1-13; København Universitetsbibliotek Lat. 19 (Anglo-Saxon minuscule s. VIII 2), fragment VII; Basel F.III.15a (Fulda Insular minuscule s. VIII ex), fols. 1v-16v; and Weimar Landesbibliothek Fragment 414a (Fulda Insular minuscule s. VIII ex). Insular exemplars appear to have been used for two other manuscripts with this version: St. Gallen Stiftsbibliothek 238 (750-70), pp. 312-84, written by Winithar; and Besançon Bibliothèque Municipale 184 (Eastern France s. IX in), fols. 1v-55v. MS Paris Bibliothèque National Lat. 10616 (Verona 796-99) fols. 1-93v, deriving from an exemplar of the "Recension courte" that had received the long version additions before being transcribed at Verona during the episcopacy of Egino, reveals no Insular traces.

59. *Rabani mogontiacensis episcopi de computo,* ed. Stevens, pp. 163-331; see pp. 190-94 for manuscripts and pp. 322-23 for glosses.

60. Abbon de Fleury, *Quaestiones Grammaticales,* ed. Anita Guerreau-Jalabert, Collection auteurs latins du moyen âge (Paris, 1982); Henry Bradley, "On the Text of Abbo of Fleury's Quaestiones Grammaticalae," *Proceedings of the British Academy* 10 (1921-23): 126-69; M. Manitius, *Geschichte der lateinischen Literatur des Mittelalters* 2 (Munich, 1923), pp. 664-72; A. Van de Vijver, "Les oeuvres inédites d'Abbon de Fleury," *Revue Bénédictine* 47 (1935): 126-69; and Ron B. Thomson, "Two Astronomical Tractates of Abbo of Fleury," in *The Light of Nature,* ed. J. D. North and J. J. Roche (Dordrecht, Netherlands, 1985), pp. 113-33. Thomson (pp. 116-18) listed eighteen manuscripts for his edition of two *Sententiae,* of which at least twelve were written in England.

61. Edited by Thomson, "Two Astronomical Tractates of Abbo of Fleury," pp. 113-33, and "Further Astronomical Material of Abbo of Fleury," *Mediaeval Studies* 50 (1988): 671-73.

62. Ludwig Traube, "Computus Helperici," *Neues Archiv* 18 (1893): 73-105, reprinted in *Vorlesungen und Abhandlungen* 3, ed. Paul Lehmann (Munich, 1920), pp. 128-56; Patrick McGurk, "Computus Helperic: Its Transmission in England in the Eleventh and Twelfth Centuries," *Medium Aevum* 43 (1974): 1-5; Arno Borst, "Computus: Zeit und Zahl im Mittelalter," *Deutsches Archiv* 44/1 (Munich, 1988): 1-81, esp. 25. W. M. Stevens, "Catalogue of Computistical Tracts, A.D. 200-1200," currently lists seventy-eight manuscripts, but more will be located.

63. MS Oxford St. John's College Library XVII (1102-10), fols. 123-37, especially fols. 127-28: " . . . fiunt XXVII. Superest in singulis signis, bisse unius horae. . . . "

64. Heiric/Helperic Autissiodurensis, *Liber de arte calculatoria*, explicits (1) " . . . de his omnibus quicquid quesierit procul dubio reperturus erit"; (2) " . . . ut his primum quasi quibusdam alphabeti characteribus inducti, illa deinceps facilius assequantur"; (3) " . . . adventum domini celebrare valebit"; (4) " . . . sequentes post se terminos in aliam feriam mutaverit"; and others. One version was prepared by Abbo of Fleury from an exemplar that used *annus praesens* 946, and the text reveals adjustments of dates from 898 to 912; see MS Biblioteca Apostolica Vaticana Vat. Lat. 3101 (1077 from Ilmunster?), fols. 42-62, cited by Traube, "Computus Helperici," p. 136. Abbo added the preface, "Me legat annotes qui vult cognoscere ciclos. . . . "

65. MS Berlin Deutsche Staatsbibliothek Phillipps 1833 [Cat. 138] (s. X ex Fleury), fols. 7v-20v; his dialectical and rhetorical justifications for arithmetic have been studied by G. R. Evans and A. M. Peden, "Natural Science and the Liberal Arts in Abbo of Fleury's Commentary on the *Calculus* of Victorius of Aquitaine," *Viator* 16 (1985): 109-27; see also Borst, "Computus," pp. 27-28.

66. MS Berlin, fols. 33v-43, ed. *PL* 90 (1850), cols. 727-60, especially cols. 757-60. This manuscript was transcribed either near the end of Abbo's life or shortly thereafter. As his computistical works have not yet found a modern editor, it is not altogether clear whether these paragraphs are directly from Abbo or from his students at Fleury. I am aware of twenty-nine manuscripts of the *Computus vulgaris* in several versions, five manuscripts of his preface to Dionysian Easter tables, seven manuscripts of his *Tabula lunaris*, seven manuscripts of his first letter to Gerald and Vitalus (1003) about altering the date of the Incarnation, and five manuscripts of another letter (ca. 1004?) to the same recipients, with further discussion of that question. No reliable editions of these works are available, and the final letter seems not to have been published.

67. S. J. Crawford, ed., *Byrhtferth's Manual*, Early English Text Society (EETS) 177 (London, 1929; repr. in EETS Original Series [Oxford, 1966], with notes and errata by N. R. Ker). The exemplar from which the Ashmole manuscript was transcribed was in disarray and so is the edition, which should be read in the following sequence: p. 2, line 1, to p. 30, line 9; 44.28 to 56.29; 30.9 to 44.27; 56.30 seq.; and passages of a folio or more are missing in several places. The manual was illustrated, but most figures have been torn out of the Ashmole manuscript in modern times. On the basis of Ashmole 328, pp. 173-74, Crawford dated the whole work to 1011, and this was repeated by Ker, *Catalogue*, no. 288, but it is difficult to see how the data would support this assertion. Various entries must have been made by the author in different years. Peter Baker has announced the preparation of a new edition.

68. Peter S. Baker, "Byrhtferth's Enchiridion and the Computus in Oxford, St. John's College 17," *Anglo-Saxon England* 10 (1982): 123-42. The earlier proposal to date that manuscript to 1081-92 by Cyril Hart, "The Ramsey Computus," *English Historical Review* 85 (1970): 29-44, depended upon an assumption that the four leaves of MS London British Library Cotton Nero C. VII (Ramsey, 1086-92), fols. 80-84, were missing folios from the Saint John's College codex, as had been proposed by N. R. Ker, "Membra disiecta," *British Museum Quarterly* 12 (1938): 130-35. Baker has demonstrated, however, that this could not have been the case and accepted the dating clause on MS fol. 3v "ad praesens tempus I. CX" or 1110. Nevertheless, the annals and the texts transcribed reveal numerous details of dating between 1102 and 1110, with a few entries extending to 1113. This has been demonstrated very well in the recent study of the manuscript by Faith E. Wallis, "Ms Oxford St John's College 17: A Mediaeval Manuscript in Its Context" (Ph.D. diss., University of Toronto, 1984), especially pp. 122-29. Therefore the transcription of computistical texts into MS St. John's College XVII occurred during the period of 1102-10.

69. *Aelfric de temporibus anni*, ed. H. Henel, Early English Text Society 213 (London, 1942), from MS Cambridge University Library Gg. III.28 (s. X ex), fols. 255-61v; this manuscript has a collection of Ælfric's early writings, made perhaps at Cerne Abbey by one of his students. See also M. R. Godden, "De temporibus anni," in *An Eleventh-Century Anglo-*

Saxon Illustrated Miscellany: British Library Cotton Tiberius B. V, part I, ed. Patrick Mc-Gurk, Early English Manuscripts in Facsimile 21 (Copenhagen, 1983), pp. 59-64.

70. MS Cambridge St. John's College Library I.15 (s. XII ex), pp. 353-402, esp. p. 396: "Unde patet lunam peragere zodiacum annum XXVII dies et octo horas." See Borst, "Computus," p. 35; on dating of Gerland's work, see idem, "Ein Forschungsbericht Hermanns des Lahmen," *Deutsches Archiv* 40 (1984): 379-477, esp. 465-66. Thus far, I have recorded only twenty-six manuscripts of the *Computus Gerlandi Bedam imitantis* in the "Catalogue of Computistical Tracts, A. D. 200-1200," mostly from the twelfth and thirteenth centuries, but there may well be at least a hundred more for this popular work, which has not been published. Cataloguers have sometimes mentioned *Tabulae Gerlandi* without determining whether the accompanying text or various texts are by the same author.

71. For example, the *Computus Cunestabuli* (Constable, s. XII Canterbury) and the work by Salomon of Canterbury (fl. 1185-98) in MS London BL Egerton 3314 (s. XII 2), fols. 1v-8v, argue strongly against Gerland. These texts have not been published. See C. H. Haskins, *Studies in the History of Medieval Science*, 2nd ed. (Cambridge, Mass., 1927), p. 87; and P. J. Willets, "A Reconstructed Astronomical Manuscript from Christ Church Library Canterbury," *British Museum Quarterly* 30 (1965-66): 22-30.

72. Robert de Losinga, *De annis domini* (1086), MS Oxford Bodleian Library Auct. F.1.9 (1120-30), fols. 2v-12v, and seven other manuscripts. The MSS Oxford BL Digby 56 (s. XII), fols. 162-219, and Vat. Regin. lat. 1281 (s. XII), fols. 1-4v, show that the *computi* of Heiric/Helperic and of Gerland and the letters of Abbo about the *aera Incarnationis* were available at Hereford while Robert was bishop; the latter manuscript contains Robert's work on fols. 4v-5. See W. H. Stevenson, "A Contemporary Description of the Domesday Survey," *English Historical Review* 22 (1907): 72-84; and A. Cordoliani, "L'activité compotistique de Robert, évêque de Hereford," in *Mélanges René Crozet* (Poitiers, 1966), pp. 333-40. The manuscripts of Robert's *computus* also give dates for movable feasts in three great (532-year) Dionysian cycles with corresponding Marian years rubricated in the right column: Dionysian year 1086 would be Marian year 1108 of the Incarnation. See Anna-Dorothea van den Brincken, "Marianus Scottus," *Deutsches Archiv* 17 (1961): 191-238; and Borst, "Computus," pp. 37-38.

73. *De naturali cursu cuiusque lunationis* is known in four manuscripts, of which the earliest is MS Cambridge Trinity College O.7.41 (s. XI 2), fols. 54v-58; it is unpublished. *Sententia Petri Ebrei cognomento Anphus de dracone quam dominus Walcerus prior malvernensis ecclesie in latinam transtulit linguam* is known in MS Oxford Bodleian Library Auct. F.1.9 (s. XII med), fols. 96-99, and MS Erfurt Wiss.-Bibl. Amploniana Q. 351 (s. XIV), fols. 15-32; ed. J. M. Millás-Vallicrosa, in "La Aportación Astronómica de Pedro Alfonso," *Sefarad* 3 (1943): 87-97; I owe this citation to Charles S. F. Burnett and J. B. Trapp of the Warburg Institute. Another work of Walcher is *De bipertita discretione horarum* (1108-12?), known in two English manuscripts but unpublished. See L. Thorndike, *A History of Magic and Experimental Science*, vol. 1 (New York, 1923), pp. 68-73, concerning Peter; and Haskins, *Studies in the History of Medieval Science*, pp. 113-17, who discussed both Walcher and Peter and quoted from the *Sententia*.

74. *Praefatio magistri Rogeri infantis in compotum* (1124, 1176?), MS Oxford BL Digby 40 (s. XII/XIII), fols. 21-50v, and Cambridge University Library Kk. I. 1 (s. XIII), fols. 222v-239, partly published by T. Wright, *Biographia Britannica Literaria* (London, 1846), 2:89-91. Roger used and corrected the works of Heiric/Helperic and Gerland. See also Roger's *Tabula astronomica* in MS London BL Arundel 377 (s. XIII), fols. 86v-87 (1120-1400), but adjusted for the meridian of Hereford in 1178 (fol. 86v apud Toletum et Herefordiam), unpublished. See C. H. Haskins, "Introduction of Arabic Science into England," *English Historical Review* 30 (1915): 65-68, and his *Studies in the History of Medieval Science*, pp. 124-26.

V

75. Concerning Robert's early life at Hereford where he may have written the *computi*, see R. W. Southern, *Robert Grosseteste: The Growth of an English Mind in Medieval Europe* (Oxford, 1986), pp. 63-82 and 127-31. Richard C. Dales, "The Computistical Works Ascribed to Robert Grosseteste," *ISIS* 80 (1989): 74-79, dates the so-called *Computus I* shortly after 1200; because of its use of Arabic sources he places the *Computus correctorius* after 1217 and probably about 1220. The *Kalendarium* is left undated by Dales, but he thinks that it "is nearly identical with that of Roger of Hereford" (citing a paper by Jennifer Moreton, to be published). The so-called *Computus minor* was found to be a post-1220 abridgment of Robert Grosseteste's *Computus I*; it refers to yet another *computus* for common rules of calendar reckoning (Heiric/Helperic, Abbo, Gerland?), but not to any of Robert's works. The *Computus correctorius* of Robert Grosseteste displaced other works, such as *De anni ratione liber* of Johannes de Sacrobosco, and brought new English computistical scholarship into the core of the *corpus astronomicum* of Oxford, Paris, and other medieval universities. See Olaf Pedersen, "The Corpus Astronomicum and the Traditions of Mediaeval Latin Astronomy," in *Colloquia Copernicana* 3: *Astronomy of Copernicus and Its Background* (Wroclaw, 1975), pp. 57-96, especially pp. 73-82.

'Sidereal Time in Anglo-Saxon England', in *Voyage to the Other World: The Legacy of Sutton Hoo*, Calvin B. Kendall and Peter S. Wells, eds. © 1992 by the Regents of the University of Minnesota. Published by the University of Minnesota Press.

VI

FULDA SCRIBES AT WORK

Bodleian Library Manuscript Canonici Miscellaneous 353

Hraban wrote a Computus during the summer of A.D. 820 while he was master of the schools and two years before he was elected abbot of the Fulda monastery. Of the sixteen manuscripts known to me, six seem to have been transcribed during the ninth century, and two of these have a special relationship with each other and with Fulda. The earliest copy of Hraban's Computus is ms S. Gallen Stiftsbibliothek 878, f. 178–240, written mostly in the hand of Walahfrid Strabo about A.D. 825 when he was a sixteen year old student at Reichenau[1]; it was taken along to Fulda when he studied with Hraban during the years A.D. 827–829. Another copy of this Computus is found in the Bodleian Library whose catalogues provide the judgment of H.O. Coxe (between 1840 and 1843) that the ms Canonici Miscellaneous 353 was written probably toward the end of the ninth century[2]. A complete collation shows that its readings agree more closely with

I am indebted to Francis S.Benjamin Jr (Emory University) for calling my attention to this and other manuscripts of Hraban's Computus, to Bernhard Bischoff (Munich) for his suggestion that I undertake proof of Fulda provenance of this manuscript, and to the Danforth Foundation and the Alexander von Humboldt-Stiftung for research grants.

1 BERNHARD BISCHOFF, *Eine Sammelhandschrift Walahfrid Strabos (Cod. Sangall. 878)*, in *Aus der Welt des Buches: Festgabe . . . Georg Leyh* (Leipzig 1950 = *Zentralblatt für Bibliothekswesen*, Beiheft 75) 30–48; rpr with a supplement in BISCHOFF's *Mittelalterliche Studien II* (Stuttgart 1967) 34–51.
My edition of HRABANI *Liber de computo* (in preparation) will show the textual agreement of these two manuscripts and explain why this is so.

2 HENRY OCTAVIUS COXE, *Catalogi codicum manuscriptorum Bibliothecae Bodleianae, Pars tertia codices Graecos et Latinos Canonicianos complectens* (Oxford 1854; Quarto Series), Canonici Misc. no. 353; »Codex membr., in 4to minori, 55 ff., sec. forsan IX exeuntis.« FALCONER MADAN, *A Summary Catalogue of Western Manuscripts in the Bodleian Library at Oxford* IV (Oxford 1897) 406. EDMUND CRASTER, *A History of the Bodleian Library, 1845–1945* (Oxford 1952) 20, 91–92.

VI

Walahfrid's copy than with any other. A more careful analysis of the Canonici manuscript is thus required.

Terminus ante quem non for ms BL Canon. Misc. 353 is provided by Hraban who gave the annus praesens of his Computus and detailed examples of how to apply it on specific days when he was writing[3]. But there is no other reference or notation in the manuscript which may be dated historically. In order to locate the scriptorium and determine more precisely the time of writing, it is necessary to rely upon criteria which were not available to Coxe. Palaeographers today would note that the scripts are continental minuscule of the Carolingian era and assume that the manuscript was copied in a German scriptorium during the first half of the ninth century, thus between A.D. 820 and ca. 850; and they might consider the possibility of provenance from the author's scriptorium.

Readers will be able to verify that the manuscript was written at Fulda during the second quarter of the ninth century from the following description of the book, delineation of the scripts, and analysis both of common elements and of practices which distinguish the scripts from each other. Comparison of these with scripts in other Fulda manuscripts[4] – more practical than reliance upon verbal description – will limit the probable period of writing to the decade before A.D. 836.

In addition to the value that this newly dated manuscript will have for analysis of Fulda script in the ninth century, there is unusual interest in the fact that several of the scribes are demonstrably immature and of non-Fulda background. They can be observed individually changing their writing habits from folio to folio and even from line to line; they were taking on a new style under the guidance of a master of the Fulda scriptorium and with the example before them of mature Fulda scribes at work on the same manuscript.

I

The leaves of ms BL Canon. Misc. 353 are thick calf-skin with many faults, a vellum such as was commonly used in the British Isles or in those continental

3 For example: in addition to repeatedly stating that the annus praesens was A.D. 820, Hraban provided his own argumentum for ferial regulars in order to demonstrate how the system could be applied (Liber de computo LXXIIII, ed. PL CVII 710 C); he showed that when it was the seventeenth moon on the kalends of August, the feria would be IV or Wednesday: that coincidence occured in A.D. 820.
4 The Fulda manuscripts cited during this discussion are listed in Appendix B.

centers which were under insular influence. Their measurements average 233x145 mm after binding; the rulings and inner bounding-lines form an average writing space of 177x97 mm, or with outer bounding-lines 177x108 mm. To guide the writing, thirty horizontal lines and two sets of double vertical bounding-lines were ruled on each folio. Normally four bifolia were laid together with each hair-side up, and thirty prickings were made along the outer vertical bounding-lines. Each bifolium was ruled separately, then the four bifolia were placed together as before and folded with hair-sides out to form a gathering. Variations in size of writing space and in pattern of rulings are discussed in Appendix C, as is also the binding.

There are six gatherings of four bifolia each (f. 1–8, 9–16, 17–24, 25–32, 39–46, 47–54) and one gathering of three bifolia (f. 33–38); the final short tract is completed on the single leaf, f. 55. Each gathering was numbered by a Roman numeral with a point on each side (·I·, ·II·, seq), placed at the lower center verso of its final folio[5].

Most of the writing was in brown ink which now has a rich brown or red-brown hue, but black ink also occurs. The scribes must have used various mixtures of brown and of black; while the ink on most folios is clearly brown in several shades or clearly black, occasionally the browns seem to have darkened not only from aging but also from a scribe's deliberate mixture of inks. An example is the writing on f. 16–17 which began with brown but darkened gradually into black; and the reverse occurred on f. 5–6 and f. 8ᵛ lines 18–30 where the scribe had begun with black ink but the last lines shaded into brown. On the other hand it appears at f. 14 lines 20–21 that a scribe using brown accidentally put his pen into the wrong ink and wrote black for two lines by mistake. The black ink on f. 12ᵛ and 13 did not adhere properly to the vellum and has flaked off to some extent, such as occasionally happened with insular scribes in continental scriptoria; a contemporary and expert corrector did not have the same difficulty[6]. In

---

5 Three gatherings also bear the more casual mark ⨍ on the recto of their initial leaves: f. 9 (top left corner), f. 25 (top right corner), and f. 33 (lower right margin at line 28).

6 E. A. Lowe, *Codices latini antiquiores* II (Oxford 1935) viii-xii, summarized the differences in methods of preparing books in the insular and continental scriptoria before A. D. 800. Unmixed black ink was used in this manuscript e. g. on f. 25ᵛ line 29 – f. 26ᵛ line 18 and on f. 35ᵛ; black initials were often made e. g. f. 2, 3, 18ᵛ; and sometimes the numeral x was black e. g. f. 8. Many corrections were made in a fresh black ink by a contemporary scribe of the same writing center; his ink did not flake off; note e. g. f. 4ᵛ line 7 *circulo* (written over an erasure).

cases of uncertainty in the attempt to differentiate scripts, the ink sometimes fails to provide an aspect by which one hand may be distinguished from the next in this manuscript. It is to the scripts themselves that we now turn.

II

Hrabanus wrote his computus at Fulda, a monastery founded by the Anglo-Saxon missionary-bishop Boniface, where both insular and continental scripts were used by scribes working side by side or even by the same scribe alternately until at least mid-ninth century. Hrabanus was from A.D. 804 a master and from 817 the director of the Fulda schools until his election as abbot in 822. He could have used notaries for dictation during the latter years of that period, but he also expressed a love of writing and may well have transcribed parts of his own works in a semi-insular style of script which has recently been identified[7]. If parts or all of a final draft of Hraban's Liber de computo had been copied out in A.D. 820 by the author, an »insular« influence even stronger than what is expected from Fulda would certainly be evident in the script. This is in fact the case: every

7 Hraban's semi-insular script was first identified from the marginal and interlinear corrections to the earliest manuscript of his Commentary on Ezechiel which was completed by Fulda scribes about A.D. 842, by HANS BUTZMANN, *Der Ezechiel-Kommentar des Hrabanus Maurus und seine älteste Handschrift*, Bibliothek und Wissenschaft I (Frankfurt am Main 1964) 1—22 with eleven reproductions from the mss Wolfenbüttel, Guelf. Weissenburg 84 and 92. This article was called to my attention by Professor Douglas A. Unfug (Emory University), editor of *Central European History*.

Butzmann's proposal, that the correcting "Hand x" is that of Hraban himself, will doubtless be validated by analysis of other manuscripts. I have already identified the same hand also in ms Wolfenbüttel, Guelf. Weissenberg 86, f.14, 25, 63ᵛ, 64ᵛ, 68 etc., as Dr. Butzmann acknowledged when I called it to his attention on 1 May 1967. This latter manuscript is itself a product of the Tours scriptorium in the earliest period (A.D. 730—750); it was brought to Fulda early in the ninth century possibly by Hraban, whence it accompanied Otfrid, a student of Hraban until ca. 830, to Weissenburg in Alsace; most of the surviving manuscripts from the Weissenburg library are now in Wolfenbüttel. See PAUL LEHMANN, *Fulda und die antike Literatur, Aus Fuldas Geistesleben*, ed. JOSEF THEELE (Fulda 1928) 21; H. BUTZMANN, *Die Weissenburger Handschriften*, vol. X of *Kataloge der Herzog-August-Bibliothek* (Frankfurt am Main 1964) 30, 248 to 250; BUTZMANN, *Der Ezechiel-Kommentar ...*, 22.

Thus the restrained but suggestive conclusions of Butzmann have established a basis for further research into the intellectual labors of Hraban. His gracious assistance as keeper of manuscripts at Wolfenbüttel is gladly acknowledged.

ninth century transcript of this computus shows insular symptoms as well as errors sufficient to prove derivation from an exemplar which had been written at least in part with scripts of insular character. And it is especially true of ms BL Canon. Misc. 353 whose continental-trained scribes on occasion mis-copied r for insular n or s, and wrote p or s in place of insular r[8]. The following insular symbols are found:

ʜ	autem	ɔ	con-
∋	eius	ꝏ	contra
ⱶ	enim	m₃	-mus
∻	est	ꝑ	per
7	et	ꞇ	-tur

Occurrence of such errors and of these symbols may be explained by an exemplar which had been written in a continental scriptorium where Anglo-Saxon script was sometimes used. Authorship of the works leads us to consider one of them, Fulda, and the scripts fulfil our expectation.

The primary scripts of ms BL Canon. Misc. 353 are characterized by rectilinear shapes and by perpendicular vertical strokes in the long letters l, h, d, p. Upright minuscule d and uncial ∂ alternate irregularly. Minuscule r predominates, but the open r often appears as ✓ or γ in ligatur with a following letter or in the -rum abbreviation. The several scripts use two common Carolingian minuscule adaptations of uncial a and of rounded semi-uncial α , but they also use the open a – not with the common shapes ∝ or ∝, but rather with the upper strokes

---

8 E. g. f. 15ᵛ line 28 *lunaris* corr ex *luraris;* f. 31ᵛ line 24 *aegyptiorum* corr ex *aegyptiosum;* f. 21 line 11 *solaris* corr ex *solapsis.*

Some of the errors are shared by other manuscripts and must derive from common exemplars or from the original: e. g. f. 48 line 24 *secte* in four manuscripts rather than *recte,* but line 10 *recte* in all manuscripts; f. 52 line 5 *tenore* corr ex *tenos ut,* but uncorrected in another manuscript.

So - called insular symptoms in continental manuscripts in most cases need not derive from an exemplar written in the British Isles. Hraban and other Fulda masters were trained at Tours before its "Regular Style" became dominant there; and Hraban's interest in grammatical studies included the ancient notae of Roman orthography. It is also the case that in Fulda insular abbreviations were not always understood. See LUD-WIG TRAUBE, *Paläographische Anzeigen* 1. II, *Neues Archiv* XXVI (1900) 228; PAUL LEH-MANN, *Enim und Autem in mittelalterlichen lateinischen Handschriften, Philologus* LXXIII (Leipzig 1916) 546—548; E. K. RAND, *A Nest of Ancient Notae, Speculum* II (1927) 174 to 175.

tending to straighten and rise to sharp points, as $\mathcal{U}$. This form was used in the "pre-Alcuinian" period of Tours script, by which E.K. Rand designated a style prevalent before Alcuin came to St Martin's at Tours and which some scribes doubtless retained after his death, even though a master of the scriptorium had instructed them to avoid both ligatures and the open a. The transition was occurring during the time that Hraban and Fredugis were students of Alcuin. Hraban seems to have returned to Fulda in A.D. 804 with a preference for the earlier style[9], rather than for the "Regular Style" which was then developing and which came to dominate Tours when Fredugis was abbot (ca. 804–834).

Usages similar to these may be found separately in the products of several continental scriptoria in which insular styles were influential. When used together by the same writers however, these practices suggest the possibility that the manuscript was written in the monastic scriptorium at Fulda during the second quarter of the ninth century. And finally, the r and the a usages are distinctive for Fulda script.

Fulda provenance may be confirmed by comparison with other Fulda manuscripts. The great libraries of the monastery, the bishop, and the city of Fulda were plundered in the fifteenth century by Renaissance (albeit humanistic!) thieves, and the remainder of their contents was almost completely dispersed during wars of the next centuries. But through the labor of many scholars several book lists of the monastic library have been found and some of the books named in them have been identified in present-day collections[10]. As yet however, no

---

9 An example of this earlier Tours style which may have affected Hraban and also other Fulda students at Tours is ms Tours 286; see facsimiles in E.K. RAND, *A Nest of Ancient Notae*, 160–176. This manuscript will be cited below concerning particular writing practices because its margins contain corrections and glosses in Fulda script from the second quarter of the ninth century.

* 10 The studies of NICOLAUS KINDLINGER (1811), C.W.M. GREIN who identified the Fulda shelf-mark (1858), ANTON RULAND (1859), PAUL CLEMEN (1890), FRANZ FALK (1902), CARL SCHERER (1902), E.H. ZIMMERMANN (1910), PAUL LEHMANN (1925, 1927), and others were summarized by KARL CHRIST, *Die Bibliothek des Klosters Fulda im 16. Jahrhundert*, Beiheft LXIV of *Zentralblatt für Bibliothekswesen* (Leipzig 1933). Christ identified the most extensive Fulda catalogue in ms Vat. Pal. lat. 1928 (s. XVI med) and correlated it with existing manuscripts thought to be of Fulda provenance. Lehmann studied many Fulda codices, but his intention to publish a full account of the school, library, and scriptorium of the monastery was frustrated when most of his notes were destroyed by the bombings of Munich in July 1944.

There is a short list of books in the early insular script of Fulda written towards the end

systematic palaeographical survey of Fulda manuscripts has been undertaken, through the need has long been recognized[11]. When he studied the eighth and ninth century scriptorium of Würzburg Cathedral however, Bernhard Bischoff discovered that a Fulda manuscript had been brought to Würzburg at the beginning of Hunbert's episcopacy (832–842) and used as a model for training scribes. This was ms BL Laud. lat. 102, written at Fulda in the first quarter of the ninth century, as demonstrated by comparison with CLM 8112 and with the "Fulda Psalter". Bischoff's description of the five hands of the Laud manuscript was the first such analysis of a Fulda manuscript. Three of the scripts were in the insular style of Fulda and a fourth was non-Fulda. But the fifth (f. 177–207) was a continental minuscule characterized by broad upright discontinuous strokes which formed the long shafts; the letter a appeared with three forms: ∝ , a, open a with two rising points; g with small closed head and long swinging lower bow; h with well-rounded bow; t usually with a short cross-bar; dotted y made from half- and quartercircles; short horizontals in a; few notable ligatures: NT and nt at line-end, open r ligature with peculiar upstroke and hook, Si ligature at beginning of sentence; abbreviations aūt autem, p̊ post, q' -que; strong punctuation with semicolon, weak with point or comma, question with rising jagged line[12].

of s. VIII, ms Basel F. III. 15a, f. 17ᵛ–18 (palimpsest), which has not been published. See PAUL LEHMANN, *Fuldaer Studien, Sitzungsberichte der Bayerischen Akademie der Wissenschaften, Phil.-hist. Klasse* (Munich 1925), 3. Abhandlung, p. 4 seq and 47 seq, with an unclear facsimile of f. 17ᵛ; KARL CHRIST 65, 168–169; E. A. LOWE, *CLA* VII (1957) 2 and 54: nos. 842–843; JOSEF HOFMANN, *Libri Sancti Kyliani* (Würzburg 1952) 141 to 148, compared it with the list on ms BL Laud. Misc. 126, f. 260. And a selection of theological books was listed by a sixteenth century visitor to the Fulda collection in ms Basel F. III. 42, discovered by the librarian of the Universitätsbibliothek at Basel, GUSTAV BINZ, *Ein bisher unbekannter Katalog der Bibliothek des Klosters Fulda, Mélanges offerts à M. Marcel Godet* (Neuchatel 1937) 97–108.

11 Cf LEHMANN, *Aufgaben und Anregungen der lateinischen Philologie des Mittelalters, Sitzungsberichte der Bayerischen Akademie der Wissenschaften, Phil.-hist. Klasse* (Munich 1918), 8. Abhandlung, p. 14–15. EDMUND E. STENGEL described the series of archival endorsements on the back of Fulda charters from ninth through twelfth centuries but noticed no relation between these and the scripts of the charters or of the Fulda Cartulary, *Untersuchungen zur Frühgeschichte des Fuldaer Klosterarchivs, Fuldensia* IV (1958), rpr *Abhandlungen und Untersuchungen zur Geschichte der Reichsabtei Fulda* (Fulda 1960) 213 to 226, 263–265, and 244 n. 134.

12 B. BISCHOFF and J. HOFMANN, *Libri Sancti Kyliani* (1952) 15–16, 55–56. The "Fulda Psalter" is ms Frankfurt am Main, Stadtbibliothek Barth. 32; see EMIL SARNOW, *Katalog der ständigen Ausstellung der Stadtbibliothek* (Frankfurt 1920) 5, no. 15; GEORG

This analysis has provided a basis for further descriptions of hitherto unnoticed products of the same scriptorium by Lieftinck (1955), Hofmann (1957), and Butzmann (1964)[13]. The manuscript here under discussion is yet another, as a detailed analysis of each of its scripts should make clear.

III

Ms BL Canon. Misc. 353 was written by at least eight and perhaps up to thirteen scribes, designated below as hands I–VIII. The following table summarizes the folios and lines written by each hand:

Hand I	f. 2 line 5 (ideo)–f. 2ᵛ	Facsimile no.1
	f. 3 line 3–f. 8ᵛ	
IIa	f. 9 line 3 (uncias)–f. 10ᵛ	2
b	f. 11–12	3
c?	f. 13 line 26 (matutinum)–f. 13ᵛ line 7 (mane)?	4
c	f. 13ᵛ line 7 (conputandos)–f. 15ᵛ line 23 (id est)	5
d	f. 15ᵛ line 23 (signiferi)–f. 16ᵛ	5
III	f. 12ᵛ–13 line 26 (levant)	4
IV	f. 17–18	
	f. 22ᵛ line 12 (caeli)–f. 24ᵛ	6
	f. 36–53	
	f. 53ᵛ?	
V	f. 18ᵛ–22ᵛ line 12 (partis)	6
	f. 25 –26 line 25 (satis)	7
	f. 26ᵛ line 19–f. 29	
	f. 29ᵛ line 6–f. 32ᵛ	

SWARZENSKI and ROSY SCHILLING, *Die illuminierten Handschriften und Einzelminiaturen des Mittelalters und der Renaissance in Frankfurter Besitz* (Frankfurt 1929) 1–2, Tafel II. The Anglo-Saxon hand is identical with the chief hand of ms BL Laud. lat. 102, f. 9 to 38ᵛ, and is similar to that of ms Erlangen, Universitätsbibliothek 9 (Fulda s. IX in), f. 20–48ᵛ, according to Bischoff 56; a facsimile of ms Erlangen 9, f. 20 may be seen in E. H. ZIMMERMANN, *Die Fuldaer Buchmalerei karolingischer und ottonischer Zeit, Kunstgeschichtliches Jahrbuch* IV (Wien 1910) 13.

13 Their contributions will be cited below; all three scholars acknowledge Professor Bischoff's heuristic advice in these discoveries.

VIa	f. 2 line 2–5 (conparari)		1
	f. 3 line 1–2		
	f. 9 line 1–3 (ergo)		2
	f. 53ᵛ (last line) – f. 54 line 2 (polos)		9
	f. 54ᵛ line 1–2 (contra)		
b	f. 26 line 25 (tibi) – f. 26ᵛ line 18 (dixerunt)		7
	f. 29ᵛ line 1–5		
VII	f. 33–35ᵛ		8
VIII	f. 54 line 2 (septentrionis) – line 30		9
	f. 54ᵛ line 2 (pedes) – f. 55		

Appendix A will indicate in sequence which of the hands wrote each folio and each gathering.

In this Fulda schoolbook the dialogue notae are always Greek $\triangle$ for Discipulus and capitalis M for Magister. But it appears that each scribe usually supplied his own rubrics (silver-grey) and initials and made his own corrections; exceptions are noted below. Although Hand VIa organized the project and initiated the text of the first two gatherings and of the final tract, there was no overall attempt to correct omissions in rubrication or textual errors. For example chapters were enumerated only in the first gathering by Hand I who nevertheless failed to supply a title for the computus at folio 2 line 1 which had been left blank for that purpose. At f. 5ᵛ line 24 he took up the rubricating pen and wrote M[agister] as usual but continued into the text with silver-grey ink for three more letters, H y s [idorus], before changing back to his regular pen and ink. Other scribes were less able to keep the procedures in mind and often skipped over chapter titles altogether or left space to add them later but failed to do so.

These scribes separated words by a brief space, save for prepositions which were often treated as prefixes and combined with a dependent substantive or intervening adjective[14]. With rare exceptions the abbreviation mark was a short horizontal stroke with the slightest wave. Common are the tall st and ri ligatures which in various forms are generally found in Carolingian manuscripts. Along with the usages listed on p. 291–294 above, the α ligature executed in a nearly

14 This is treated as a common practice by PAUL LEHMANN, *Einzelheiten und Eigenheiten des Schrift- und Buchwesens, Sitzungsberichte der Bayerischen Akademie der Wissenschaften, Philos.-hist. Klasse* (Munich 1940), Heft 9, p. 23; B. BISCHOFF, *Paläographie*, in *Deutsche Philologie im Aufriss*, ed. W. Stammler, I (2 ed. Berlin 1957) 436.

perfect semi-circle may be expected of an experienced Fulda scribe. It was smoothly done by Hand I for example but not always by the others. In fact Hand IIc could scarcely manage it; that he made the attempt may indicate that a Fulda master was imposing his style upon a novice.

The major writing on f. 2–8ᵛ of the first gathering is designated Hand I (Facsimile No. 1) and is unmistakably a Fulda script within the above description. The upright strokes, rectilinear letters, and spaced words of Hand I include the symbols ⁓ or ·ē· est and ·ēē· esse (with or without points), & et (as conjunction or syllable, e.g. c& cetera), ủ vero, ł vèl. For nomina sacra the contractions are conventional with Deus, Dominus, and their declensions; note x̄p̄o Christo and the related words s̄c̄orum sanctorum, f̄r̄ frater, fr̄ī fratri. Other abbreviations are tñ tamen, sic̄ sicut, ꝑ per, p̄ pre, ꝓ pro; t̄ -ter but not t'; I-longa and N only at beginnings of words; ꝏ -orum twice at line-ends with a shape not found in the other hands.

Abbreviation of some words takes alternate forms: dix̄ and dixī dixit, ñ and n̄ and N̄ either non or nam, r̊q and r̊iq and rel̊iq reliqua. More striking is the variation of abbreviation mark in an' ante, sem' semel, quiqm' quiquam, and the numeral .c̊. centum milia; the snakelike mark was used on these four occasions by Hand I but nowhere else in the manuscript. It may be related to his use of the half-moon apostrophe for -us in the words primus and minus; both ' and ꝰ may have been used in the exemplar and occasionally transcribed here.

Concern for proper spacing is evident throughout the first gathering; more than enough room was left for the epistolary "Prologus", while the list of capitula was begun at the top of the next folio; full lines were used for short chapter titles; extra space was allowed for oversize initials[15]; and on f. 8ᵛ line 26–27 a complete two line sequence of Roman numerals with corresponding Greek letters was erased and recopied on line 27–28, leaving line 26 blank simply in order to avoid the appearance of a crowded page.

For punctuation Hand I used the single point on or slightly above the line for all stops. A few exceptions occur where a comma above the line is used once for a weak stop and three times possibly intending a strong stop. Once: appears for a weak stop and once; for a strong stop. The point is sometimes used after a question, but more often questions are left unpunctuated. His own nota for correcting

15 Some of these initials were capitalis quadrata in black ink; an example is the first title in the list of capitula at f. 3 line 2 (I. De numerorum potentia) where D was written in black ink and the scribe could not resist retouching the line with his black pen in hand.

an omission was the chryphia Ͻ as used at f. 7 line 12 within the text and in the lower margin where the missing words were supplied[16]. Corrections of spelling and punctuation have been made by various hands. One reader of the first gathering erratically dotted some of the i's, as well as each minim of one u, and extended the vertical stroke of t upwards.

The scribes who followed on f. 9–16ᵛ, second gathering, seem to have been novices working under the example of Hand I, changing and improving in style as they wrote. I have called the script Hand II, even though there were probably four scribes at work. The stroke of Hand IIa is thinner and less firm at first than the previous script, and it leans slightly to the right; with hands IIb, c, d the stroke tends to broaden. They used more contractions and suspensions than Hand I: characteristic is suprascript iota in p' pri- and q' qui. They often failed   *
to provide chapter titles or even space for them, made many textual errors, and four times left uncertain words to be filled in later. Spaces within the initials of each script are occasionally touched with the silver-grey rubricating ink[17]. While

16 HRABANI *Liber de computo* V 8 Ͻ *et* X *anteponitur* C *quando nonaginta significant* .... . This chryphia acquired a history when it was taken into the text of the computus in two manuscripts; the late ninth century CLM 14221 has the correct text but also has the marginal addition inserted again nine lines later; ms London BM Addit. 10801 (s. XVII) omitted the words where they belong but added them to the text where it had ended on f. 7 on the Canonici, taking both nota and marginal addition as just one more line of text – evidence that Addit. 10801 may have been transcribed from this manuscript. The same nota has been remarked in other Fulda manuscripts: Kassel Astron. F 2 (s. IX²) and Astron. Q 1 (s. IX med), Wolfenbüttel Guelf. Weissenburg 73 (s. IX med) and 84 (ca. A.D. 842); in Walafrid Strabo's copybook during the period following his study at Fulda: ms. S. Gallen 878 p. 278; and often in works copied or corrected by another Fulda student, Lupus of Ferrières: mss Bern 366, Vat. Regin. lat 597, and London BM Harley 2736. In the latter manuscript of Cicero's De oratore Lupus used the chryphia to indicate textual problems which were to be investigated (see f. 17ᵛ Ͻ ; f. 51ᵛ, 53ᵛ etc. ᴂ), although corrections were not always made; see the facsimile edition by C.H. BEESON, *Lupus of Ferrières as Scribe and Text Critic* (Cambridge 1930) 27; the manuscripts of Lupus are conveniently listed by ROBERT J. GARIÉPY JR, *Lupus of Ferrières: Carolingian Scribe and Text Critic, Mediaeval Studies* XXX (Toronto 1968) 90–105.

17 Spaces were left in text and later completed at f. 10 line 17–23, f. 10ᵛ line 1, f. 10ᵛ line 17–18, f. 14ᵛ line 7–8. Three times a corrector-rubricator of this gathering used the nota ⫮ to indicate either a suprascript insertion (f. 9ᵛ line 15) or placement in the text of an omitted chapter title, added in the margin (f. 9ᵛ line 22, f. 15 line 26). The identical nota was used by Lupus of Ferrières; BEESON loc cit.

it may be that many differences in Hand II derive from changes of pen or ink, variations in hair-side and flesh-side of the thick vellum, interruptions, fatigue, or occasional laxity in absence of the writing master, nevertheless the distinctions should be noted.

In f. 9 line 3 (quae uncias)–f. 10ᵛ Hand IIa (Facsimile No. 2) used the symbols ·e· or ē or ⌁ est, & et (conjunction or syllable), q̣ quia and ꝗ quod, q; and q;· -que, q̄q̄ quoque, ꝑ per, ꝓ pro, b; -bus, t̄ -ter (but not t'), ů vero, initial I-longa, NT at line-end; contractions dr̄ dicitur, m̄ tamen. In addition to the usual alternating forms of a, d, and r, the diphthong ae takes four forms within four lines on f. 9: conputistae, dispersę, quᴇ, q;. quae, and another form on f. 10ᵛ: q: quae. Monosyllables are accented with an arc over the vowel: tê, sê, âs. Punctuation is predominately a semi-colon for strong pause and reversed semi-colon for weak, but a medial point may serve for either.

Insular usages were transcribed from the exemplar: ꝯ con-, ⌁est, ł vel, ⁊ et, and the single instances of m; -mus and q; quae. Certain errors also support the assumption that the exemplar had been written partly in an insular minuscule of Fulda or in Hraban's semi-insular minuscule, scripts which Hand IIa was not accustomed to read[18].

Hand IIb (Facsimile No. 3) copied f. 11–12 with a firmer stroke and fewer words per line. Most of the same symbols were used and fewer abbreviations, save that suprascript iota is not found; notable are ⧾ enim[19], ƫ -tur (but not

18 Four errors derive from misreading of insular usages:

> f. 9ᵛ line 6 (partiri): *partisi* corr ex *partis et* [r/s; 7 et]
>
> line 17: *a temperamento* corr ex *aut̄ peramento* [at̄ autem]
>
> line 24: *decurrit* corr ex *decussit* [r/s]
>
> f. 10ᵛ line 20: *incipiebamus* corr ex *incipiebami* [m, or m; -mus]

An unusually clear example of how the rr/ss error in f. 9ᵛ line 24 could derive from a Fulda exemplar may be seen in the Fulda book-list from before ca. A. D. 850 in ms Vat. Pal. lat. 1877, f. 36 *seorsum*, where a very steady non-insular minuscule script suddenly switched to a cursive open r and s in ligature. A facsimile is in PAUL LEHMANN, *Quot et quorum libri fuerint in libraria Fuldensi, Bok- och Bibliothekshistorika Studier tillagnade Isak Collijn* (Uppsala 1925), facing p. 52. For a similar rs ligature at Bobbio, see ms Milan, Ambros. I. 1 parte superiore (s. VIII/IX); facsimile in FRANZ STEFFENS, *Paläographische Studien* (2 ed Leipzig 1929), Tafel 45(3), line I. 7 *universa*.

19 At f. 11ᵛ line 21 Hand IIb omitted the words *sic supra ostendimus* which a contemporary corrector supplied in the right margin. A few words later the scribe also omitted the preposition *in;* then he placed the mark ⧾ above the place. Later a second corrector added the preposition below the same place. This line of text was transcribed in CLM

-tus), tall rt ligature[20]; monosyllables were accented either with an arc or with a rising stroke; NT is found once at mid-line. Punctuation is a semi-colon for strong and reversed semi-colon for weak pause; but the medial point is introduced on f. 11^v, and on f. 12 it is used exclusively for all purposes.

The scribe of f. 13^v line 7 (conputandos)–f. 15^v line 23 (zodiaci id est) has been designated Hand IIc (Facsimile No. 5) because its stroke and shape of letters are similar to those of hands IIa and IIb, especially as he gained confidence on his last folio; all of them were trying to come to terms with the characteristics of Fulda style. Hand IIc was uncertain of his stroke in the ct ligature and the swinging lower bow of g; along with the previous three hands he formed x with a long right to left stroke curved below the line, but occasionally that stroke became stubby and turned back to the right, as found with Hand IIb; the two curving strokes of ẏ meet well above the ruling; and he produced initials ↲ U and ⅄ Q, unique within this manuscript. Insular symbols were transcribed: 7 et, ⱨ autem, ɔ con-, and ꝱ contra. He used the suspensions τ' -tur but not -tus, ꝙ quod, b; -bus and q'b; quibus at line-end, oꞃ and then oꞃ -orum; contractions dr̄ dicitur, qñd quando and qs̄i quasi, r̊q reliqua, scďm secundum; initial I-longa; NT at line-end and once at mid-line; t for d once. He usually disregarded the right bounding-line, but in the last few lines of f. 13^v, 14, and 15 he kept it neatly. Punctuation is a medial point.

It is possible that this hand began at f. 13 line 26 (matutinum) in a very shaky script (Facsimile No. 4) learned elsewhere. If so, we are able to observe even more clearly how the script was being shaped according to the Fulda style, as Hand IIc attempted the ct ligature, altered the shape of his -orum suspension, and abandoned subscript iota and the ro and rt ligatures; note also ⱨ autem, qs̄i, and the initials.

Hand IId (Facsimile No. 5) on f. 15^v line 23 (signiferi)–16^v has the general appearance of its predecessors in the gathering, and most usages are almost indistinguishable from theirs: q̣ quia, τ̄ ter, m̄ tamen, initial I-longa; NT at line-end. Distinctive are τ' -tur in which the apostrophe is no longer similar to

14221 as *sicut supra ostendimus; quicquid enim in corporibus* etc – probably from an exemplar which it shared with the Canonici manuscript. That common exemplar may have been Hraban's working copy of A. D. 820, from which the first fair copy had been made and sent to Marcharius.

20 The same rt ligature was used by Hand VI. Cf also the ninth century Fulda glossator of ms Tours 286 (s. IXin), facsimiles in E. K. RAND, *A Nest of Ancient Notae*, op cit (note 8).

a half-moon but has become a hooked-9; uncial a with exaggerated vertical stroke and the bow tending to flatten out, but no form of open a; e with a subsumed in diphthong; ƕ with the bow ending in a point, as seen hitherto only in the autem symbol; k preferred to c as a hard sound at beginning of some words; tall thin initial Q capitalis rustica; foreign terms for Hebrew and Egyptian months were over-lined in chapters XXIX–XXX. Punctuation is a medial point, doubled in the series of Hebrew months.

I have so far passed over the script of f.12ᵛ–13 line 26 (levant) which shall be designated Hand III (Facsimile No.4). There is a transition in the first lines of f.12ᵛ which prevents a sharp distinction from being made, but perhaps by line 3 and certainly by line 8 a heavier and broader stroke leaves the impression of a crowded and not very neat page and writing which at times is careless. The scribe used the Fulda open a, alternate d's, initial I-longa, and sparingly the abbreviations common to Hand II: dɼ dicitur, noṁ nomen, usq:, usque, ƕ autem (rather than ƕ), ꝫ eius – the insular symbol awkwardly transcribed from the exemplar. A medial point was used for all punctuation. The scribe's ink darkened gradually from brown to black, and on f.13 the black ink did not adhere well to poorly prepared vellum.

The third gathering was begun and ended by Hand IV (Facsimile No.6) who wrote f.17–18 and f.22ᵛ line 12 (per caeli)–f.24ᵛ; he also copied the last half of the short fifth gathering and all of gatherings six and seven to the end of the computus, f.36–53. This scribe was mature and his writing practices varied little as he worked. The letters are small, neat, and generally rectilinear, though less so than in hands I and V. Distinctive is uncial a, often with elongated vertical stroke and compressed bow; other forms of a were consistently excluded, and the open a occurs only once but was later begun by mistake and emphatically rejected three times[21]. There was a tendency to open the r slightly and to peak the upper stroke, a tendency fully expressed only in ꝝ -rum which recurs frequently[22]. The upright minuscule d is always used. At first g took the form

---

[21] See f.17 line 5 *intercalatum;* f.37 line 25 *epactas,* line 28 *epactae;* f.46 line 5 *tertia.* All instances reflect the exemplar. The scribe mistook s/r four times in errors that were corrected by a contemporary; examples were given above in note 8, and cf those in note 18.

[22] The identical -rum suspension is also found in ms Milan Ambros.I.1ˈsup.(s. VIII/IX), facsimile in STEFFENS, loc cit (note 18), who called it "die alte Form". The ro ligature and other cursive ligatures are also found in that manuscript. Cf further Hand A in mss Weiss. 84 and 92, facsimiles in BUTZMANN, *Der Ezechiel - Kommentar ...,* op cit (note 7).

of two almost closed circles but later stretched out into the swinging style of
Fulda; yet Hand IV does not attain the grace of the g developed by Lupus during
his decade in the Fulda scriptorium. I longa occurred once in mid-word (quIbus),
made with three short strokes[23]; y with its half-circle stroke to the left is small
and above the ruling; k is preferred to hard c at the beginning of words; τ' -tur
(but not -tus); alternate forms ꝗ and qđ quod, q· and q; and q;. -que, p̄p̄ and
ꝑp̄r̄ propter. Insular usages transcribed from the exemplar are 7 et, m; -mus,
p̊ post. Monosyllables were accented with an arc.

Within this manuscript Hand IV is unique in the use of h̄r̄ huius, ·C·vies
cenvies, apr̄e aprelis, b̄ -bet (cuiuslibet), fēr feria, ur̄is vestris, cursive ro ligature
(cf Hand VI), and sa ligature several times. Overlarge capitalis rustica initials
occasionally have inner spaces filled with oblongs of bright red ink on f.17; notable
are the forms of A, F, S, U, two forms of Q and of I, enlarged half-uncial e, uncial
m, and uncial d with an unusually straight upper stroke ending in a thin hook.
Initials and numerals of the nineteen verses on f.46ᵛ were rubricated silver-grey.

The final minim of numerals was often lengthened and hooked slightly to the
left; it was only in this hand that declination of ordinals was indicated. Punctuation was a point on or above the line for all stops; but the semicolon occurs for
full stop three times on f.22ᵛ and on the average of once per folio throughout this
script. A corrector has supplied the semicolon for full stop several times, especially
at chapter ends, and ⸜ once for a weak stop. An omitted line of text was copied
in the upper margin and its place in the text indicated by the ⨉ nota[24].

23 I longa is found in this manuscript at the beginning of words in hands I, IIc, IV, and V.
However the medial I longa occurs rarely and not so much for the semi-vocal sound
(e.g. *Ius, huIus*) as because it sometimes appeared to be an initial following a prefix (e.g.
preIudicare, deInde) — analogous to the practice throughout the manuscript of joining prepositions with their objects. But the occurance of *quIbus* at f.38 line 23 is anomalous.

24 This omission-insertion nota has been seen many times but rarely remarked upon in
print; see E. A. LOWE, *The Oldest Omission Signs in Latin Manuscripts: their Origin and
Significance, Miscellanea Giovanni Mercati* VI (Studi e Testi 126; Vatican 1946) 36—79,
esp. p. 75 (ms Paris BN lat. 11709, s. IX Visigothic?) and p. 70 (ms Paris BN n.a. 2171,
A. D. 1067 Visigothic). I believe that Lowe's wish to make this a Visigothic nota was
mistaken, since it tends to occur in manuscripts from the Germanic regions; but this
question will have to be pursued elsewhere.
Another corrector has twice indicated interlinear insertions by two points, like a colon;
that nota was used to correct Hand V; again cf Lupus in ms London BM Harley 2736,
f.78; BEESON loc cit. More common are the notae ./ in text and ./· in margin which
were used by another corrector of both scripts.

Hand V alternated with Hand IV in the third gathering, f.18–22ᵛ line 12 (partis); the fourth gathering, f.25–32ᵛ, seems to have been rubricated by Hand IV, but Hand V wrote almost all of the text[25].

The broad and even vertical stroke and thin horizontal stroke of Hand V (Facsimile No.6) give an impression of upright rectilinear script with well-spaced letters and words, such as we have seen in the first hand. Actually the level of lower letters tends to rise and fall, the long strokes often lean right with a slight bow, and the right bounding-line is little observed. The pointed Fulda open a alternates with miniscule a in which the stubby vertical stroke bends upright or back to the right. The upper stroke of r is peaked, but the letter does not open as in Hand IV; the upright minuscule d is always used, save for initials; g is open in semi-insular fashion and the downward vertical stroke arches slightly left to a point before making its lower bow which often does not close; the strokes of x also swing more freely than in earlier hands; two strokes of y meet well above the line and in some cases the short stroke on the left is low, horizontal, and almost without curve. Numerals were often not set off by points on each side, and v was used for the numeral rather than the usual u. A single monosyllable has an acute accent.

There are relatively few abbreviations in this script: ·ēē· esse and ·e&̄· esset (cf Hand VII), dt̄ dicit and dr̄ dicitur, q quia (rarely), ꝑ and qꝺ quod, q: or q; -que carefully distinguished from quę and q;· quae (all were spelled out by a corrector), ꝓꝑter propter, N̄ non. Found only with Hand V are d/t occasionally reversed[26], initial I longa two-thirds the height of a tall letter, and the suspensions epīsc episcopus[27], ꝺꝝ -orum (cf Hand I), lib· libro, verḡ Vergilius.

25 Two brief sections of the fourth gathering were copied by Hand VIb which will be discussed below.

It should be noted that Hand V usually wrote with a broad point and a brown ink with a slight red tint. But at f.25ᵛ line 29 the same scribe apparently picked up a different pen and used black ink, continuing to f.26 line 25 (satis), after which Hand VIb supplied a few lines. With that change of pen and ink Hand V's control seems to have been shaken, as may be noted in the letter g which henceforth varies each time as it occurs.

26 An exception is found in Hand IIc; this novice appears to have confused d/t once or may have transcribed the reversal from his exemplar; see p. 299.

27 The usual suspension at Fulda would have been eps̄; but epīsc had been known in Latin manuscripts at least since Victor, Bishop of Capua, corrected the Gospel Harmony in ms Fulda, Hessische Landesbibliothek, Codex Bonifatianus 1, about A.D. 546; E.A. LOWE, CLA VIII (1959) 1196. This is the oldest and most valuable manuscript known to have been in the Fulda library and must have attracted Hraban's antiquarian interest.

Initials are in two patterns: a handsome capitalis rustica, and a set of black and graceless capitalis quadrata extending over the space of two lines[28]. On several folios the spaces closed by one or more initial of either set were partly filled with red ink (by Hand IV?).

Punctuation is by means of a point slightly above the line, but ; will sometimes indicate a full stop at the end of a section or of a chapter. The scribe noted one omission with an apostrophe in the left margin of f.27 line 25 *aquila* and made the correction himself. Three further corrections illustrate his rare problems with the text: *ait*, used only twice in the computus, occurs at f.20 line 21; the scribe wrote *icit* but hastily and clumsily corrected it to *ait*. On f.21 line 11 is *solaris* corr ex *solapsis*, and on f.30ᵛ line 26 is *possem* corr ex *posrem* – errors deriving from misreading insular r.

At widely separated places in ms BL Canon. Misc. 353 similar fugitive scripts appear for a few lines. They share open a, insular e raised above adjacent letters, open r which forms ligatures ra re ro not found in the other hands, and rt ligature with strokes rising to double height. It is likely that these scripts are from the same Hand VI (Facsimiles No. 1, 2, 7, 9). It is also probable that this is the hand of a writing master who organized the production of this book. He copied the first four lines of the Prologus, the rubric for the list of capitula and first title of the list, as well as the first and part of the third lines for the second gathering,

The same suspension was used in a Salzburg manuscript towards the end of the eighth century, ms Salzburg, Stiftsarchiv St. Peter a. IX. 16; *CLA* X (1963) 1462. On the use of Greek H and C in abbreviations for nomina dei and then for other words with sacral associations, see LUDWIG TRAUBE, *Nomina Sacra* (Munich 1907) 5–7, 18, 161 to 166, 255; *Lehre und Geschichte der Abkürzungen, Vorlesungen und Abhandlungen* I (Munich 1909) 136, 145–152; W. M. LINDSAY, *Notae Latinae* (Cambridge 1915) 424–425; E. K. RAND, *Studies in the Script of Tours* I (Cambridge, Mass., 1929) 26, 54–55; PAUL LEHMANN, *Sammlungen und Erörterungen lateinischer Abkürzungen im Altertum und Mittelalter, Abhandlungen der Bayerischen Akademie der Wissenschaften, Philol.-hist. Abteilung*, Neue Folge, 4. Abh. (Munich 1929) 17–22, esp. p. 19.

28 The best known example of Fulda capitals which developed from Roman capitalis   ✳ rustica is the copy of Hraban's De laudibus sanctae crucis in ms Wien NB lat. 652 (ca. A. D. 840); the folios on which the text in capitals is superimposed upon colour drawings of Louis the Pious and of Christ are reproduced by ZIMMERMANN, op cit (note 12), figures 35 and 36; see also plate V in *Die Ausstellung Karl der Grosse – Werk und Wirkung* (Aachen 1965) no. 497: Louis the Pious as a Christian knight. Other copies produced at about the same time are mss Amiens 223, Paris BN lat. 2423, and Vat. Regin. lat. 124.

leaving the title of chapter VIII and the first △[iscipulus] for rubrication on lines 2–3, which was never supplied. However the main work of transcription appears to have been assigned by him to other scribes who may have been copying the first and second gatherings simultaneously. When the computus was finished similar controls were exercized for the addition of other items: the same master transcribed the initial line on f.54 of the short tract *Duos sunt extremi vertices mundi* ... (again not rubricated)[29] and two lines of its text at the top of f.54ᵛ, leaving most of the text to another. It may be the same hand, here designated VIb, which twice relieved Hand V for a few lines in the third gathering.

The script of f.33–35ᵛ, first half of the fifth gathering, will be described as a single Hand VII (Facsimile No.8)[30]. It shares a certain style and many usages with Hand II and may be considered together with it, but the overall appearance of the page is quite different in Hand VII. The scribe used a short choppy stroke evident especially in g and the ct ligature, a stroke nevertheless which was strong and assured. The letters are small and tended to be rectilinear (after the example of Hand I), but this was only partly successful as the vertical shafts lean first one way then the other.

In this hand we see the Fulda practice of alternate forms for a and d as well as the open r several times: the most common r is distinguished by a slight peak at midpoint of the upper stroke, with an ơ2 ligature (exordium) and suspension ơ2-orum; on the last third of f.35ᵛ however, this practice is alternated with ơ𝖞 -orum and α𝗒x -arum. The tall rt ligature also reappears; insular m𝟑 -mus was transcribed from the exemplar, but n' -nus shows the scribe's own usage;

29 The first words of this tract were copied onto a line at the bottom of f.53ᵛ by Hand VIa, leaving enough space for illustration or title, before he decided to begin again at the top of the next folio.

With Hand VIa cf ms Paris BN lat. 2423 (Fulda A.D. 844?) f.9; facsimile in *Catalogue des manuscrits en écriture latine* II 117 and plate VI. If it is the same hand, the Paris manuscript shows none of the uncertainties of ms BL Canon. Misc. 353 and must have been copied after the scribe had gained further years of experience. Cf also ms Wolfenbüttel, Guelf. Weissenburg 92 (Fulda 842) f.70.

With Hand VIb on f.26ᵛ, note ⤳ est, m, and x with three distinct strokes; cf ms Wolfenbüttel, Guelf. Weissenburg 84 (Fulda 842) f.103 lines 13–23.

30 Here it is assumed that differences between the script of successive portions, which in some cases are coterminous with shades of ink, may better be explained by external factors rather than by change of scribe. However the following succession of scripts may be worth further analysis: f.33 lines 1–28, f.33 line 29–f.33ᵛ line 28 (suppleri), f.33ᵛ line 28 (videamus)–f.35 line 2 (est), f.35 line 2 (quia)–f.35ᵛ.

p'us prius and q'b; quibus at line-end occur as in Hand II. Note also aūt autem, *
ē& esset, q· -que, NŌ Nonas. Rare is the ancient nota p' post which must have
been transcribed from the exemplar. For Hand VII had abandoned the apostrophe
as a suspension with τ' -tur and used it only once with τ' -tus (again from the
exemplar?), while repeatedly and consistently using the symbol τ² -tur. It is the
only hand to do so in this manuscript. Finally, h lacks the foot, x is either long
or short, and the two strokes of y meet above the line; note also the capitalis
rustica initial A and the distinctively thin and elongated initial P.

Folio 54 line 2 (septentrionis)–f.55 completes the tract *Duos sunt extremi
vertices mundi* . . . in a new Hand VIII (Facsimile No.9). The script is similar to
Hand IV, though much more erratic; it is distinguished by the angle of strokes
in s and st ligature, the rt ligature, open g under insular influence, m, -mus
and b, -bus.

In summary thus far: these scribes were copying an exemplar whose scripts
had many insular usages and which had probably been produced at Fulda; that
exemplar may have been a working draft of the author or a draft which included
at least his personal corrections. Hands I, V, and VI show the characteristics of
mature Fulda script, and Hand VI appears to have supervised the project. Hands
IV and perhaps VII were equally experienced but wrote in non-Fulda styles, and
Hand IV rejected some elements of Fulda script from the exemplar. Hands IIa,
b, c, d, III, and VIII were inexperienced and unsure of themselves; under instruction they tended to give up previously acquired writing habits in favour of Fulda
usages, for which Hand I was the prime example. Therefore I conclude that ms
BL Canon. Misc. 353 was written at Fulda[31].

IV

Although the same general characteristics make Fulda provenance certain,
ms BL Canon. Misc. 353 cannot have been written so early as were the Fulda
manuscripts cited by Bischoff from the first quarter of the ninth century.
Hraban's handbook-computus was composed during the summer of A.D. 820;

---

31 The possibility should be mentioned that the manuscript might have been written in
some other scriptorium which was being developed under the guidance of a Fulda
writing master or a former Fulda student, such as occurred at Würzburg Cathedral
during the episcopacy of Hunbert. I have pointed out several bits of evidence which
might allow such an interpretation, but the weight of evidence supports direct Fulda
provenance.

how much later was this copy made? By comparing it with other dated Fulda manuscripts[32] we may narrow the probable period of writing.

The earliest copy of Hraban's Commentary on Ezechiel was completed by Fulda scribes about A.D. 842, the year in which the author had been forced by Ludwig the German to give up his influential position of abbot and retire to the nearby Petersberg. This commentary with corrections in the author's hand is now at Wolfenbüttel in two volumes, mss Guelf. Weissenburg 92 and 84, and has been studied by Hans Butzmann[33]. There are many similarities between the Canonici scripts and those of the Weissenburg manuscripts: the common stroke of Canonici Hand I and Weissenburg Hand E, the rectilinear shapes and perpendicular vertical strokes in all hands, the a and d and r usages[34], semi-circular ct ligature, occasional insular symbols, and many particular writing practices indicate a continuity of habit in the same scriptorium. Inspection of these manuscripts has convinced me that the scripts of the Ezechiel commentary reflect a greater uniformity of style and consistency of usage, both in letter form and in abbreviation, than do those of the computus, and that the latter may have been written a decade or more earlier than 842.

Comparison with four fragments of Hraban's commentaries on Genesis, Leviticus, and Ecclesiastes leads to the same judgement. They have been dated between 838 and 840 by Josef Hofmann[35], and the non-insular scripts not only

32 While subjectivity is a bane for some palaeographical analysis, ultimately manuscripts must be set side by side and comparative judgments made — in part by use of clear and large reproductions. My opinions about the relative ages of these Fulda manuscripts are recorded here in expectation that their accuracy will be tested, refined, and perhaps corrected by other scholars.

33 *Der Ezechiel-Kommentar des Hrabanus Maurus ...*, op cit (note 7); he discussed ten of the sixteen hands which wrote the two volumes and provided excellent facsimiles. In addition to the hands A, E, and F which show similarities with the Canonici scripts, it is possible that the hand which copied ms Weiss. 84, f. 60 and the corrections on f. 25 to 42ᵛ etc is the later work of a matured Hand IIa, but this is a script not seen in Butzmann's facsimiles.

34 Canonici Hand IV had rejected the Fulda open a; likewise the Weissenburg Hand C does not use the open a and other peculiarly Fulda practices; both worked in the Fulda scriptorium nevertheless, though at different times.

35 *Fragmente von Bibelkommentaren des Hrabanus Maurus, Würzburger Diözesangeschichtsblätter* 18./19.Jahrgang (1956/57) 5—19, facsimiles 1,3,4. He thought that the Würzburg fragments of Hraban's commentary on Genesis had been written in the same script in both insular and noninsular styles as the primary hand that worked on the Fulda Cartu-

show Fulda usages in general but also many of the specific practices noted above in Canonici hands I, V, and VIa. These fragments however were written later than the scripts here being analyzed.

The Noctes Atticae of Aulus Gellius allows a further step to be taken. When Lupus was a student at Fulda in A.D. 829/30 he requested this work on loan from Einhard, then abbot of Seligenstadt. He did receive it and still had it in 836, for in that year he wrote to Einhard that Hraban was using the book and was having it copied for the Fulda library. The copy ordered by Hraban was discovered by G. I. Lieftinck[36] in ms no. 55 of the Bibliothèque Provinciale de Frise at Leeuwarden. Its Fulda scripts show even more certainly the same characteristics under discussion, and the parallels with hands I and V are greater than in the foregoing comparisons. However it is still probable that the Canonici scripts were written earlier than these.

In these manuscripts the 2-symbol was normal for -tur, while -tus and -mur were spelled out, and the apostrophe was used for -mus, -nus, eius. Two exceptions are shown in the available facsimiles: Hraban was still using τ' -tur in corrections of his Commentary on Ezechiel about A.D. 842; and the Hand VI of ms Leeuwarden B.A.Fr. 55 did the same once on f. 104^v in mid 836. Apparently a uniformity of practice had developed at Fulda before 836–842 in use of the apostrophe and the 2-symbol which paralleled other uniformities noted above. The Canonici manuscript however does not reflect that pattern; on the contrary it must have been written during an earlier transitional period. Its scribes dealt variously with the ambiguous apostrophe: hands I, IIa, III, V, VI, VIII did not use any symbol for either -tus or -tur but wrote them out; hands IIb, c, d, IV used the apostrophe to form -tur but wrote out -tus; and Hand VII consistently wrote τ^2 -tur. Hands I and VII used the apostrophe to form -mus and -nus, although Hand VII also transcribed τ' -tus once and mγ -mus once from the exemplar; Hand VIII used m, -mus, and the other hands wrote m; -mus.

lary ca. 828 or later, and perhaps also in a separate fragment of the Cartulary dated ca. 835 by PAUL LEHMANN, cited below.

36 *Le MS. d'Aulu-Gelle à Leeuwarden exécuté à Fulda en 836, Bulletino dell' Archivio Paleografico Italiano* N. S. I (1955) 11–17, especially facsimiles VI, IX, and XI. Lieftinck thought that Lupus' manuscript of the work, Vat. Regin. lat. 497, had been corrected from Einhard's during the Fulda loan. More recently P. K. MARSHALL has argued that the omissions and faulty readings which were not corrected in Vatican manuscript make this a doubtful assumption; see his Praefatio, *A. Gellii Noctes Atticae* (Oxford: Clarendon Press, 1968) p. xi-xiv.

At the time that this manuscript was transcribed therefore, the ambiguity of the apostrophe which all continental scriptoria faced during the second half of the ninth century was being temporarily resolved in Fulda by always writing out -tus, while taking one of two alternatives: τ' was not used at all and m' -mus was retained (e.g. hands I, V), or τ' -tur was retained with m; -mus (e.g. Hand IV). These were the examples before the students, so far as the evidence of this manuscript allows: two student-scribes (hands IIa, III) responded to the first alternative and used m' -mus but wrote out -tus and -tur; and three student-scribes (hands IIb, c, d) took the second alternative to use τ' -tur and m; -mus. No confusion is evident, and each scribe maintained one of these patterns consistently. However the use of τ^2 -tur in differentiation from -tus by Hand VII was tolerated at the same time.

It is uncertain when the pattern of Hand VII came to dominate Fulda practice. After leaving Fulda in ca. A.D. 836 Lupus of Ferrières used the apostrophe for -tur only twice in ms London BM Harley 2776, and he corrected τ' -tur to τ^2 -tur in the Tours ms Vat. Regin. lat. 1484. On the other hand the apostrophe was still used for -tur in the Fulda catalogue of ms Vat. Pal. lat 1877, f.35–43, from A.D. 815–836. In any case the "sure criterion" of apostrophe and 2-symbol practices by which W.M.Lindsay attempted to date manuscripts early or late in the period 700–850 and before or after ca. 820 may not be reinstated, especially not for a scriptorium in which insular script was written until after mid-century[37].

---

37 On the usages of Lupus see BEESON op cit (note 16) 20–21. The Fulda catalogue was dated not later than ca. 850, perhaps earlier, by LEHMANN, *Quot et quorum libri fuerint in libraria Fuldensi*, op cit (note 18) 51–52; but LINDSAY thought that it must be early ninth century because it has the apostrophe for -tur and corrections by an insular hand, *The Early Lorsch Scriptorium, Palaeographica Latina* III (1924) 14. Nevertheless the two Fulda book-lists "de cella paugolfi" on f.35ᵛ of ms Vat. Pal. lat. 1877 were probably written between the years 815 when former abbot Baugulf died at his Hammelburg retreat and 836 when Hraban and Rudolf visited the place.

The data from which Lindsay drew his criterion were presented in all their variety with appropriate demurs about certainty, *Notae Latinae* (Cambridge 1915) 376–378; yet he allowed his opinion to become simplified into an assertion of -tur usage as a "sure criterion". See his remarks in *Palaeographica Latina* III (1924) 13–14 and V (1927) 29–30, 35–36; his Forword to DORIS BAINS, *A Supplement to Notae Latinae* (Cambridge 1936); and Miss Bains' note on the Vatican Terence, *Classical Review* XLVI (1932) 153–154. Unfortunately Lindsay's test case, the dating of two book-lists from Lorsch in ms Vat. Pal. lat. 1877, actually failed: in conversation BISCHOFF has shown on independent grounds that f.44–79 which uses the 2-symbol must be dated earlier than

Rather it is sufficient to point out that the tendency towards a uniform differentiation between τ^2 -tur and τ' -tus, found in the Fulda scripts of 836–842, was used by none of the mature hands in ms BL Canon. Misc. 353 and by only one of the hands which were apparently under instruction. This reinforces my judgment that the Canonici manuscript should be dated earlier than those with which it has so far been compared.

One of the best known manuscripts from this period is the Fulda Cartulary, ms Marburg Staatsarchiv K 424. It was begun during the first few years of Hraban's abbacy (822–842) and was supervised by Rudolf. Intended to be an orderly record of the properties of the monastery, the Cartulary is a transcription of charters listed in fifteen regions and by abbatial periods within each region. The several volumes seem to have been dispersed after the twelfth century and most were lost. A surviving section of the original collection is f. 10^v–70^v of the Marburg manuscript, written in several insular Fulda scripts. When the primary work in this section had been completed a scribe added Roman numerals I– CLXXVIII in the margins, and within this enumerated sequence the latest date given is 3 October 828 in the next to last charter. However the final charter was copied in non-insular script, and the plan of organization was not maintained after the initial effort; further non-insular entries were made on f. 8 and 50^v before midcentury without regard for region or period. Two of these noninsular transcripts are of special interest.

A charter which concerned Rheingau during Hraban's abbacy was entered without enumeration by a non-insular Fulda hand on the unused verso of f. 50 (last leaf of former gathering ·VI·), following the Wormsgau charters from Baugulf's time (779–802); it represents the final settlement of two years of litigation[38] and was probably copied into the Cartulary soon after its date, 30 July 836.

f. 1–33 which uses the apostrophe for -tur; hopefully this demonstration will be included in the papers celebrating the 1200 th anniversary of Lorsch, now being printed. Further significant exceptions to Lindsay's criterion were given by E. K. RAND, *On the Symbols of Abbreviations for -tur, Speculum* II (1927) 52—65, and *Studies in the Script of Tours* I (1929) 26.

38 According to the enumeration written into the manuscript during the early eighteenth century by J. F. Schannat, this is charter no. 110; it superseded no. 217 (30 August 834) which was recorded out of place on f. 82 towards the end of s. IX. E. F. J. DRONKE, *Codex Diplomaticus Fuldensis* (Kassel 1850; rpr Aalen 1962), printed no. 217 (Dronke no. 487) but did not print no. 110 which he considered to be a duplicate.

The Wormsgau charters which made up the original contents of f.10ᵛ–70ᵛ ended with those from 823, 824, 825, 828, and 822 during Hraban's abbacy, the first four of which were copied in an insular script. The charter dated 3 October 828 has been used by all scholars to estimate the Cartulary's date of writing. Immediately following on f.70–70ᵛ is a Wormsgau charter from 822 in non-insular Fulda script. The date and script of this final charter suggest that it may have been briefly overlooked, though strict chronological order was not kept in any of the abbatial periods. But its enumeration in the primary collection indicates that it was transcribed shortly after the preceding charter, perhaps within the same year[39].

Contents of the two charters are the same, though the formulae vary; both bore the name of Rudolf as responsible for the writing, but the later version explicitly invoked the abbot's authority as well: "Hruodolfus indignus presbiter et monachus iussu domni Habani abbatis sui scripsit." (The slip which resulted in misspelling the abbot's name may reflect the tension behind this agreement, but it was also the copiest's tendency to squeeze letters together for ligatures with r.) Both versions have Tironian notes at the end. Transcripts of both charters were printed in parallel columns by EDUARD HEYDENREICH, *Das älteste Fuldaer Cartular im Staatsarchiv zu Marburg* (Leipzig 1899) 59, and discussed on p. 36, 49–52, with a facsimile of f.50ᵛ facing the title page. As I have not yet been able to see the manuscript in Marburg, the staff of the Staatsarchiv generously provided a microfilm copy of it. See also EDMUND E. STENGEL, *Über die karlingischen Cartulare des Klosters Fulda*, Fuldensia II (1921), rpr op cit (note 11) 156–159; and his *Urkundenbuch des Klosters Fulda* I 2 (Marburg 1956) xviii-xxi.
A further bifolium of the lost Thuringian section of the Fulda Cartulary, written in both insular and non-insular Fulda scripts ca. A.D. 835, was discovered by Paul Lehmann in the Benedictine monastery at Sarnen (Switzerland); his facsimiles are reduced in size and do not show the non-insular hands clearly enough for comparison. See LEHMANN, *Mitteilungen aus Handschriften, IX: Zu Hrabanus Maurus und Fulda*, Sitzungsberichte der Bayerischen Akademie der Wissenschaften, Philos.-hist. Klasse (Munich 1950), 9. Abhandlung, p. 6–18; and STENGEL, *Fragmente der verschollenen Cartulare des Hrabanus Maurus*, Fuldensia III (1956), rpr op cit 195–197.

39 The final charter of the primary collection, ms Marburg Staatsarchiv K 424, f.70–70ᵛ, non-insular minuscule, has in the margin a Roman numeral CLXXVIII (charter no.192 [Schannat], Dronke no.403). The scribe placed his initial letter within the double bounding-lines and used the inner line as guide for the text — normal practice which the insular hands had ignored. For further argument on including this charter within the primary Wormsgau collection, see STENGEL, *Über die karlingischen Cartulare des Klosters Fulda*, op cit 158 n. 56.
The insular script of the 828 charter is called Hand a by Dronke and Heydenreich; see HEYDENREICH op cit 40–43 and his facsimile of f.70 (facing p. 41) which shows both scripts.

The similarities of these two non-insular minuscule scripts with the Canonici hands I and V are again striking. It is even possible that on f.50^v of the Fulda Cartulary we see the script of Hand V after some years of experience. The irregular stroke on f.70–70^v however, with its changing letter shapes and inability to keep the vertical shafts upright, the tall st and rt ligatures, a cursive tendency to ligature & with t or i, and the lack of open a are very strong reminders of usages in Canonici hands II, III, and IV. In the development of Fulda scripts which we are beginning to visualize, it appears that the mature Canonici scripts had not yet taken on the style of mid 836 and that the practices of immature Canonici scribes are consonant with a hand at work in Fulda during late 828 or soon thereafter.

V

Analysis of the scripts of ms BL Canon. Misc. 353 has confirmed the impression that this book was made in the Fulda scriptorium during the second quarter of the ninth century. A terminus no later than mid 836 has been established by comparing these scripts with those in dated Fulda manuscripts; and a Fulda charter transcribed towards the end of 828 or early 829 is not too early for correlation with some of the Canonici scripts. Consequently this manuscript was probably written during the decade before A.D. 836.

Together with the other manuscripts cited, the Canonici will allow development of a composite picture of the Fulda scriptorium during the first half of the ninth century, with no appreciable lacunae and including the work of more than fifty scribes writing in both insular and continental styles. A collection of facsimiles from these few Fulda manuscripts would allow identification of many others which are still of uncertain origin. It would eventually make possible a systematic analysis of the Fulda scriptorium and a more adequate knowledge of the magnificent collections which once composed the medieval Fulda library.

Such a study should also include the manuscripts of those who were trained elsewhere but who worked for a time in the Fulda scriptorium. The Laud manuscript analyzed by Bischoff and the Weissenburg manuscript analyzed by Butzmann show insular Fulda, non-insular Fulda, and non-Fulda hands in the same books. And in the Canonici manuscript the largest portion of the writing was done by Fulda Hand V alternating with a non-Fulda Hand IV. The latter script not only lacked distinguishing Fulda characteristics but also positively rejected such practices when they faced him in the exemplar. Yet he too wrote at Fulda.

As the scripts of ms BL Canon. Misc. 353 were being analyzed, the realization slowly emerged that the book had been made under the direction of Fulda Hand VI. This writing master used the project to instruct immature scribes (IIa, b, c, d, III) whose earlier non-Fulda training was yielding to new usages, as well as mature scribes (IV, VII) whose non-Fulda styles were less adaptable. Throughout the second gathering the students neglected to include numerals, titles, and dialogue marks for some chapters; they rubricated erratically, allowed pens to dull and inks to darken, used many abbreviations, transcribed occasional insular symbols awkwardly, and made numerous textual errors. Hands IIa, b, c, d and VII had brought with them a distinctive suprascript iota in suspensions p' pri and q' qui, suspensions for -rum, and ligatures which did not satisfy the Fulda master.

His instruction that quae and -que be clearly distinguished must have resulted in some initial confusion, for we find three forms of the conjunction and six forms of the ae diphthong in hands IIa, d and IV. Attempts to form rectilinear letters and upright verticals also had no consistent results, nor did efforts at making free swinging lower strokes of g and x, though the lower circle of Hand IV's g tended to open and lengthen more gracefully as he wrote. Hand IV was least affected by writing at Fulda, but there was some change in his -rum suspension with open r. More evident is the adaptation in Hand VII; and the same direction of change in forming -rum is very pronounced in novice Hand IIc. Before them are the effective examples of mature Fulda Hands I and V: under this influence Hand IIc may also be seen to abandon subscript iota and the ro rt ligatures.

The open a does not appear in hands IId and IV who always used uncial a with exaggerated vertical stroke and flattened bow; they also preferred K to initial c in some words. These are practices which must derive from elsewhere than Fulda. But the shibboleth is the ct ligature: the Fulda writing master required every scribe to try and all had difficulty; Hand IIc could not make it at all, and Hand VII did it with short choppy strokes. The ct ligature executed in a nearly perfect semi-circle is found in the mature Fulda hands of each manuscript cited in this essay and should be included in the combination of writing practices which indicate Fulda provenance.

While medieval scholars have long known in general that writing was an essential part of Carolingian education, ms BL Canon. Misc. 353 allows us an opportunity to catch Fulda script in the very process of being taught and learned during the decade before A.D. 836.

Appendix A: Sequence of folios

Blank	f. 1–2 line 1 [probationes pennae]	
Hand VIa	f. 2 line 2–5 (conparari)	
I	f. 2 line 5 (ideo) – f. 2ᵛ	
VIa	f. 3 line 1–2	
I	f. 3 line 3 – f. 8ᵛ	Gathering ·I·
VIa	f. 9 line 1–3 (ergo)	
IIa	f. 9 line 3 (uncias) – f. 10ᵛ	
b	f. 11–12	
III	f. 12ᵛ–13 line 26 (levant)	
IIc?	f. 13 line 26 (matutinum) – f. 13ᵛ line 7 (mane)	
IIc	f. 13ᵛ line 7 (conputandos) – f. 15ᵛ line 23 (id est)	
d	f. 15ᵛ line 23 (signiferi) – f. 16ᵛ	·II·
IV	f. 17–18	
V	f. 18ᵛ–22ᵛ line 12 (partis)	
IV	f. 22ᵛ line 12 (caeli) – f. 24ᵛ	·III·
V	f. 25–26 line 25 (satis)	
VIb	f. 26 line 25 (tibi) – f. 26ᵛ line 18 (dixerunt)	
V	f. 26ᵛ line 19 – f. 29	
VIb	f. 29ᵛ line 1–5	
V	f. 29ᵛ line 6 – f. 32ᵛ	·IV·
VII	f. 33–35ᵛ	
IV	f. 36–38ᵛ	·V·
IV	f. 39–46ᵛ	·VI·
IV	f. 47–53	
IV	f. 53ᵛ?	
VIa	f. 53ᵛ (last line) – f. 54 line 2 (polos)	
VIII	f. 54 line 2 (septentrionis) – line 30	
VIa	f. 54ᵛ line 1–2 (contra)	
VIII	f. 54ᵛ line 2 (pedes) – line 30	·VII·
VIII	f. 55	
Blank	f. 55ᵛ	

Appendix B: Dated Fulda Manuscripts

s. IX in	Erlangen, Universitätsbibliothek 9
s. IX early first quarter	Munich, Bayrische Staatsbibliothek, CLM 8112
s. IX first quarter	Frankfurt-am-Main, Stadtbibliothek, Barth. 32
s. IX first quarter	⌐Oxford, Bodleian Library, Laud lat. 102
815–836	Vat. Pal. lat. 1877, f. 35–43
826–836	Oxford, Bodleian Library, Canon. Misc. 353
828–836	Marburg, Staatsarchiv K 424, f. 10^v–70^v
* 836 med	Leeuwarden, Bibliothèque de Frise, B.A.Fr. 55
838–840	Fragments from Würzburg, Weissenburg i.B., and Windsheim
842	Wolfenbüttel, Bibl. Guelf., Weiss. 92 and 84

Appendix C: Prickings, Rulings, Bindings

Gatherings ·II·, ·III·, ·IV·, and ·V· were made in the same way. Four bifolia with hair-sides up received thirty prickings simultaneously in two vertical lines, perhaps guided by a straight-edge parallel to the left and right edges of the bifolia. This may be observed in the second gathering, in which folios 9, 10, 11, 12 were pricked on the right and folios 16^v, 15^v, 14^v, 13^v on the left; movement of an awl may be followed through identical holes of decreasing proportions from top to bottom bifolia. Visualized on the schema of Figure 1, the prickings were done first between positions LL. A second set of prickings was made at II, and two more sets 1–5 mm above the top line and 1–3 mm below the bottom line at CF. On the hair-side of each separate bifolium the bounding-lines CC, FF, II, LL were ruled with a dry stylus; and between CC and LL thirty horizontal rulings were done; due to carelessness some had to be ruled a second time. The four bifolia were again placed together as before and folded with hair-sides out.

So far as I know, this practice has not been described before, despite a considerable modern literature on manuscript prickings. A microfilm copy of the Fulda ms Marburg, Staatsarchiv K 424 shows that it may also have been prepared in this manner.

Bifolia of the first gathering were pricked along the bounding-lines LL and at II, FF, CC as described above; but differences of pricking and ruling suggest that each bifolium was drawn separately from a common store where they had lain in varying stages of readiness. The sixth and seventh gatherings also show variations: bifolium 39–46^v was ruled in the same manner but has a larger writing space 177x100 mm, or 177x113 mm with outer bounding-lines. The other bifolia use prickings at F'F' and JJ (5–8 mm above and below EE and II) to guide bounding-lines which limit the writing space 177x101 mm; prickings along the unruled BB and LL served to limit extension of horizontal rulings, and the writing commonly runs over JJ to the right. In these two gatherings the vellum is thicker (except f.42–43^v), and the hair-side rulings are not easy to see, even though each bifolium was ruled separately.

Figure 1 is modelled on the schema provided by L.W.Jones, *Where are the Prickings? Transactions of the American Philological Association* LXXV (Philadelphia 1944) 72, which he used to illustrate several other systems of pricking and ruling. The figures in E.K.Rand, *Prickings in a Manuscript of Orleans*, ibid. LXX (1939) 329–337, are also very helpful in visualizing the steps in putting together a number of bifolia to form a gathering.

The present binding allows only thread-holes of earlier bindings to be observed on the inner fold of each gathering, and some of these are obscured by re-use of old holes and overlapping of new threads. A pair of the oldest darkened holes 25 mm apart is 34/59 mm form the top of the first gathering; within a tolerance of 5–6 mm the same pair occurs on each gathering. Another pair of the oldest thread-holes 21 mm apart is about 120 mm below the first pair, 180/201 mm from the top (or 39/32 mm from the bottom) of gatherings ·I·, ·II·, ·III· (probably), ·VII· and of the final folio. The lowest of these holes at 201 mm from the top occurs also in gatherings ·IV· and ·V·, for which the second hole of the pair is perhaps at 194 mm, but not in gathering ·VI· where the pair seems to have shifted to 176/194 mm. Each of the two oldest pair had a separate thread passed through its two holes and tied on the outside of the gathering; threads of seven gatherings and the final folio were then tied together. But there is now no evidence of a cover (vellum or board) or of reinforcement along the spine for ties.

To reinforce the older threading, a later binder added two more pair of holes: the upper pair 13 mm apart is 111/124 mm from the top, and the lower pair 9 mm apart is at 158/167 mm in gatherings ·I·, ·II·, and the final folio. The other gatherings also show two pair of holes between the oldest ones, perhaps added by the same later binder, but they are spaced irregularly. It was apparently the Soranzo binder who removed all materials of the original and of the later bindings, made new holes, and threaded them afresh; but in several gatherings he re-used a single old thread-hole at 32 or 39 or 56 mm from the bottom.

Such limited observations of the original binding technique are not sufficient to include ms BL Canon. Misc. 353 with certainty among the manuscripts described by BERTHE VAN REGEMORTER, *La reliure souple des manuscrits carolingiens de Fulda*, *Scriptorium* XI (1957) 249–257, viz p. 252: "On trouve toujours la reunion des deux techniques: ais souples et deux fils." It may be noted also that the manuscripts studied by Van Regemorter were attributed to Fulda on the basis of sixteenth century shelf-marks and their presence in the seventeenth century collection of Remi Faesch; date and provenance of their scripts are quite varied and for the most part still uncertain, and so may be the origins of their bindings.

LIST OF FACSIMILES OF BL CANON. MISC. 353

The facsimiles are reproduced in a slightly smaller scale than the original Ms. For original size cf. p. 289.

Facsimile no. 1: folio 2^r – hands VIa and I
Facsimile no. 2: folio 9^r – hands VIa and IIa
Facsimile no. 3: folio 11^r – hand IIb
Facsimile no. 4: folio 13^r – hands III and IIc (?)
Facsimile no. 5: folio 15^v – hands IIc and IId
Facsimile no. 6: folio 22^v – hands V and IV
Facsimile no. 7: folio 26^r – hands V and VIb
Facsimile no. 8: folio 33^r – hand VII
Facsimile no. 9: folio 54^r – hands VIa and VIII

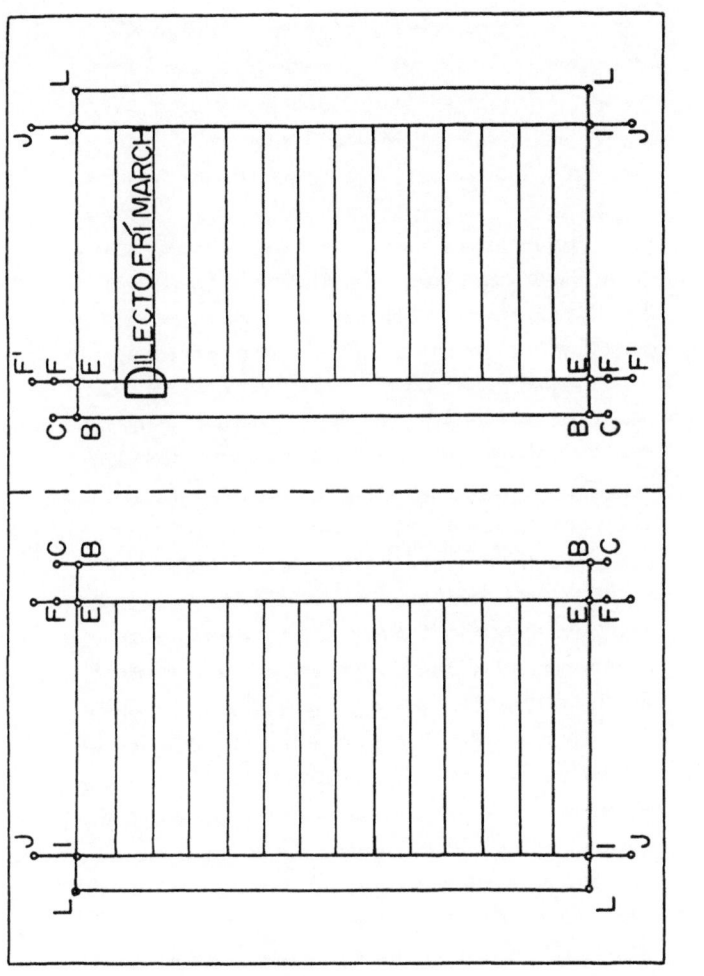

Figure 1

Dilecto fri macchario monacho hirubanus
peccator inxpo sal̄. legimus scriptum Inpro
uerbis melior e sapientia cunctis prosissimis. &
omne desiderabile a nonpotest comparari. Ideo
fr dilectissime gratias ago dō qui tibi eius amorē
Inspirauit. cuius possessio mundi diuitias contēp
nente. Teq; reddidit suofulgore decoratum. & p
·imis tuis pfectuosum· pācebas ergo utquibus
dam decōputo propositionib; earumque min
perfectis responsionib; quas mihi protuleras
nescio aquibus confectas Pilum adhiberem. eáq;
tibi ludidio resredderem· feci quantum potui sed
non eo ordine quo ibi positas repperi· quia con
fusa series uim cognoscendi abstulit. & tędium
lectionis inuexit· pleraque ergo quae mihi
magis necessaria uidebantur addidi. & ordinē
Inipsis rebus disponere contendi. Inde quoque eue
nit dum breuitati studerem. & tamen ipsarum re
rum ueritatem patefacere uellem. quod unius li
bri quantitatem deuitare nonpossem; Conposui
quidem exnumero & temporum articulis quendā
dialogum. & nomini tuo ipsum dicaui, Inquo quę
necessaria mihi uidebantur interrogandi discipu
li nomine. & quae respondendi magistri uocabulo
pnotaui. & nonhæc tantummodo propriis ratiōci
nationibus. sed etiam ex antiquorum dictis & scorum
patrum sententiis enodare curaui. Ideo fr karis
sime hæclegens· scias me non difficultatem uerboʳ
aut obscuritatem sententiarum studuisse. Immo

+ ...enıſ quoq; parte hıſden ꝓ fıgere ſcıunt

Debıſ ergo qın eıaſ coꝑutaſ eae appellant rogout
edıſſere? Unaarū qq diuiſione qua non mınuſ
temporibuſ rebuſ ue aliıſ quam num miſſ ap eıcon
putandıſ non ıgnobılıſ ın tertıo ÷ quıbuſ quıddıſ
pote paſſım hıſtorıae ꝙ ı̄p̄ra ſacra ſcrıptura
uteatur nomına partet ꝙ fıguraſ eorum paucaſ
adfıgere curauımuſ; lıbra uel aſſıuſ aſſıſ xıı unciae;
deunx ſıue labuſ xı unciae;
deaınx uel dextanſ x unciae;
dodranſ ſıue dorıſ uıı unciae;
beſ ſıue bıſſe uıı unciae;
ſeptunx ſıue ſeptuſ
uı unciae.
ſemıſ ſex unciae ·   ſeun ...utter wıelt
quıncunx ſıue ancuſ unciae · ꝑ
trıenſ ſıue ertaſ ıııı unciae ·
quadranſ ſıue quadraſ treſ unciae;
ſextanſ ſıue ſex teſ duo unciae;
ſeſcunx ſıue ſeſean eıa ·
ı unciad ꝙ ſemıſ· ueorıſ ergo eu dıeıſ ...ı quam
unciad xxıııı ſcrıpulı · ꝑe.........
ſem unciad xıı
ducae ſextulae ſıue ſerelae uıı· ıd eert ıa paſ unciae;
ſtalicuſ uı
ſextula ſıue ſerele ıııı·
dımıdıa ſextulae ſıue ſerele ıj·
ſcrıpuluſ uı· ſılıquae·
haec ınquam ponderū uocabula t charae te reſ

reddit integram; A. unde dicitur momentum. M.
a motu siderum celerrimo cum aliquid tibi breuiss-
mis in spatiis cedere atq; succedere sentitur;
capit ergo unum momentum. ostenta & dimidiu;
a domos qq. d. lx. IIII; de partib; A. quid nomi-
natur in computo partes? M. partes a partic-
one circuli zodiaci uocantur; quæ tricenis diebus
per m̅ ses singulos findunt; recipiunt h̅ singulæ
partes mom̅ tꝛ. II. & duas partes unius mom̅tei
ostentæ IIII. & atomos. i. d. IIII; de minuto. A. quid
est minutum; M. decima pars horæ. A. unde dicitur
minutum; M. a minore licet ualló qm̅ minus mom̅
tu; q minus numeretur q̅ magis im̅ plet; habet
ergo minutum partem unam & dimidiam mom̅tea
IIII. ostentæ uii. atomos. ii. ccl. uii; de puncto.
A. quid est punctus; M. quarta pars unius horæ;
A. unde dicitur est punctus; M. a paruo puncti transcensu q̅ fit in horologio; punctus quip-
pe a pungendo dicitur ÷. eo q̅ quibusdam punc-
tionib; certæ designationis in horologiis desig-
netur; punctus autem habet minuta duo & di-
mediu; partes IIII. momentu x̅. ostentæ xu.
atomos. u. d. c. xl. quattuor ergo punctum unam
horam faciunt; de hora. quid ÷ hora. M. certus
est minus temporis; si qui de hora duo decimæ
pars est diei. ihō ad restante qui a̅c; nonne
xii. horæ sunt diei. si quis ambulauerit in die
non offendit; ubi quam uir allegorice redie.
discipulor̅ u̅ q̅ ait in lustran diem erant; horas cap
pellucerit; solito cū more humanæ reputationis.

Facs. no. 3: Hand IIb (Ms. Canon. Misc. 353, Bodleian Library Oxford, fol. 11ʳ)

denoctem solis absentia terrarum umbra
con dita donec aboccasu redeat adexor
tum · unde dicta ē nox · quod noceat
aspectibus t negotiis humanis siue quod in
ea fures t latrones nocendi alius occasionē
nanciscantur · ob quam causam facta
ē nox · praetemper uncta laboris huma
ni ut corporu requiem haberent & ut
animalibus quibusdam solem non feren
tibus uictum que stando daretur occasio
de partibus noctis noctis partes quot
sunt aquem crepusculum uespere
uesperum conticinium Intempestum
galli cinium matutinum diluculum ·
horum proprietates per singulas species dic
crepusculum ÷ dubia lux nam creperu
dubium dicimus · hoc inter tenebras
& lucem · uesperum ab lappurente stel
la eiusdem nominis dicitur dequa poeta
ait ante diem clauso componet uesp
olimpo · conticinium quo omnia con
ticescunt ide silent Intempestum me
dia nox ide qsi Inactuosum qnd om
nibus sopore qui&is nihilo perandi tē
pus ÷ gallicinium qnd galli cantum
leuant matutinum Inter abscessum tenebrar
& lucis re aduentum diluculum qsi iam Incipiens
paruadies lux hec & aurora pangere usq ad so
lis exortum · Harum ergo partium natura
lis dies quae precedit Intempore utru nox seu dies

Facs. no 4: Hands III and II c (?) (Ms. Canon. Misc. 353, Bodleian Library Oxford, fol. 13ʳ)

feriam quartam quintam & sextam de suo adnectentur:
sabbatum excudens scriptura recinuit · Δ unde dr̄ feria · M · a fando scilicet & ideo dominicus
dies In quo dixit dr̄ fiat lux prima feria dici potest · Deinde céterae feriae a prima numerentur·
Δ ebdomada ergo quot horas continet · M · c·lx
uiii Δ quot punctorum · M · dc·lx·ii · Δ quot minutiae·
M · l·dc·lxxx · Δ quot partes · M · ii·d·xx·Δ quot momenta · M · uidcc·xx · Δ quot ostentae · M · x̄·l·xx ·
Δ quot atomos · M · trea milia milium·&dcc·xc·dccc·
de mensibus · Δ · mensis quid est · · M · lunaris
per zodiacum circuitus · Δ · unde dr̄ mensis · M ·
a luna quae greco sermone mane dr̄ nam &a
pude or menses uo carctur menses nc & apud
hebreor hieronimo teste luna quecon lare no-
minante mensib; nomen dedit · unde & iursibus
strahe de luna loquentibus · mensis secundum
nomen ± eius · Δ · utrum ad solem an lunam m̄r
pertinet · · M · ad utrosque sed magis ad lunam
ideoq; rectius tac difiniendum puto q̄m sint
luminis lunaris circuitus a credintegratio de
noua ad nouam · Solaris h̄ m̄sis digressio é solis
per duodecimam partem zodiaci idest signiferi
circuli quae xxx diebus & x semis horis Inpl&·
xx·duobus uidelicet horis aedimidia lunari m̄e
per ductior siquidem solaris mensis aequa diui-
sione totius anno p·xii· partita continet dies
xxx·x·horas desemissem · lunaris uero continet naturaliter dies xx·uiiii· & xii horas
sed quhaec proputatio difficilis fuit antiqui

Facs. no. 5: Hands IIc and IId (Ms. Canon. Misc. 353, Bodleian Library Oxford, fol. 15ᵛ)

pcedunt · statiua dicuntur quia dū stella
semp mouetur tamen ĭnaliquibus locis sta
re uidetur · Deplurieū uū cursup signiferū aenū ı
Sedquia audi utredicertem planetarū uan signi
fercum esse ĭnzodiaco ipsius zodiaci ū pprietā fer
con a cursū eorū ĭneorogo ut exponas ·
Macium zodiacus uel signifer · ē · circulusobli
quus; xii · signis constans pquem errantes stel
le feruntur · Hec aliud habitatur ĭnterris
quā quod illi subiaca · reliqua apolis squalent ·
ē uit signifer · cccbcv · partabus ā quadran
te unius partis p cæli ambitū longus · & xii · par
tib; latus · p transuersū · ueniens tamen stella ex
cedit eumbinis partib; · luna qq ptotā latitu
diné eius uagatur · Sed omnino ñ excedens eum
abhis mercurii stella laxissime e · ut tam ēduo
denis partab; tot sunt eni latitudinis ut dicimus
ñ amplius octonas p errā; · Neq; has equalir sed
duas medio eius & supra quattuor ĭnfraduas
sed ĭndimedio feré ine duas partes flexuoso dra
conū meatu inæqualis · marts stella quattuor
medus · iouis media & supra duab; · saturni dua
bus ut sol; · De · xii · signis
Signa aut zodiaci uellē ut ediceres unde traxe
rint nomina ex causis annalib; t agestaliū
fabulis nomina sumserunt signa xii · nam
ariete marto mense pp ammoné ioue tri
buunt · unde & meus simulacro arietis cor
nua singunt · taurū aprili ppé eundé ioue
ep ĭnboue sic fabulose conuertit · castore &

M Suus quidem cuique color ē. saturno candi
dus. Iouis clarus. mars igneus. lucifero gaudens
uesper orefulgens. mercurio radicus. lunae
blandus. sol lucet oritur ardens postea splendens
sed colores ratio altitudinem temperat. Si qui
demearum similitudinē trahunt. In quarū aerē
uenere subeundo. tingiturq; adppinquantes u
tralibo aliena meatu circulus frigidior In pal
lorem. ardentior In ruborem. uertuosus In
horrorē sol atque commisurae apsidum ex
tremeq; orbitae atram In obscuritatem.

A Quo autem In loco uel In quibus signis sunt in
psentia. M Modo autem id ē. anno dominicae
incarnationis. dcccxx. mense iulio. nona die
mensis est sol In uicesima tertia parte cancri.
luna in nona parte tauri. stella saturni In
signo arietis. Iouis In librae. martis In pisciū
ueneris quoque stella. & mercurii quia iuxta
solen In luce diurna modo sunt non appa
ret In quo signo morentur. A Cetera quoq;
sidera quae extra zodiacum sunt pauco uere
uter enumeres atque ad extra parte ipsius
zodiaci quae uea sinistra sint differes.

M Si de cursu planetarum tantum contentus
esset. Iam satis tibi dictum. ē. putarem. sed
quia nimiū curiosus ad cetera signa quaep
toui sperae caelestis ambriu consistunt. Inqui
rendo pgrederis. necessariū reor uel psius
sperae situ & naturā prius exponā secundū
maiorū sententiā breuiter. & sic quae licebit

Sed uel In ipso aequinoctio hoc e̅ xii k̅l̅ n̅ apl̄i̅
die. uel eo transgresso legitima procedat.
embolismorum autem sic dionisius ait ista
ratio probatur existere q̅ annor̅ communium
uideatur damna supplere. quatinus ad solare
tempus lunaris exequatur excursio quamuis
enim solis annuu̅ eidem p̅ singulos m̅ses luna
circum eat. tn̅ eius p̅fectione̅ xii suis mensibus
Implere n̅ ualdt. deniq; In annis communi
bus ad ratione̅ solaris anni xi dier̅ lune deesse
cernuntur. In embolismis uero xl iiii dier̅ bꝰ
eunde̅ annu̅ uideatur solare̅ lunatranscendere.
Xebrei quidem qui solos lunares in lege nouera̅t
Xobseruabant menses. iuxta naturalem lunae
cursum tricenis un detricenisq; diebus communiu̅
annoru̅ m̅ses duodenos explicabant. Xe aiot secun
do ubi debebat. Anno e̅ tu̅ decimu̅ In fine anni m̅ se̅
xxx dier̅ adponebant embolismu̅. porro roma
ni quidis pares habent m̅ses. Non uno quolib̅ & in loco
embolismos co̅putando In e̅ponere uoluerunt. Sed
potius ubi libd̅ mediis anni temporibʒ uacuu̅ con
gruumq; In k̅alendas locu̅ In uenire potuissent.
p̅rima̅ igitur embolismorum luna̅ iiii No̅ decem
bris. Secunda̅ iiii No̅ septe̅bris. Tertia̅ ii No̅ mar
tias. Quartam pridie No̅ decembris. Quintam
iiii No̅ nouembris. Sextam iiii No̅ augustas.
Septima̅ iiii No̅ martias nasci dixerunt.
 	De ordo ade &en deca de̅.
A quia ordo adis &en decadis ratione̅ breuiter
superius fecisti p̅ co ut planius dicas ob qua̅ utilita̅te̅

Duo sunt extremi uertices mundi quos appellant
polos septentrionis & austri · quoru alter a nobis
semp uidetur alter nunquam · In eo quia nobis cernitur
tria sunt signa constructa · duo scilicet arcturi & ser
pens circu · Atq; inter illos in more fluminis means ·
Helice arcturos maior · cynosora minor appellat ·
Que diuersu quide aspiciunt · Nam dorsa earu sibi ma
mutuo auertuntur · auersis huc atq; illuc pedibus· ser
pens uero cauda angit helicen · Cetero circet cynosura
ita tamen ut easdem medius inter labens separat cuius
caput ad dextru pedem eius qui in geniculo stat quem
hercule dicunt · dextro pede caput primit serpentis
capite ad austru conuerso · humeris suis coronae tangit
confinia · Ipsa aute corona post tergu herculis sita ·
capiti serpentis que serpentarius tenet adppinquat ·
Serpentarius uero quia graecis ephiancus uocatur sub her
cule ponitur ad austru uersis pedibus scorpionem
calcat serpente pcinctus que utraq; manu tenet·
Qui plurima longitudine protentus usq; ad coronam
extenditur · iuxta huius serpentis fligi uo fa uolumi
na · nullo alio inter ueniente signo bootes post tergu
arcturi maioris uidetur pedibus ad uirgine uersis·
Nam uirgo sub pedibus bootis est constructa ·
Contra guttur uero & pedes anteriores ursae ma
ioris In commissura zodiaci atq; lactei circulorum
gemini sunt locata · habenas alteuo latere a gracatore
cu edulis duobus qui contra caput helias capite uer
ro pede dextro · sinistru tauri cornu tangere uide
tur · iuxta geminos quia zodiacus alcissime erigitur ·
Contra uentre ursae maioris cancer situs est habens

Facs. no. 9: Hands VIa and VIII (Ms. Canon. Misc. 353, Bodleian Library Oxford, fol. 54r)

A NINTH-CENTURY MANUSCRIPT FROM FULDA

MS. CANONICI MISCELLANEOUS 353

INTEREST in the early medieval computus led me to notice that in Volume IV of the *Summary Catalogue* Falconer Madan listed a manuscript as:

19829. = Qu. Catal. III (Canon.), Misc., no. 353 (Rabanus Maurus).[1]

The Canonici manuscripts in the Bodleian Library had been catalogued between 1840 and 1843 by Henry Octavius Coxe in the Quarto Series, Part III, where this description of Canon. Misc. 353 appears:

Codex membr., in 4to minori, 55 ff., sec. forsan IX exeuntis.[2] It is one of the manuscripts purchased in 1817 for the Bodleian from the heirs of Matteo Luigi Canonici (d. 1803) who had received it from the Venetian collector, Jacopo Soranzo (d. 1761). It had been the fourth item under no. 744 in Soranzo's library

---

[1] *A Summary Catalogue of Western Manuscripts in the Bodleian Library at Oxford*, iv (1897), 406.

[2] Coxe added a brief list of contents and the three primary incipits; see *Catalogi codicum manuscriptorum Bibliothecæ Bodleianæ*, Pars tertia codices Græcos et Latinos Canonicianos complectens (1854): Canonici Misc. no. 353; and Edmund Craster, *A History of the Bodleian Library, 1845–1945* (1952), 20, 91–2.

In addition to the Bodleian's acquisition of 2047 Canonici manuscripts, another 915 were purchased from the heirs in 1835 by the Revd. Walter Sneyd who sold a part of these in the following year to the British Museum (Additional MSS. nos. 10629–10919); the remainder of Sneyd's purchase was dispersed in a series of sales; see Craster, 20 and Mitchell (cited note 3), 133–5. One of the Sneyd–Canonici manuscripts is now B.M. Addit. 10801, a seventeenth-century transcript which is identical in contents with MS. Canon. Misc. 353. Canonici manuscripts in the libraries of Florence have been discussed by the directrix of the Laurentian library, Irma Merolle, *L'abate Matteo Luigi Canonici e la sua biblioteca; i manoscritti Canonici e Canonici-Soranzo delle biblioteche fiorentine* (Florence, 1958).

My thanks are due to Francis S. Benjamin and to Bernhard Bischoff for their generous assistance and encouragement in computistical and palaeographical studies, as well as to the Danforth Foundation and the Alexander von Humboldt-Stiftung for research aid.

vol. iii in his catalogue of quarto manuscripts), and traces of the shelfmark are still visible on the lower spine.[1]

Contents of this manuscript may be described as follows:

Folio 1–1ᵛ   Originally blank, this cover folio may have been intended to receive on its verso a rubricated title of the primary work, but that was not done. A few scribblings were added later on the recto.[2]

2–53   Hrabani *Liber de computo* (untitled)
Incipit: Dilecto fratri Marchario monacho Hrabanus
         peccator in Christo salutem. . . .

This work was written at Fulda in A.D. 820 by Hraban, who was then director of the schools and two years later was elected abbot of the monastery. It is a handbook in dialogue form for use in Carolingian schools and exists now in sixteen manuscripts, several adaptations, and many fragments and excerpts. The first edition was published by Étienne Baluze, *Miscellaneorum Liber Primus* (Paris, 1678), 1–92, with the text based upon Paris, B.N. MS. lat. 4860 (s. IX²) f. 119ᵛ–135ᵛ, which had been in the possession of Jean Baptiste Colbert in the period during which Baluze served as Colbert's librarian (1667–1700). It was reprinted in the second edition of Baluze's *Miscellanea, Tomus Secundus continens Monumenta Sacra* (Lucca, 1761), 62–85. It is from this second edition

---

[1] Characteristics of former Soranzo manuscripts have been summarized usefully by J. B. Mitchell, 'Trevisan and Soranzo: Some Canonici Manuscripts from Two Eighteenth Century Venetian Collections', *Bodleian Library Record*, VIII (1969), 127–30. Manuscripts of the Soranzo catalogue are in Venice, Biblioteca de San Marco, Ital. X. cl. 137–9, photocopies of which are available at the Bodleian Library: MS. Facs. c. 40/7, fol. 90ᵛ. Mr. Mitchell has deposited a card index to the Canonici manuscripts from the Trevisan and Soranzo collections in the office of the Keeper of Manuscripts, and I wish to thank Dr. J. J. G. Alexander for making it available to me.

[2] At the extreme upper left corner of f. 1 is a suspension for the word spiritum, perhaps written in s. IX². During the next century someone took this folio in hand upside down and scribbled on mid-page:

         a b c [e d a r i']
         a u t
         audivi quoniam
         audivi dominum loquentem [m e g h o]
         dicentem finem [mumn r h a]

My brackets indicate uncertainty and include strokes which may simply be iterated minims. But the last group of strokes suggests an abbreviation for the name, rha[banus], as it was spelled occasionally toward the end of the ninth century and thereafter. Such *probationes pennae* were partially recalled from scripture, especially the Psalms, memorized in early school years: cf. Ps. 30: 13, 29: 12, Ezekiel 2: 1–2, 43: 6, etc. In general see B. Bischoff, 'Elementarunterricht und Probationes Pennae in der ersten Hälfte des Mittelalters', *Mittelalterliche Studien*, I (1966), 74–87.

A Ninth-Century Manuscript from Fulda

of Baluze that the work was reprinted and is readily accessible in Migne, *Pat. Lat.* cvii, cols. 669–728.

Scholars have been fortunate that Baluze had at hand a relatively early and reliable copy of Hraban's computus; nevertheless it has interpolations, omissions, and tables for a period later than the date of writing. The first critical edition is now in preparation by the present writer.

53ᵛ   Incipit martius v, aprilis i, maius iii, . . .

Lines 1–2 are a list of solar regulars, sometimes called ferial regulars; cf. Hrabani *Liber de computo* lxxiii. (= Bedae *Liber de temporum ratione* xxi. 5–7, ed. C. W. Jones). They give the number of the feria (I Sunday, II Monday, seq.)[1] for the kalends of each month in a year for which seven concurrents are prescribed in the Dionysian–Bedan Easter Table, such as the table found in this manuscript: Hrabani *Liber de computo* lxxxiii.

Concurrents repeat in a twenty-eight year Solar Cycle, taking leap year into account. It happened that Hraban published his computus in a year with seven concurrents, A.D. 820; other years with seven concurrents were 826, 831, 839, etc. Adjustment to another year could be made by adding the lunar regulars for each month of that year; and following a leap year, a further addition of one must be made for the months of January and February in order to determine their feria. This is reflected in the list here under consideration, which is probably a copy of notes made for and during the period March 820–February 821.

53ᵛ   Ianuarius, augustus et december iiii nonas habent, xviiii kalendas post idus, et dies xxxi; . . .

Lines 3–19 provide simple instructions for beginners about nones, ides, and kalends, numbers of days in the months, and how to divide months into weeks. It assumes that the user normally counts the days forward from the kalends rather than the reverse, as in the Roman system which was being taught.

Such material recurs often in manuscripts and has been printed e.g. in *Pat. Lat.* xc. 799–800 and cxxix. 1281. Lynn Thorndike and Pearl Kibre, *A Catalogue of Incipits of Mediaeval Scientific*

---

[1] The origin of this system of solar or ferial regulars is unknown; Bede seems to have repeated it from an earlier source, according to C. W. Jones, *Bedae Opera de temporibus* (Cambridge, Mass., 1943), 157–8. It was incorporated into many medieval computi and has been printed e.g. in *Pat. Lat.* xc. 705, 800–1, and cxxix. 1288–9.

Hraban provided his own *argumentum* for ferial regulars (end of lxxiiii) in order to apply the system to his *annus praesens*; he showed that when it was the seventeenth moon on the kalends of August, the feria would be IV (Wednesday). That coincidence occurred in A.D. 820.

Writings in Latin (rev. ed., London 1963) 652, list MS. Berne 441 (s. IX) f. 1ᵛ as the earliest occurrence of this instruction.

53ᵛ   Regulares ad kalendas inveniendas quota sit feria

Lines 19–20 introduce another *argumentum*, of which only the title was transcribed; cf. Hrabani, *Liber de computo* LXXIIII; Thorndike & Kibre 1346.

53ᵛ—55   Duo sunt extremi vertices mundi, quos appellant polos . . . (untitled)

This tract has sometimes been called *Excerptum de astrologia Arati*, or attributed to Hyginus; cf. *Pat. Lat.* XC. 368–9 (gloss to Bede) and M. Manitius, 'Das sogenannte Fragment Hygins', *Rheinisches Museum für Philologie*, Neue Folge LIII (1898), 393–8. It appears in seven manuscripts together with Hraban's computus and often with other works in manuscripts of the ninth century and later; the earliest may be the computistical collection in CLM. 210 (Salzburg 818?), f. 74ᵛ–76ᵛ; see Thorndike & Kibre 473 and 1689.

In order to make proper use of MS. Canon. Misc. 353 in preparing a new edition of Hraban's computus, it became necessary to reconsider Coxe's dating. The scripts are continental minuscule of the Carolingian era and, to judge from the style of writing, must have been written between A.D. 820 and *c.* 850 in a German scriptorium. Elsewhere[1] I have offered a more detailed palaeographical analysis which establishes Fulda provenance during the decade before A.D. 836; and that argument may be summarized here.

There are 55 folios measuring 233 × 145 mm. after binding. The writing area is 177 × 97 mm., enclosed by double bounding-lines, with thirty lines ruled on the flesh-side of each bifolium before folding; prickings with an awl had been made along outer bounding-lines of four bifolia simultaneously with hair-sides up. Six gatherings were formed from four bifolia and one from three, folded with hair-sides out and signed at lower centre of the last folio verso with a Roman numeral between two medial points; gatherings ·VI· and ·VII· show significant variations, particularly wider writing areas of 177 × 110–113 mm. Thread holes for the original binding may be observed: first pair 25 mm. apart, at present 34 and 59 mm. from top; second pair 21 mm. apart, at

---

[1] 'Fulda scribes at work', *Bibliothek und Wissenschaft*, IX (1972), 287–316 and facs.

A Ninth-Century Manuscript from Fulda

present 180 and 201 mm. from top, with some variation for the latter pair.

The parchment is a thick calf-skin with many faults, such as was commonly used in continental centres under insular influence. Ink is mostly a rich brown, but some black; mixtures of inks were used in some cases which darkened progressively into black.

The scripts differentiate themselves into three mature Fulda hands (I, V, VI; plates III (a) and IV), two mature non-Fulda hands (IV, VII), and six novices (IIa, b, c, d, III, VIII; plate III (b) shows hand IIc):

Hand   I   f. 2 line 5 (ideo)–f. 2^v
               f. 3 line 3–f. 8^v
      IIa   f. 9 line 3 (uncias)–f. 10^v
        b   f. 11–12
        c   f. 13 line 26 (matutinum)–f. 15^v line 23 (id est)
        d   f. 15^v line 23 (signiferi)–f. 16^v
    III   f. 12^v–13 line 26 (levant)
    IV   f. 17–18
               f. 22^v line 12 (caeli)–f. 24^v
               f. 36–53^v
     V   f. 18^v–22^v line 12 (partis)
               f. 25–26 line 25 (satis)
               f. 26^v line 19–f. 29
               f. 29^v line 6–f. 32^v
    VI   f. 2 line 2–5 (conparari)
               f. 3 line 1–2
               f. 9 line 1–3 (ergo)
               f. 26 line 25 (tibi)–f. 26^v line 18 (dixerunt)
               f. 29^v line 1–5
               f. 53^v last line–f. 54 line 2 (polos)
               f. 54^v line 1–2 (contra)
   VII   f. 33–35^v
   VIII   f. 54 line 2 (septentrionis)–line 30
               f. 54^v line 2 (pedes)–f. 55

The writing master who laid out the project was Hand VI; he began the computus' prologue and its capitula, initiated the second gathering, twice filled in a few lines, and wrote the first lines of what were intended to be the last two folios (though an extra leaf had to be attached to complete the final tract). Hand I copied the remainder of the first gathering; novice hands were trained on the second gathering; hands IV, V, and VII alternated through five more gatherings to complete the computus; and novice hand VIII was assigned an addendum under supervision of the master.

The Fulda scripts are characterized by rectilinear shapes and perpendicular verticals in the long letters; open 'a' with upper strokes rising to sharp points alternates with the more common minuscule adaptations of uncial 'a' and rounded semi-uncial 'a'; both open 'r' and minuscule 'r' with the upper stroke occasionally peaked; both upright minuscule 'd' and uncial 'd'; 'g' with small head closed or opening slightly and a long swinging lower bow; strokes of 'x' also swinging freely; undotted 'y' with semi-circular strokes meeting well above the line, but the short left stroke in Hand V is sometimes horizontal and uncurved; tall 'si', 'ri', 'st', 'ct' ligatures, the latter executed in a nearly perfect semi-circle. Especially prominent in the script of Hand VI are insular 'e' raised above adjacent letters, open 'r' which also forms 'ra', 're', 'ro' ligatures, 'rt' ligature rising to double height, and distinctive strokes for 'm' and 'x'. The abbreviation mark was normally a short horizontal stroke, save for special symbols. Fulda allowed numerous alternative abbreviations, of which only a few will be noted here: ∼ and ∻ and ·ē· est, ñ and p̄ and N̄ either non or nam, ꝗ and qđ quod but ꝗ. quia rarely, q: and q; -que carefully distinguished from qu ꝼ and q;· quae (all spelled out by a corrector). The -rum suspension is contrived from sinuous strokes rather than from open r; ꞇ ter and m' mus n' nus occur in Hand I, but -tus -tur -mur were always written out by these three Fulda scribes; abbreviations for Nomina Sacra are conventional with the exception of x̄p̄ō Christo and epīsc episcopus which may derive from Hraban's original notations; insular symbols survive only for est, i̇ vel, ů vero, but Hand V misread 'p' for insular 'r' and 'r' for insular 's' from the exemplar—not surprising in a scriptorium in which both insular and continental scripts were regularly used for several more decades.

Chapter titles were rubricated in silver-grey ink which now often shows a bright orange hue. Initials are a handsome capitalis rustica and a set of graceless black capitalis quadrata of two lines height; the spaces of a few of these were filled with red ink by some other hand. Spacing was carefully observed throughout, the

A Ninth-Century Manuscript from Fulda

list of capitula begins at the top of a folio, short chapter titles receive full lines, and words are separated from each other. Punctuation is normally a medial point for the various pauses, but a comma and rarely a colon sometimes appear for weak pause, semi-colon for strong pause, especially at the end of a section or a chapter, and questions were usually left unmarked. Hand I used a chryphia in text and in margin to indicate an omission, and Hand V used an apostrophe in margin for this purpose.

Evidence of the instruction which the novice and non-Fulda hands in the manuscript were receiving is very interesting but will have to be elaborated elsewhere. Hand IV positively rejected the Fulda open 'a' when it appeared in the exemplar, though he and others somewhat grudgingly yielded to the example of hands I and V on other usages, such as the -rum suspension. Hand IIc (plate IIIb) gave up the use of subscript iota and his ligatures 'ro' and 'rt' in learning the Fulda style. The 'ct' ligature in a semi-circle was new to them, and all had difficulty with it.

When the Fulda scripts in MS. Canon. Misc. 353 are compared with other Fulda manuscripts it is possible to narrow the period of writing. From the above description it may be seen that these scripts have the same general characteristics as the continental minuscule of f. 177–207 in Bodleian MS. Laud Lat. 102 which Bernard Bischoff identified as a Fulda script of the first quarter of the ninth century.[1] But Hraban's computus dates only from A.D. 820, and this is not the original; how much later was it copied? Lieftinck, Hofmann, and Butzmann have identified later manuscripts of Fulda provenance which show both the general usages and many of the specific practices we have noted in Canonici hands I, V, and VI, but they seem to reflect later developments.[2] Rectilinear letters and perpendicular verticals,

[1] B. Bischoff and J. Hofmann, *Libri Sancti Kyliani* (Würzburg, 1952), 15–16, 55–6; he dated the Laud manuscript by comparing it with CLM. 8112, Frankfurt-am-Main Stadtbibliothek Barth. 32, and Erlangen Universitätsbibliothek 9. From having seen the Laud and Munich manuscripts and facsimiles of the others I concur in this dating.

[2] G. I. Lieftinck, 'Le MS. d'Aulu-Gelle à Leeuwarden exécuté à Fulda en 836', *Bullettino dell'Archivio Paleografico Italiano*, N.S. I (1955), 11–17; Josef Hofmann, 'Fragmente von Bibelkommentaren des Hrabanus Maurus', *Würzburger Diözesangeschichtsblätter*, 18./19.Jahrgang (1956–7), 5–19; Hans Butzmann, 'Der Ezechiel-Kommentar des Hrabanus Maurus und seine älteste Handschrift', *Bibliothek und Wissenschaft*, I (1964), 1–22; facsimiles with each article are excellent.

'a' and 'd' and 'r' usages, semi-circular 'ct' ligature, and occasional insular symbols indicate continuity with the same scriptorium and perhaps even development of some of the same hands over several years. But inspection of the earliest manuscripts of Hraban's Commentary on Ezekiel and analysis of facsimiles of fragments of his commentaries on Genesis, Leviticus, and Ecclesiastes and of his copy of Aulus Gellius' Noctes Atticae have convinced me that these are later products of Fulda. Written between 836 and 842, they reflect a greater uniformity of style and consistency of usage than may be found even in the most mature scripts of the Canonici.

The Fulda Cartulary was begun during the first few years of Hraban's abbacy; a section written mostly in insular scripts of Fulda survives in MS. Marburg Staatsarchiv. K 424, f. 10ᵛ–70ᵛ. The latest charter bore the date 3 October 828 on f. 70, and the next one to it on f. 70–70ᵛ was written in non-insular Fulda minuscule. Both of these charters were enumerated in the original sequence and were probably transcribed toward the end of 828 or in 829, and the latter script shows similarities with usages of Canonici hands II, III, and IV which I have described as being under instruction. A later addition to the Cartulary was probably copied on f. 50 soon after its date, 30 July 836, and has similarities with Fulda Hand V.[1]

Such correlation of scripts establishes a high probability that MS. Canon. Misc. 353 was written at Fulda during the decade before A.D. 836.

Fulda was pre-eminent among those ninth-century centres which were transcribing and creating Latin literature, as well as educating teachers, providing books, and training scribes for other monasteries and cathedrals. But surprisingly few Fulda manuscripts have been carefully dated. The Canonici manuscript together with the others cited will now allow us to see a composite picture of that scriptorium during its most creative half-century.

---

[1] Charters no. 192 (f. 70; Dronke no. 403) and no. 110 (f. 50; cf. no. 217, Dronke no. 487) are discussed by E. Heydenreich, *Das älteste Fuldaer Cartular im Staatsarchiv zu Marburg* (Leipzig, 1899), 36, 49–52, 59, with facsimiles; E. E. Stengel, 'Über die karlingischen Cartulare des Klosters Fulda', *Fuldensia*, II (1921), repr. *Abhandlungen und Untersuchungen zur Geschichte der Reichsabtei Fulda* (1960), 156–9; Stengel, *Urkundenbuch des Klosters Fulda*, I.2 (Marburg, 1956), xviii–xxi.

VII

PLATE III

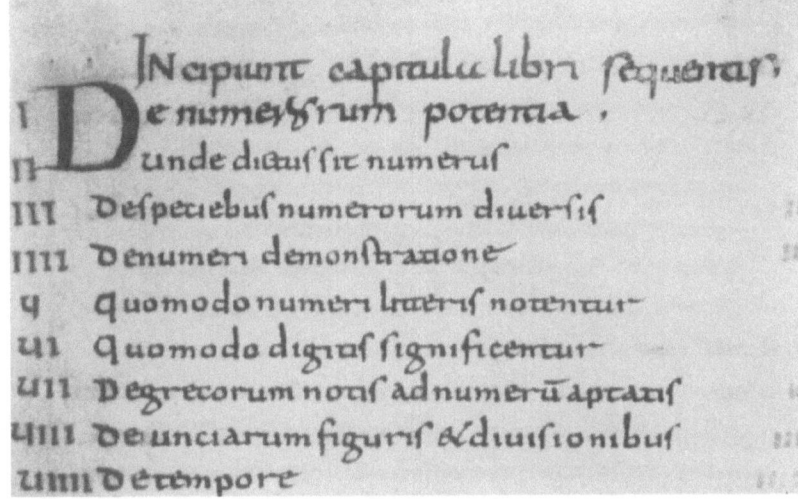

(a) Fol. 3ʳ, showing Hands VI and I

MS. Canonici Misc. 353

PLATE IV

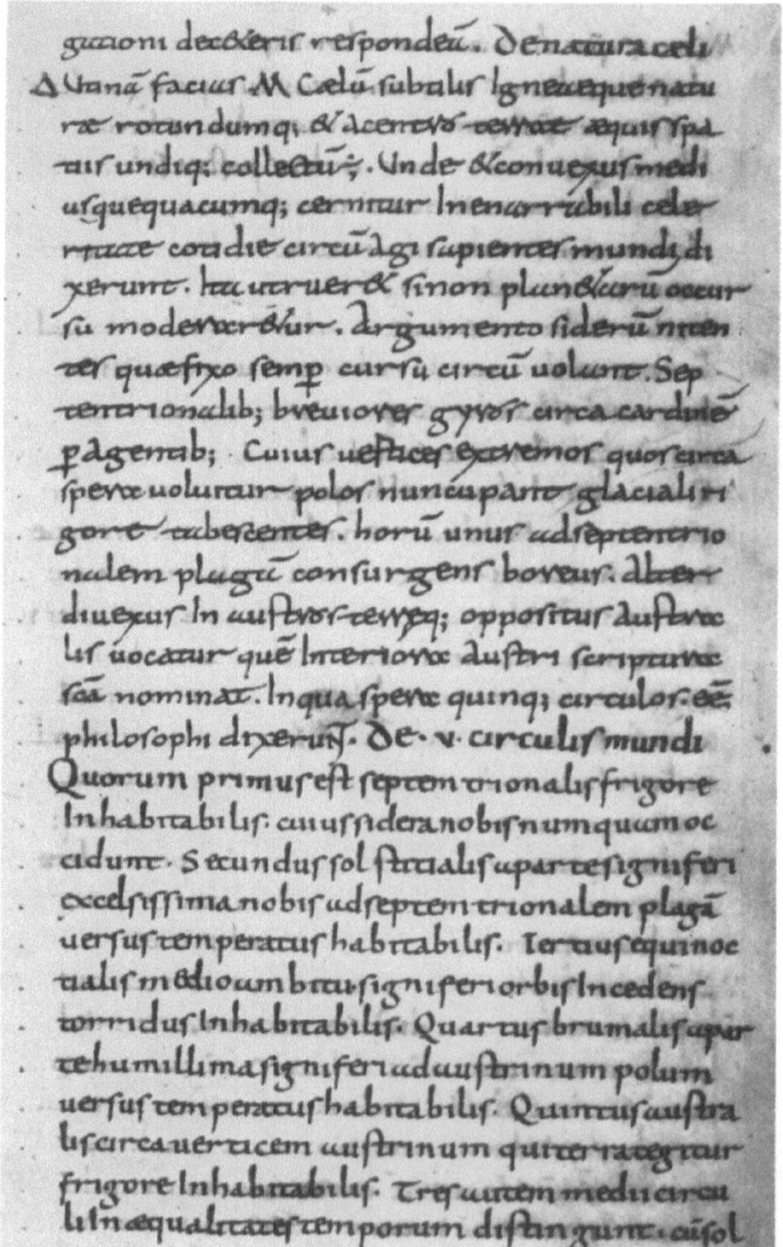

gicaoni decdérit respondeu. de natura celi
Δ Vana facict M Celu subtilis Igneæque natu
ræ rotundumq; & acentus certat æquis spa
tiis undiq; colleccui;. Unde & conuexus medi
usquequacumq; cernitur In encorribili cele
ritate cotidie circu agi superiores mundi di
xerunt. hac utruer& sinon plu n&ceru occur
su moderaretur. argumento siderũ ita en
ter quæfixo semp cursu circu uoluntr. Sep
tentrionalib; breuiores gyros circa cardiné
pagentib; Cuius uertices extremos quor circa
sperc uoluitur polos nuncupato glaciali ri
gore acbescentes. horũ unus ad septentrio
nalem plagã consurgent boreus. alter
diuersus In austros tenq; opposit aust ræ
lis uocatur quẽ Interiore austrũ scriptura
sca nominat. Inqua spere quinq; circulos. ee
philosophi dixerunt. de· v· circulis mundi
Quorum primus est septentrionalis frigore
Inhabitabilis. cuius sidera nobis numquam oc
cidunt. Secundus sol stitialis a parte signiferi
excedissima nobis ad septentrionalem plagã
uersus temperatus habitabilis. tertius equinoc
tialis medium breui signiferi orbis Incedens
torridus Inhabitabilis. Quartus brumalis a par
te humillima signiferi ad austrinum polum
uersus temperatus habitabilis. Quintus austra
lis circa uerticem austrinum qui certate agitur
frigore Inhabitabilis. Tres autem medii circu
li Inæqualitates temporum distinguunt. cuisol

MS. Canonici Misc. 353, fol. 26ᵛ
showing Hands VI and V

VIII

INTRODUCTION TO HRABANI *DE COMPUTO LIBER*

Hraban Maur is remembered not only as an abbot and later as a bishop active in ninth century councils, but especially as a teacher and author. (¹) What he taught however has not been made clear. (²) One could assume that Gottschalk came from Saxony, Walahfrid Strabo from Reichenau, and Servatus Lupus from Ferrières to Fulda to study the Scriptures with a pious theologian. Yet Hraban was not yet known for his biblical commentaries, and only Gottschalk was preoccupied with theological questions. All three enhanced their poetic abilities and training in metrics at Fulda, but the numerous manuscripts annotated or copied by Lupus demonstrate a controlling interest in the pagan Latin classics and in literary criticism. (³) Walahfrid's *vademecum* shows that over half of his entries were items of natural history, including Bedae *De temporibus* and *De natura rerum* copied at Fulda as well as the earliest extant copy of Hrabani *De computo* studied and transcribed at Reichenau and other computistical *argumenta* found at Fulda, Weissenburg, and Aachen. (⁴)

This evidence is consistent with the development of monastic schools. The bookish drills of Roman schoolmasters in Hellenistic lore and grammatical forms of Latin traditions survived the

---

(1) Well-written popular accounts were published for the anniversary of Hraban's death (A.D. 856) by Stephen HILPISCH, *Der Heilige Rabanus Maurus, Abt des Klosters Fulda und Erzbischof von Mainz* (Fulda 1956), 55 pages; and by Theodor SCHIEFFER, *Hrabanus Maurus (zum 1100. Todestag am 4. Februar 1956), Archiv für mittelrheinische Kirchengeschichte* VIII (Mainz 1956) 9-20. See further Ernst DÜMMLER, *Hrabanstudien, Sitzungsberichte der preussischen Akademie der Wissenschaften*, Philol.-hist. Klasse (Berlin 1898) IV, p. 24-42; Albert HAUCK, *Kirchengeschichte Deutschlands* II (Leipzig 1903, rpr 1954) p. 630-659, 669-673; Max MANITIUS, *Geschichte der Lateinischen Literatur des Mittelalters* I (München 1911), p. 693-647; Paul LEHMANN, *Fuldaer Studien, Sitzungsberichte der Bayerischen Akademie der Wissenschaften*, Philos.-philol.-hist. Klasse (München 1925), 3. Abh.; IDEM, *Fuldaer Studien, Neue Folge, ibid.* (1927), 2. Abh.; and portions of many other essays by Lehmann.

(2) Cf. for example R. STACHNIK, *Die Bildung des Weltklerus im Frankenreich* (Paderborn 1926), p. 92-96; LEHMANN, *Zu Hrabans geistiger Bedeutung*, in *St. Bonifatius. Gedenkgabe zum 1200jährigen Todestag* (Fulda 1954), rpr *Erforschung des Mittelalters* IV (Stuttgart 1961) 198-212; E. VON SEVERUS, *Hrabanus Maurus und die Fuldaer Schultradition, Fuldaer Geschichtsblätter* XXXII (1956) 113-124. Professor Raymund Kottje expects to prepare a new biography of Hraban.

(3) These manuscripts were conveniently listed by R.J. GARIÉPY Jr., *Lupus of Ferrières: Carolingian Scribe and Text Critic, Mediaeval Studies* XXX (Toronto 1968) 90-105

(4) B. BISCHOFF, *Eine Sammelhandschrift Walahfrid Strabos (Cod. Sangall. 878)*, in *Aus der Welt des Buches: Festgabe Georg Leyh* (Leipzig 1950), rpr with additions in his *Mittelalterliche Studien* II (Stuttgart 1967) 34-51; W.M. STEVENS, *Walahfrid Strabo — a student at Fulda, Historical Papers 1971 of the Canadian Historical Association*, ed. J. Atherton (Ottawa 1972) 13-20.

166   INTRODUCTION

Roman empire only in fragments, as Italian and Gallo-Roman patricians practiced economic restraint in education from the third century onwards, walled their cities, and abandoned lesser peoples to bear the brunt of economic chaos. Yet the tradition of *artes liberales* was continued by episcopal churches to some extent. The records of those churches and their schools are not continuous, yet there was no time when schools could not be found during the early middle ages. By the seventh century in Visigothic Spain and by the eighth century in Francia on both sides of the Rhein, endowments and personnel were again available for many centers of learning, old and new. Costly books were provided, and disciplines of advanced study were taking new forms for the most part in monasteries.

The new *artes et scientiae uel disciplinae* did not further the Roman social forms of caste: training in the exercise of power through rhetoric and law was abandoned in favour of a different vocation, *opus Dei* — service to God. Both work and prayer within the monastic shape of the Christian life were conceived to be egalitarian, while new or young members of the community were introduced to the new way of life as apprentices to more experienced elders who looked after their learning of letters initially by means of the Lord's Prayer, the Creed, the Psalms, the Liturgy and their moral well-being and maturity one by one. *Sapientia* was approached through liturgy and Scripture rather than through knowledge of the world and its ways. But this new goal and new approach required *grammar* in order to gain understanding in *litteris et bonis moribus ;* and it implied both *cantus* and *computus* in order to carry forward the *opus Dei* with regularity and good form. Grammar, song, and reckoning the times were essential to the monastic schooling, as Charles W. Jones has emphasized, (5) in contrast with the Hellenistic and Roman pedagogy.

The *deus* who was served in episcopal and monastic schools was not a *dea creatrix* (one among many in confusion) or a *Juppiter* whose priests neglected their calendar and let the seasons slip by unnoticed, but rather the *Dominus* of all the earth who *omnia in mensura et numero et pondere disposuit* (cf. *Sapientia* 11, 21). Augustine's emphasis both on the Creation of nature and time together and on the historical Incarnation constantly informed the hearing of the Psalms by canons and monks who repeated them at least weekly. This provided a cultural language in which the objects

---

(5) Charles W. JONES, Preface, *Bedae venerabilis Opera*, pars I *Opera didascalica*, *Corpus Christianorum Series Latina* CXXIII A (Turnhout: Brepols, 1975), p. v-vii, xii-xv; IDEM, *Carolingian aesthetics: Why modular verse? Viator* VI (1975) 309-340.

and forces of nature were brought to mind and their order considered, especially the regular movement of sun and moon through the stars *ut sint in signa et tempora et dies et annos* (Gen. I, 14). ([6]) The reckoning of times was thus not only necessary for the common life but also appropriate for its meaning and purpose in the *opus Dei*.

Computus is a term denoting such reckoning at all levels from initial learning of numerals, practice of arithmetic, use of tables of dates, mastery of methods for calculating the dates themselves, explanation of the significance of such data and of the whole order of the cosmos — theologically as well as mathematically. *Computus* could also denote a tract or collection of *argumenta* or an essay or a codex in which such contents formed a large part, though mixed with other subjects. Time-reckoning required some manipulation of numbers, recognition of the parts of the zodiac and the simple astronomy of solar and lunar motions, along with some history and theology. In the search for understanding however, *computus* tended to attract related and sophisticated materials far beyond calendarial requirements. This development may be observed by comparing the *Calculus* of Victorius (s.V), the cycles and argumenta of Dionysius Exiguus (s.VI), and the more broadly conceived works of Beda (s. VIII) ([7]). Computistical codices of the eighth, ninth, and tenth centuries bring together for example *mappaemundi* displaying the globe of the earth with three continents (or four) along with descriptions of peoples and places ([8]);

---

(6) Isidori *Etym*. III, 4, 1; Bedae *DTR* II, 23/24: "Sed et errantia sidera suis quaeque spatiis zodiaco circumferri, quae natura non iuxta ethnicorum dementiam dea creatrix una de pluribus sed ab uno uero Deo creata est, quando sideribus caelo inditis praecepit ut sint in signa et tempora et dies et annos"; Hrabani *Comp*. I, passim.

(7) See the survey of computistical literature from ca. s.II until the time of Bede by C.W. Jones, *Bedae Opera de temporibus* (Cambridge, Mass., 1943), p. 6-122, with citations of the literature in abundance; this has been supplemented by other essays by Jones and by *Clavis* 2249-2323. My *Catalogue of computistical tracts* (sub prelo) will list all Latin computistical materials thus far identified prior to the 1582 alteration of the calendar system, with incipits, manuscripts, editions, and a selection of modern literature for each item.

(8) Cf. mss Karlsruhe Landesbibliothek, Reichenau 167 (A.D. 834-848), Irish materials from ca. A.D. 700; Sankt Gallen Stiftsbibliothek 248 (s.IX, X, XI), partly copied from Karlsruhe 167 but continuously adding to contents; Paris BN lat. 7530 (Monte Cassino s.VIII²); London BL Cotton Caligula A.XV (s.VIII² northern France) f. 65-119 from a Visigothic exemplar but with Anglo-Saxon glosses; Paris BN lat. 4860 (s.IX²) with school-dialogues and added notes; Bern 417 (s.IX med) and 610 (s.X) and 250 (s.IX, X ex), a series of materials from the region of Auxerre, Paris, and Fleury where many teachers were active with computistical materials. See further *Monumenta cartographica vetustioris aevi*, A.D. 1200-1500, ed Robert Almagia and Marcel Destombes, vol I: Mappemondes (Amsterdam 1964), sections 1-25 in which M. Destombes included lists of manuscripts containing mappaemundi from before A.D. 1200.

planisphaeria with stereographic projections of stars and constellations for both the northern and the southern hemispheres [9] ;
alternate systems of planetary motion (epicyclic, eccentric) and
diagrams of the motions both descriptive and logical (i.e., saving
the appearances) [10] ; and multiple *argumenta* on the same computistical topic providing not only alternate explanations but often
irreconciliable data to be analyzed. [11] Thus *computus* was a part
of the monastic curriculum which met the needs of the *opus Dei* on
the one hand but on the other provided a basis for further
explorations of nature and for critical thinking about the phenomena for their own sake.

COMPVTVS

Computistical writings from the Julian reform (45 B.C.) to the
essays of Beda (A.D. 703, 725) have been studied separately by
many scholars. Charles W. Jones, the most recent editor of Beda, [12]
has introduced the texts by an excellent history of the literature and
analysis of problems which arose and were solved in the tradition of
the Western calendar from the second to the early eighth century.
On the other hand interest in later calendar reforms had led to the
cursory study of Hermann the Lame of Reichenau, Robert of
Hereford, Robert Grosseteste of Oxford, and Campanus of Novara

---

(9) There are two traditions of medieval planispheres; their earliest exemplars are from Salzburg: Wien 387 (A.D. 810-816) with its twin CLM 210,
copied from the same exemplar of astronomical-computistical contents and
illustrations; and ms Basle Universitätsbibliothek A.N.IV.18 (s.IX med) from
the Fulda scriptorium. See Patrick McGurk, *Germanici Caesaris Aratea cum
scholiis, a new illustrated witness from Wales*, The National Library of Wales Journal
XVIII (1973) 197-216, esp. p. 200-201.

(10) There are three kinds of illustrations in early medieval manuscripts
which demonstrate critical study of planetary motion during the ninth and tenth
centuries but which have not yet received serious attention: 1) motion of Venus
and Mercury around the Sun; 2) eccentric motions; and 3) rectilinear graphs.
All of these appear in computistical manuscripts of our period.

(11) For example Walahfrid in his ms Sankt Gallen 878, p. 176-177, 284-302
which were entered after he had left Fulda in the Spring of A.D. 829 and
probably when he was either visiting Weissenburg or working at Aachen during
829-838. See W. STEVENS, *Walahfrid Strabo — a student at Fulda*, p. 17.

(12) C.W. JONES, *Bedae Opera de temporibus* (Cambridge, Mass., Mediaeval
Academy of America, 1943), p. 3-122; see also *Bedae venerabilis opera, Opera
didascalica, Corpus Christianorum, Series latina* CXXIII (Turnhout: Brepols, 1975
seq), for another printing of DTR, and a new edition of DNR with other
teaching manuals of Bede.

as precursors of the Gregorian Calendar (1582). ([13]) Hraban stands in an intermediate period of three centuries (eighth, ninth, and tenth) during which interest in the calendar was high.

The "Christian era", numbering the years successively after the birth of Jesus Christ, became acceptable in Europe partly through the influence of Bedae *Historia ecclesiastica gentis Anglorum* and his essays on time. This "Christian era" was based upon the 95-year table and explanations written about A.D. 525 by Dionysius Exiguus for (African?) Bishop Petronius and then for Roman archdeacon Hilarius. ([14]) Dionysius explained the nineteen-year *
cycle as used in the Cyrillan table by the Church of Alexandria and more generally by Greek-speaking churches after the Council of Nicaea had rejected the quartodecimans. ([15]) The Alexandrian usage had been introduced in northern and southern Italy and some churches in Roman Africa by 400. Dionysius' table applied to the years A.D. 532-626 inclusive and was intended to replace a

---

(13) Hermann the Lame, Reichenau (A.D. 1013-1054), has received attention primarily for his essays on the abacus and the astrolabe, rather than for his interest in the calendar; on his computistical calculations see Gabriel MEIER, *Die sieben freien Künste im Mittelalter*, Second Part (Einsiedeln 1887), p. 17-19, 34-36. Robert, bishop of Hereford (1079-1095), tried in A.D. 1086 to introduce into England the theory of Marianus Scotus (1028-1082, lived and wrote at Fulda) that the Dionysian era placed the Incarnation twenty-two years too late; for both see W.H. STEVENSON, *A Contemporary Description of the Domesday Survey*, *English Historical Review* XXII (1907) 72-84, esp. p. 73-75. Little has been added by A. CORDOLIANI, *L'activité computistique de Robert, évêque de Hereford*, in *Mélanges offerts à René Crozet*, eds. P. Gallais, Y.-J. Riou (Poitiers 1966) I 333-340. Robert Grosseteste wrote three *Computi*; see S.H. THOMPSON, *The Writings of Robert Grosseteste, Bishop of Lincoln, 1235-1253* (Cambridge 1940), p. 94-97, 106, 115; A.C. CROMBIE, *Robert Grosseteste and the Origins of Experimental Science, 1100-1700* (Oxford 1953), p. 45-48, 97 n. 3 (bibliography). Campanus of Novara wrote his *Computus* about A.D. 1264; cf. Francis S. BENJAMIN's comparative text, *John of Gmunden and Campanus of Novara*, *Osiris* XI (1954) 221-246. Most known computi have not been published or are available only in misleading texts, even for such interesting reformers as Florus of Lyon (s.IX), Abbo of Fleury (s.X), and Marianus Scotus of Fulda (s.XI).

(14) Bruno KRUSCH, *Studien zur christlich-mittelalterlichen Chronologie: Die Entstehung unserer heutigen Zeitrechnung*, Abhandlungen der Preussischen Akademie, 1937, Philol.-hist. Klasse, Nr. 8 (Berlin 1938), p. 59-86. Krusch failed to distinguish between *argumenta* which were authentic and those which Dionysius could not have prepared; in this regard a more reliable edition is by J.W. Jan (1718), rpr *PL* LXVII 497-508. The best discussion is JONES *Bedae Opera* 68-73 passim.

(15) Cyril C. RICHARDSON, *The Quartodecimans and the Synoptic Chronology*, *Harvard Theological Review* XXXIII (1940) 177-190; and especially *A New Solution to the Quartodeciman Riddle*, *Journal of Theological Studies*, New Series XXIV (1973) 74-84. He explained divergent sources as originating from quite distant groups who were attempting to keep old customs or were making different adjustments to new contexts.

similar cycle by Victorius of Aquitaine (A.D. 457) which required too many exceptions in its application to Roman usage and which was not really cyclic. ([16])

Although the work of Dionysius was prefaced by the letter from archdeacon Hilarius and appeared to gain stronger sanction when Hilarius became bishop of Rome, the new era was not applied by the Roman curia for official documents until the tenth century. It found advocates however in Rome before A.D. 550, Kent ca. 597, southern Ireland 632, southern England s.VII, northern Ireland and Northumbria s.VII², but only s.IX in Wales. ([17]) A misplaced stone in the wall of a church at Périgueux dating from A.D. 631 shows a table of data for reckoning Easter which was based upon the Victorian table, and that usage continued even into the eighth century in various parts of Francia. ([18]) But the Dionysian system had found acceptance as early as A.D. 541 at the Council of Orléans. Two chapters of a book of instructions on how to use it in A.D. 737 have been published, ([19]) and there are many short directions for particular calculations or *argumenta* which bear an eighth century *annus praesens*, though copied at later times. Examples of how these *argumenta* were collected and augmented in the ninth century are the *Compotus sancti Augustini, sancti Isidori, sancti Dionysii, sancti Quirilli Greciae et ceterorum*, and the *Computus grecorum siue latinorum*. ([20]) A handbook written in Salzburg about A.D. 818 drew not only upon such *argumenta* but also on Bedae *De temporum ratione* and Plinii *Historia naturalis* at a time when the latter was known at the Carolingian court but was rarely available in the eastern parts of the Carolingian empire. ([21]) The most popular collection of all during the ninth

---

(16) The table prepared by Victorius was edited by KRUSCH, *Studien* II (1938) 16-52; see also JONES, *Bedae Opera*, p. 61-68.

(17) JONES, *A Legend of St Pachomius, Speculum* XVIII (1943) 200-203.

(18) CORDOLIANI, *La Table pascale de Périgueux, Cahiers de civilisation médiévale* IV (Poitiers 1961) 57-60 with a photograph. Further evidence of continued influence of Victorius may be seen in mss Bern Stadtbibliothek 611, and Milan B. Ambros. H.150; the former was dated A.D. 727 by Krusch, *Studien* II (1938) · 53-57. See also CORDOLIANI, *Contribution à la littérature du comput ecclésiastique au moyen âge, Studi medievali*, Third Series II (Torino 1961) 169-171.

(19) Ms Berlin, Deutsche Staatsbibliothek, Philipps 1831 [Cat. 128], f. 138-142, ed. KRUSCH, *Das älteste fränkische Lehrbuch der dionysianischen Zeitrechnung, Mélanges à Émile Chatelain* (Paris 1910), p. 232-242.

(20) An exemple of the first is CLM 14456, an ninth century manuscript which retained *argumenta* from as early as A.D. 719. An eighth century version of the second is ms Paris BN lat. 7569, f. 86-125. Other collections of these types were cited by CORDOLIANI, *Contribution...* (1961) 174.

(21) Ms Wien NB lat. 387; its Regensburg twin, CLM 210, has been studied by Karl RÜCK, *Auszüge aus der Naturgeschichte des C. Plinius Secundus in einem astronomischen Sammelwerke des VIII. Jahrhunderts, Programm des königlichen Ludwigs-Gymnasiums... 1887/88* (Munich 1888); IDEM *Die Naturalis Historia des*

and tenth centuries was sometimes entitled, *Sententiae sancti Agustini et Isidori in laude compoti,* and exists in several versions. ([22])

Letters and verse which inquire about or express various aspects of this science are abundant. In addition to the well-known correspondence of Charlemagne and Alcuin, there are extant an anonymous letter addressed in A.D 809 to Agnardus (Einhard?) on the calculation of Easter, ([23]) another written in A.D 809 by Adalbert of Corbie, ([24]) and Dungal's answer to Charlemagne about the eclipse of the sun in A.D. 810. ([25]) Computistical verses had long been used as aids to the memory, such as *Ecloga* XVII of Ausonius which was used by Beda and Hraban (see XXXVIIII 27-28) or the verse *Nonae aprilis norunt quinos/...* which seems to have been known to Isidore of Sevilla and was later at Fleury attributed to an angel's revelation to Pachomius, thence to be repeated by Hraban (LXXXIIII 6-25) without reference to angels or the founder of cenobite monasticism. ([26]) An entire booklet of anonymous verses accompanied some computi, appearing with Hraban's work for example in mss *ADEPV,* and another all-purpose

---

Plinius im Mittelalter, Sitzungsberichte der Bayerischen Akademie, Philos.-Philol. Klasse (Munich 1908), p. 203-318.

([22]) One version is printed as *De computo dialogus* in PL XC 647-664; see CORDOLIANI, *Une encyclopédie carolingienne de comput: les "Sententiae in laude compoti", Bibliothèque de l'Ecole des Chartes* CIV (1943) 237-243; and C.W. JONES, *The Lost Sirmond Manuscript of Bede's Computus, English Historical Review* LII (1937) 213-214: items no. 4 and 5; Jones thinks that it was a cathecism for use in Irish schools of s. VII and a source for Bede. Several versions are discussed by JONES, *Bedae Pseudepigrapha* (Ithaca 1939) 48-51.

([23]) Einsiedeln, Stiftsbibliothek 321 (s.IX), p. 145-156.

([24]) Ms Paris BN lat. 2796 (A.D. 813-815?) f. 98-101; ed E. DÜMMLER, *MGH. Epp.* III (1934) 569-572.

([25]) Ms London BL Regius 13.A.XI (s.XI-XII) f. 120-126ᵛ; ed *MGH. Epp.* IV 570-577; *PL* CV 447-458. Dungal's reputation as a student of astronomy and chronology is further enhanced by the identification of his hand as Corrector Saxonicus, the first corrector and annotator of the earliest existing manuscript of Lucretius, Leiden BPL Voss. F.30; the text itself had been transcribed shortly after A.D. 800 in northeastern France by a scribe who may have been trained in the court school, according to B. BISCHOFF, cited by Virginia Brown, *The Insular Intermediary in the Tradition of Lucretius, Harvard Studies in Classical Philology* LXXII (1967) 307-308.

([26]) Ausonii, *Ecloga* XVII, was quoted in full by Bedae *DTR* XVI 7-18, but only the first two lines were provided by Hraban who assumed that his students had memorized the verse. For *Nonae aprilis norunt quinos/...* see C.W. JONES, *A Legend of St Pachomius* 33. Other computistical memory verses were discussed by BISCHOFF, *Ostertagtexte und Intervalltexten,* rpr *Mittelalterlichen Studien* II (Stuttgart 1967) 192-227.

set travelled under the rubric *Carmina Salisburgensa*. [27] Such verses were written as well by Paul the Deacon (A.D. 763), Walahfrid Strabo (before A.D. 849), Wandalbert of Prüm (A.D. 813 - c. 870), Agius of Corvey (865-888), and Erchempert of Monte Cassino (A.D. 904). [28]

The computistical dialogue found in the manuscript of Basel, Universitätsbibliothek, F. III. 15^k (Benediktbeuren, A.D. 820-840) f. 24^v-36, was derived from Hraban's work. But there was also a·completely independent essay by the Irish grammarian, Dicuil, during the years A.D. 814-816, composed at court and presented to the emperor Louis the Pious, chapter by chapter: *Liber de astronomia et computo*. [29] After s.IX med. a book *De computo lunae* was published by Heiric of Auxerre [30] and from Lyons came a *Tractatus computi* possibly by Florus (d. ca. A.D. 860). [31] At the turn of the century computistical works were circulated by Notker the Stammerer (840-912) who was choirmaster-computist of Sankt Gallen, and by Wichram of the same abbey

---

*

(27) The booklet in mss *ADEPV* has ten items concerning the cycle of twelve months and twelve signs of the year, the nineteen year cycle, and related materials which have often been noted. *Carmina Salisburgensia* was edited from CLM 14743 (s.IX) f. 157-184 by E. Dümmler, *MGH. Poetae latini* II 644-646.

(28) Paul the Deacon addressed his acrostic poem on the computus to the wife of Desiderius, *Adelperga pia*, ed. E. Dümmler, *MGH, Poetae latini* I, no. 35; Manitius I, p. 257, 259. Walahfrid: ed E. Dümmler, *MGH, Poetae latini* II, p. 422-423; ed. E. Baluze, *Miscellanea Sacra* I (Paris 1678), p. 491. Wandalbert: ed. Dümmler, p. 576-578, 604-618; Cordoliani, p. 67-68. Agius of Corvey: ed. K. Strecker, *MGH, Poetae latini* IV (1914), p. 937-943. Erchempert of Monte Cassino: see Cordoliani, *Un manuscrit de comput ecclésiastique mal connu de la Bibliothèque Nationale de Madrid* [no. 19], *Revista de Archivos, Bibliotecas y Museos* LVII (1951) 11-12; Manitius I 709-710.

(29) Dicuil: see p. 175 infra.

(30) This *computus*, also called *Ars calculatoria*, should be attributed to Heiric of Auxerre (A.D. 841-876), not Helperic of St. Gallen (c. A.D. 980), according to Ludwig Traube, *Compotus Helperici, Neues Archiv* XVIII (1893) 73-105. However the recent study by Patrick McGurk tends to reinstate the name of Helperic for this work; see *Computus Helperici: its transmission in England in the eleventh and twelfth centuries, Medium Aevum* XLIII (1974) 1-5.

(31) Ms Montpellier Faculté de Médécine 157 (s.IX²), cited by C. Charlier in *Traditio* VIII (1952) 94.

(32) De quattuor questionibus compoti, addressed by Notker to his student, Erkenhart; ed. G. Meier, *Die sieben freien Künste im Mittelalter* (Einsiedeln 1887), p. 31-34; Manitius I 362; Cordoliani, *Les Traités du comput...*, p. 64-65; TKr 1125. Wichrammi Tractatus de compoto Bedae, in CLM 14221, f. 20^v-22 (s.X fragment); see also ms Sankt Gallen 459 (ca. 900) p. 347-66. Note the *Chronicle* of world history completed in A.D. 906 by Regino of Prüm (d. 915, Trier) who discussed at the end of his first book the difficulties in coordinating imperial and papal years with the Dionysian Era that he was using (ed. F. Kurze, 1890, p. 40); Manitius I 699.

(s.IX ex). ([32]) A limit will be set for this sketch of three centuries of studies with two writers in the last decades of s.X. who reopened the question of dating the Incarnation and the Passion. Notable are a letter to Hugo by Heriger of Lobbes ([33]) and two letters and a manual by Abbo of Fleury ([34]) who perhaps also introduced Hraban's *Computus* to the English. ([35])

Within the more immediate intellectual climate in which Hraban studied at Aachen and perhaps also at Tours, taught at Fulda, and administered first the monastery at Fulda and later the diocese of Mainz, computus was also considered to be of primary importance. The conversations which Charlemagne encouraged and shared with his intellectual friends included astronomical and computistical questions, as may be found in the correspondence of Alcuin. ([36]) When he joined Charlemagne's court Alcuin probably brought from York a collection, *Libellus annalis,* from which he taught De bissexto, De cursu et saltu lunae, and Calculatio. ([37]) This was entirely consistent with the concern for education by the Pepinides which Boniface met in Carloman whose capitulary from

---

(33) *PL* CXXXIX 1129-1154; A. CORDOLIANI, *Abbon de Fleury, Hériger de Lobbes, et Gerland de Besançon sur l'ère de l'Incarnation de Denys le Petit, Revue d'histoire ecclésiastique* XLIV (1949) 480-484.

(34) "Epistola prima ad Geraldum et Vitalem" (A.D. 1003), "Epistola secunda...," *Computus vulgaris qui dicitur ephemeris;* see A. CORDOLIANI, *Abbon de Fleury...,* p. 463-487.

(35) Excerpts from Hrabani Liber de computo in Fleury ms Berlin Staatsbibliothek Phillips 1833, f. 3ᵛ by a scribe perhaps trained at Angers and contemporary with Abbo's entries on f. 5 seq demonstrate that Hraban was known in Fleury during Abbo's lifetime. Abbo was at Ramsey abbey and travelled at least to Bury St Edmunds and Canterbury, teaching grammar and computus during A.D. 986-988. The earliest English transcription of Hraban's work and of Isidori *De natura rerum* is ms Exeter 3507 (s.X²) which was written after A.D. 960 and probably after 986. Its exemplar was wrapped with two bifolia containing Abbo's *Sententiae,* including those composed at Ramsey, when these contents were transcribed into ms London BL Cotton Vitellius A.XII (s.XI ex). In the meantime Abbo's student, Byrhtferth of Ramsey, used both the *Sententiae* and Hraban's work to compose his own computus about A.D. 1000. It thus appears that Abbo had brought Hraban's handbook with him to revive learning among the English (cf. Oswald's letter).

(36) Concerning these exchanges, see Manitius I 248, 285, and M. L. W. LAISTNER, *Thought and Letters in western Europe, A.D. 500 to 900* (2ed Ithaca 1957), p. 202.

(37) The monumental title, *LIBELLVS ANNALIS BEDAE,* appears in ms Vat. Pal. lat. 1449 (Lorsch s.IX¹) f. 2ᵛ at the head of a great collection of computistical tables, argumenta, and later Bedae DTR, transcribed in the court library, according to B. BISCHOFF, *Lorsch im Spiegel seiner Handschriften* (München 1974), p. 43-44 and 81 n. 5. It includes all those items which were attributed to Alcuin in Froben's edition on the basis that the *Calculatio* bears Alcuin's name

A.D. 742 provided a model for bishops to make annual tours of their dioceses to confirm and to instruct the people ; furthermore the education of priests required a better than average ability to read and to instruct : some should leave their duties in order to gain prerequisite skills ; and any who stubbornly remain ignorant shall be removed, "For those who are ignorant of God's law cannot proclaim and preach it to others." [38] In his *Admonitio generalis* of A.D. 789, Charlemagne's demand and legal support for more and better schools included specifically instruction in "Psalmos, notas, cantus, compotum, grammaticum per singula monasteria uel episcopia..." [39] Not only was this program carried forward by Charlemagne's well known *Epistola de litteris colendis,* a mandatum composed with Alcuin's assistance between 794 and 800 and circulated further by Baugulf of Fulda and other abbots and bishops in their areas of influence ; [40] but also after Alcuin's death the emperor's emissaries were directed in A.D. 805 to inquire "De lectionibus, 2. De cantu, 3. De scribis ut non uitiose scribant, 4. De

---

in this manuscript f. 11^v-12^v. Such a collection under Alcuin's name is also attested by the rubric *EXCERPTVM DE LIBRO ALBINI MAGISTRI* which Walahfrid placed on his ms p. 277 at the beginning of an *Adbreviatio Chronicae* (ms p. 278-283) in ms St Gallen 878, probably at Aachen between A.D. 829 and 834 ; it is followed by numerous computistical argumenta on ms p. 284-302. JONES, *The lost Sirmonds Manuscript* 104-106, has shown that *De bissexto* (ed *PL* CI 993-999) and *De cursu et saltu lunae* (981-993) in the Lorsch manuscript were known to Bede and perhaps Irish computists and cannot be attributed to Alcuin's authorship ; but they probably were brought from York to the continent by Alcuin when he joined Charlemagne's court.

(38) Carloman's capitulary was edited by Alfred BORETIUS, *MGH,* 2. *Leges In Quarto,* B. Legum Sectio II : Capitularia Regum Francorum I (Hannover 1883), no. XIX, cap. 15 : "et populum confirmare et plebos docere et investigare..." ; cap. 16 : "Quia ignorantes legem Dei eam aliis annuntiare et praedicare non possunt."

(39) Admonitio generalis, cap. 72, *ibid.* p. 54-62. This chapter was "the decisive legal formulation of educational goals," according to Josef FLECKEN-STEIN, *Die Bildungsreform Karls des Grossen* (Freiburg im Breisgau 1953), p. 74. Cf. also the Capitulare primum of A.D. 769-774, though doubt about date and attribution was expressed by F. LOT, *Le premier capitulaire de Charlemagne,* 1924-1925, repr. *Recueil des travaux historiques de Ferdinand Lot,* II (Paris 1970), p. 317-323.

(40) *Epistola de litteris colendis :* ed Luitpold WALLACH, *Alcuin and Charlemagne* (Ithaca, N.Y., 1959), p. 201-207 with excellent notes. Only a single late transcription of this existed until recently in ms Metz Stadtbibliothek 226 (s.XI/XII, destroyed during 1942-1945). An early Fulda transcript of Baugulf's circular based upon Charlemagne's mandatum survives in ms Oxford BL Laud. misc. 126 (s.IX^1), identified and edited by Paul LEHMANN, *Fuldaer Studien, Neue Folge, Sitzungsberichte der Bayerischen Akademie der Wissenschaften,* Phil.-hist. Klasse (München 1927), Abhandlung 2, p. 6-13 ; Eng. trans. M. L. W. LAISTNER, *Thought and Letters in Western Europe* (2ed, Ithaca, N.Y., 1957), p. 196-197.

notariis, 5. De caeteris disciplinis, 6. De compoto, 7. De medicina-
lia arte,...'' Evidence of more such inquiries ([41]) makes certain that
this was an on-going Carolingian policy which did not change
during Hraban's lifetime but persisted through the reigns of Louis
the·Pious and his quarelling sons — all of whom sponsored schools
and maintained scholars, libraries, and scriptoria at their
courts. ([42])

Instructors in computus nevertheless had difficulty with its
complexities, and there were numerous *argumenta* which not only
promised to shorten the Bedan explanations but might derive
from Victorian, ps Anatolian, or some other system. The Aachen
conference of A.D. 809 was intended to clear up some of the resulting confusion ([43]) but Hraban was not in attendance, prevent-
ed even from access to his pens and personal books by the Fulda
abbot Ratgar (802-817) who required all to work on his new
basilica and reduced hours of prayer as well as hours of study for
this purpose. ([44]) Hraban may also not have known of the Irish
grammarian Dicuil who had gained temporary favour at court and
presented the first chapter of his *Liber de astronomia et computo* to
Louis in May 814; apparently Louis and his advisors raised
questions about it and rebuffed the second chapter. Dicuil's work
was completed in 818 and gained only limited circulation. ([45])

---

(41) *Capitulare missorum in Theodonis villa*, A.D. 805, *ibid.*, no. XLIII (p. 121);
see further p. 235 (end of A.D. 805), and p. 363 (A.D. 813).

(42) Cf. Louis' capitulary of A.D. 827, *ibid.*, p. 403. He also took a special
interest in astronomy, according to *Vita Hludowici*, cap. 58, ed *MGH. Scriptores
In Folio* II (Hannover 1829), p. 642-643. See Bernhard BISCHOFF, *Die Hofbibliothek unter Ludwig dem Frommen*, in *Medieval Learning and Literature : Essays
presented to Richard W. Hunt* (Oxford 1976), p. 3-22.

(43) *Capitula de quibus convocati compotistae interrogati fuerint*, ed. E. DÜMMLER,
MGH. Epp. IV, p. 565-567; rpr with English translation by C.W. JONES, *An
early medieval licensing examination, History of Education Quarterly* III (1963) 19-29.

(44) Appeals by Fulda monks in 809 and 812 failed to obtain help from
Charlemagne, but Ratgar was removed from office in 817 by Louis; these
unusual appeals and responses are discussed by John McCULLOH, above, p.
XIII-XIV. See further D. HELLER, *Die ältesten Geschichtschreiber des Klosters
Fulda* (Fulda 1952), p. 31-34, 79-82; and J. SEMMLER, *Fuldaer Klosterporteien,
Zeitschrift für Kirchengeschichte* LXIX (1958) 268-298. Ratgar was abbot 802 to
817 and died 820.

(45) The manuscripts are Valenciennes 404 (St Amand s.IX ex) f. 66-118,
Tours 803, II (St Martin's s.IX ex) f. 58-103^v, and Paris BN n.a. lat. 1645 (St
Martins s.IX ex) f. 1-14^v; ed M. ESPOSITO, *An unpublished astronomical treatise by
the Irish Monk Dicuil, Proceedings of the Royal Irish Academy* XXVI (1907) Section
C, p. 378-445, and A. CORDOLIANI, *Le comput de Dicuil, Cahiers de civilisation
médiévale* III (1960) 325-337. Cf. Esposito's revisions in *Modern Philology* XVIII
(1920) 177-188; A. VAN DE VIJVER, *Dicuil et Micon de Saint Riquier, Revue Belge
de Philologie et d'Histoire* XIV (1935) 25-47; JAMES J. TIERNEY, *Introduction,*

COMPOSITION AND SOURCES

The contents of Hraban's *Computus* reflect his previous seventeen years of teaching straightforwardly from the classic sources of the Dionysian-Bedan tradition.

Probably in A.D. 819 Hraban received a letter from Marcharius setting out a series of computistical problems and asking that Hraban relieve his confusion. The Fulda master replied that he did not know the source of these difficulties or of the inadequate solutions sent to him; and we cannot detect the particular situation out of which Marcharius wrote. Hraban knew him to be a teacher of good report, ([46]) but thus far we have been unable to determine anything else about Marcharius. ([47])

The magister scholae of Fulda attempted to answer the problems presented by the calendar and to add enough other information so that an adequate understanding could be gained of reckoning the moveable feasts of the Christian year. He gave these materials an organization which he felt would be appropriate for the student: that is, a series of ninety-six chapters through which a teacher could lead a student step by step from the most elementary arithmetic and questions which anyone could answer, through more complex matters of the nature of the heavens and the earth and observation of planetary motion, then into the real complications of the Easter Table and the historical-theological assumptions on which it is founded. The table data were given a new form by which Lent and Rogation could easily be reckoned without expert knowledge. Finally the meaning of time itself, often referred to, was given a brief theological explanation in contrast to the earlier chapters from which most of the theology had been pared away from his sources.

Dicuili Liber de mensura orbis terrae (Dublin 1967), p. 12-15.

(46) "... teque reddidit suo fulgore decoratum et proximis tuis profectuosum," Hrabani *Comp.*, Prologus 7-8.

(47) The earliest manuscripts support the spelling Marcharius-Marchari-Marchar (mss *BCFHIMN* lac *GV*) rather than Macharius. In the latter form one could recall Greek saints from the Martyrologies as did Alcuin when he applied a nickname Makarius to Ricbod (bishop of Trier, d. 804); or Macarius Scottus whose ideas were opposed by Ratramnus of Corbey half a century later; or some otherwise unrecorded Irish Man-chén or a Breton Marcan-Marcar-Machar. See note to the text.

The dialogue form of his textbook on reckoning was a series of questions leading to needed information; it provided methods of applying the data appropriate for technical subject matter rather than for personal evaluation. Even as a Biblical scholar Hraban would know not only the question-posing rhetoric of Job but also several dialogue forms, both argumentative and didactic, in the Gospels which at the time of their composition were common in Hellenistic literature and continued in early Christian writings. (48)

Sapienter interrogare, docere est : these are evidently the words of Charlemagne to which Alcuin replied (49) about A.D. 798. In the early portion of *De rhetorica et uirtutibus* Alcuin had asked and Charlemagne answered, but later the emperor posed questions and the teacher responded. (50) Hraban had experienced this pattern of intellectual give and take at Aachen and doubtless practiced it at Fulda. In his *Computus* Hraban urged the "Disciple" to overcome his inertia and to learn the science of number because it is the key to elucidating many mysteries. But what the "Master" teaches through dialogue is how to solve the problem of determining the date of Easter this year and any year.

Hraban's sources were primarily Bedae *De temporibus, De natura rerum*, and *De temporum ratione;* he also drew upon Isidori *De natura rerum* and *Origines siue Etymologiae*, as well as one or more collections which included excerpts and paraphrases from these works; and he used *argumenta* and the nineteen year table created by Dionysius Exiguus. At various times during his life Hraban had access to some of the works of Ambrose, Augustine, Jerome, Donatus, Priscian, as well as Vergil and Josephus. He could well have known Boethii *Arithmetica* cited in cap. I, Plinii Secundi *Naturalis historia* used in cap. XXII, and Hygini *Astronomicon* used in cap. XXVIII — though such passages found in this *Computus* could also have been drawn from the more general computistical collections discussed below.

Hraban certainly expended enormous energy and funds to build up the Fulda library into one of the best centers of study for both

---

(48) C.H. Dodd, *The dialogue form in the gospels, Bull. of the John Rylands Library* XXXVII (1954) 54-67, esp. p. 63-65 ; Manfred Hoffman, *Der Dialog bei den christlichen Schriftstellern der ersten vier Jahrhunderte (Texte und Untersuchungen zur Geschichte der altchristlichen Literatur* XCVI ; Berlin 1966), esp. p. 5 n. 4.

(49) Alcuini *Ep.*136, ed E. Dümmler, *MGH. Epp* IV, p. 205 ; Paul Lehmann, *Das literarische Bild Karls des Grossen vornehmlich im lateinischen Schrifttum des Mittelalters* (1934), rpr *Erforschung des Mittelalters* I (Stuttgart 1959), p. 156 passim ; M.L.W. Laistner, *Thought and Letters*, p. 199-200 ; J. Fleckenstein, *Die Bildungsreform Karls des Grossen*, p. 34-36.

(50) Alcuini *Dialogus de rhetorica et de uirtutibus*, ed. PL CI, 919-976; cf. the *Disputatio Pippini cum Albino*, et PL CI 975-978. Concerning *Aliae propositiones ad acuendos iuvenos*, ed PL CI 1143-1160, which is sometimes attributed to Alcuin, see Jones *Bedae Ps* 51-52.

178   INTRODUCTION

secular and sacred literature in the Carolingian empire. ([51]) The authors and works which he used in the *Computus* could have been explicitly identified by Hraban but were not, even though the reader's attention was called to the fact that the discussions had been supplemented by citations of ancient authorities (Prologus 17-18)

It is well known that in their theological essays and biblical commentaries Beda, Alcuin, and Hraban customarily cited the authors whom they quoted or paraphrased by placing reference notae in margins. For example in Hrabani *Commentarii in Genesim* (CLM 6260, Freising s.IX) the sources were identified AG (Augustinus), H (Hieronymus), and so on. ([52]) Handbooks for school use however are a different sort of literature then and now. ([53]) Neither Hraban nor his literary mentors was so meticulous about sources for these, and the relatively elementary materials of the school texts were made available for all without technical apparatus.

The *Computus* of Hraban is organized into five distinct sections, cap. I-VIII, cap. IX-XXXVI, cap. XXXVII-LIII, cap. LIV-XCII and cap. XCIII-XCVI. His sources have been outlined only once before: Maria Rissell, *Rezeption antiker und patrischer Wissenschaft bei Hrabanus Maurus* (Frankfurt-am-Main 1976), p. 30-40. She divided the work into three parts and attempted to indicate a source for each chapter; however apparently due to haste she omitted some of the chapters, misnumbered some of them, named a source for a chapter which applied only to a few lines, and failed to distinguish quotations from paraphrases or simply related ideas. In the first set of annotations to the text offered below, Hraban's sources will be indicated more precisely and more completely. Rissell's discussion of the overall contents (41-75) is quite useful and will rarely need to be revised here, save for misinterpretations due to her use of the printed edition rather than manuscripts.

---

(51) See the essays of Karl CHRIST and Paul LEHMANN in the memorial volume, *Aus Fuldas Geistesleben*, ed Josef THEELE (Fulda 1928): LEHMANN, *Quot et quorum libri fuerint in libraria Fuldensi, Bok-och Bibliotekshistoriska Studier tillägrade Isak Collijn* (Uppsala 1925), p. 47-57; and CHRIST, *Die Bibliothek des Klosters Fulda im 16. Jahrhundert. Die Handschriften-Verzeichnisse* (Zentralblatt für Bibliothekswesen, Beiheft 64; Leipzig 1933, rpr Wiesbaden 1968).

(52) Paul LEHMANN, *Zu Hrabans geistiger Bedeutung, St Bonifatius: Gedenkgabe zum 1200 jährigen Todestag* (Fulda 1954), p. 473-487, rpr *Erforschung des Mittelalters* III (Stuttgart 1960), p. 198-212, esp. p. 203-205 and two facsimiles. For Bede's practice see C.W. JONES, *The manuscripts of Bede's "De natura rerum"*, Isis XXVIII (1937) 436-438, and the introduction to his new edition of this work in *Corpus Christianorum, Series Latina* CXXIII A (1975), p. 184, 187.

(53) William H. STAHL, *The systematic handbook in antiquity and the early middle ages*, *Latomus* XXIII (1964) 311-321, without mention of Hraban.

Cap. I-VIII. First Hraban discussed number, numerals, what they signify, and how to use them; for the science of number constat omnium disciplinarum esse magistram (1.3/4). His thoughts parallel those of Alcuin who explained to Charlemagne that scholars had discovered mathematics, whereas God has created it; (54) and Hraban is sure that this is evident from Scripture (I,4/8). (55) Without number how could we comprehend evening and morning, the succession of night and day from one to seven, the sun and moon, the signs and seasons and days and years? More proximate understanding of number was sought in the etymology of numerus and in definitions offered by Isidore, Donatus, and Augustine. (56)

Varieties of applications of number were indicated first by their grammatical forms, but terms for weights and measures and many types of multiples were also supplied. (57) Three sorts of number could be indicated by Roman numerals, by Greek letters, or by showing the fingers in certain positions. His explanation of finger-reckoning (VI) is partly paralleled by Beda and partly by another version commonly available. (58) In outline these eight chapters are an abbreviated introduction to arithmetic as taught in Fulda (cf Rissell op.cit. 41-48).

Many of the materials used thus far by Hraban will be found in computistical collections called Sententiae sancti Agustini et Isidori in laude compoti, almost every version of which added, deleted, and rearranged the contents. (59) Lacking an edition of any of these, I have indicated various parts which have been published by Herwagen, Noviomagus, and thus Patrologia Latina XC; they are not sources, but they contain material similar to Hraban's discussions: De arithmeticis numeris liber (ed PL XC 641-648), De

---

(54) Alcuini Ep. 144, ed MGH. Epp. IV, p. 239.

(55) He quoted Isidori Etym. III, 4, 1; this is not the same as Alcuin's point that arithmetic is necessary for understanding Holy Scripture.

(56) Quotations from Boethius, Isidore, Augustine, and Donatus in cap. I-II may be found in the Sententiae discussed below.

(57) Citation of Varro and Vergil in cap. III also may be found in the Sententiae discussed below.

(58) The tract with which Hraban begins his discussion of finger-reckoning (VI, 4/15), inc: Tres digiti in sinistra manu..., is found in many manuscripts of the early ninth century; see JONES Bedae Ps 54 and TKr 1583. There are several versions and a large bibliography. Later Hraban added in cap. LXXVI Bede's application of this system to the observation of planetary motions.

(59) A. CORDOLIANI, Une encyclopédie carolingienne de comput: les "Sententiae in laude compoti", Bibliothèque de l'Ecole des Chartes CIV (1943) 1-7; for an even wider range of versions, see JONES, Sirmonds Ms 204-219; IDEM Bedae Ps 48-51; IDEM Bedae Opera 105-113; and Bedae Opera didascalia 1, Corpus Christianorum Series Latina CXXIII A (Turnhout 1975), p. xii-xiv.

computo dialogus (647-652), *De diuisionibus temporum liber* (652-664), and *De argumentis lunae libellus* (701-728). Successive passages may be found in ms Vat. Pal. lat. 1447, written about A.D. 813 with later provenance from the monastery of Saint Alban where Hraban was buried and then Mainz Cathedral; cf also ms Vat. Rossi lat. 247 (s.XI) whose exemplar used the date A.D. 820. In dialogue form some of the material occurs in the Benediktbeuern ms Basel F.III.15k (A.D. 820-840) f. 21-57 with *argumenta* for 789 and for 820. Without a critical study of such collections however it is not yet possible to say which of them might have been available for Hraban's use.

Cap. VIIII -XXXVI. Tempus est mundi instabilis motus rerum-que labentium cursus. With his own poetic definition (VIIII, 5/6) Hraban introduced the discussion of time itself and its terminology, deceptively simple at first but becoming increasingly complex. Bede is quoted on the common derivation of tempus from tempe-ramentum, or time as the measure and divider of any period and of the whole life span(8/11). The terms are from nature, human custom, and divine or human decision, and the student must understand them for various periods and their proportional relationships: atom, ostentum, momentum, partes, minutum, punctus, hora, quadrans, and dies — the number of each being calculated separately for each successively large category, provi-ding occasion for students to practice multiplication. ([60]) Discus-sion of day, night, and their several parts (XVIII, 7 to XXIII, 19) is rather full and appears to be derived not only from Isidore and Bede but also from the *Sententiae*. Ebdomada was a rubric for the seven day week but also seven weeks (and a day) of pentecost, seven months, and seven years. Names of Hebrew days and those derived from planets led to the innovation of feria for the seven-day week in the Roman calendar, attributed to Silvester, bishop of Rome, as Beda taught.

Hraban summarized the necessary information about the month in the Roman calendar with a few lines (XXVIII, 20/24), but the invention of months was attributed to the Chaldeans, ancient master astronomers from whom according to Josephus Abraham learned about the movement of stars and taught the Egyptians (XXVIIII-XXXII) — a conceptual mythos which

---

(60) The arithmetical ratios were not simply copied out by the scribes who were usually very careful about transcribing numerals in surviving manuscripts of this work. In four places student scribes at various times introduced errors in chapter XXVII by practicing multiplication; one of these was in the school at Reichenau ca. 825, later corrected: ms *G*.

provided a rationale for much of Isidore's and Beda's intellectual endeavors. (⁶¹) Hraban added that despite the silence of scripture it ought not be supposed that the Hebrew patriarchs before the deluge lacked such knowledge (XXVIII, 41/46). Names and application of months by Hebrews, Egyptians, Greeks, and Romans were given from Beda, while Hraban added a further explanation of Roman usages which derived from their divinities, historical events, rulers, or terms of numerical order. It was his own composition (XXXII, 28/89) from information supplied by Isidore and Beda, though not from similar material in Macrobius whose works were not known in Fulda at this time. Roman terms Kalendae, Nonae, Idus were explained but actual application of them was assumed, apparently on the basis that students were drilled on such things in their classes ; such classroom exercises are indeed found to follow this computus in the earliest mss CG. (⁶²)

In this section Hraban's authorship is strongly in evidence for cap. XI-XII, XVIII-XVIIII, and long portions of XIIII, XXVII, XXVIII, XXXII. Only a few sentences of De divisionibus temporum liber (cited by Rissel 32-34) may be found in these chapters, though there are parallels of content and closer parallels of wording in other chapters ; and his use of Isidore and Beda does not appear to derive from any of the *Sententiae* items. It is a well-organized discussion for teaching purposes which is briefer but perhaps more coherent than his sources.

Cap. XXXVII-LIII. In his third integral section of materials for instruction in computus Hraban introduces celestial phenomena : sun, moon, planets, stars, and their relative motions. He named the seven planets, told how to recognize them, and mentioned some characteristics of their movements in the heavens. Then he led his students into an elementary acquaintance with the zodiac, describing the movement of all the planets but especially the sun and the moon through various constellations within and without the zodiacal circle. The unusual phenomena of eclipses and comets were described, but more important were the regular solstices and equinoxes ; and he supplied the general concepts of the heavenly sphere and the projected regions into which two spheres of heaven and earth are divided. There is no reasonable doubt that Hraban

---

(61) For this tradition see E.R. CURTIUS, *Europäische Literatur und lateinisches Mittelalter* (Bern 1948), p. 216, 449-450 ; for evidence prior to Josephus, Louis H. FELDMAN, *Abraham the Greek Philosopher in Josephus, Transactions of the American Philological Association* IC (1968) 143-156.

(62) These examples have been explained in STEVENS, *Walahfrid Strabo — a student at Fulda*, p. 13-20.

shared with Isidore, Beda, and all medieval scholars who have written in Latin on this subject the globular concept of both earth and heaven. [63]

This was largely derived from Beda's early work *De natura rerum* liber V, VIIII, XV-XX, XXI (epitome), XXII-XXIIII; see the edition by C. W. Jones, *Bedae opera didascalica* 1, Corpus Christianorum Series Latina CXXIII A (1975), p. 189-234. A comparison of variant readings has indicated that ms Paris BN 4860 (s.IX² Diocese of Konstanz?) f. 108-111ᵛ provides a text of Bedae DNR closest to the one used by Hraban. To these chapters he added selections from Bedae DT and especially DTR, and he cited Arati *Phenomenae* from a version which has not yet been identified. Similar to cap. LI is De duodecim signis liber, ed Ernest Maass, *Commentariorum in Aratum Reliquiae* (Berlin 1898), p.582-594; but cf also Isidori *Etym.* III, 71, 4/15, Bedae DNR XVIII which is being supplemented here, and the tract Duo sunt extremi uertices mundi... which accompanied Hraban's *Computus* in mss *ABCDEV*. Hraban supplemented his earlier description of the zodiac by reference to the popular memory verse, *Ecloga* XVII of Ausonius, and by a few lines to indicate that the signs are of such great distance that at least two hours may be required to observe their motion (XXXVIIII 28/30). Pointing out the necessity of direct observation and instruction by someone expert in these things (XL), Hraban then supplied Beda's method for determining the positions of the moon and sun in the zodiac on any day (XLI-XLII), adding that the extra ten and a half hours in any month are neglected in this calculation but are accounted for in the twelve month period to be discussed later. The use of instruments and a star chart is implied in these chapters; drawings of

---

(63) The relevant Latin texts have been reviewed by F.S. Betten, *Knowledge of the sphericity of the earth during the earlier middle ages, Catholic Historical Review*, New Series III (1923) 74-90; J.K. Wright, *Early Christian Belief in a Flat Earth*, in his *Geographical Lore of the Time of the Crusades* (1925), p. 53-54, who nevertheless suspected that some believed the earth to be flat like a disk rather than spherical; and Charles W. Jones, *The Flat Earth, Thought* IX (1934) 296-307, whose copious references to the earth's sphericity in medieval literature of all centuries left no such doubts. See also Rissel 54-57 who attempts to relate Hraban's phraseology in biblical commentaries and *De rerum natura* to this matter, even though she had noted that he had omitted an astronomical section entirely from the latter in view of its presence in the *Liber de computo*.

Concerning the purely modern mythology that there was a doctrine of the Church condemning the idea that there might be inhabitants of the southern temperate zone or of the antipodes, see Stevens in *Speculum* LI (1976) 752-755: a review of K. Hillkowitz, *Zur Kosmographie des Aethicus* (1973), and related literature.

Fulda and Salzburg-Regensburg planispheres survive from this period, as does a moon-finder. ([64]) Finally Hraban added a few lines on the positions of the seven planets at the time of writing:

Right now, that is 9 July 820, the sun is in the twenty-third part *
of cancer, the moon in the ninth part of taurus, saturn in the sign
of aries, jupiter in libra, mars in pisces, while it does not appear
in which sign venus and mercury tarry because they are now near
the sun during daylight (XLVIII, 16/19).

Again, the discussion of this astronomical section by Rissel 49-58 is quite useful, though her identification of sources is undependable.

Cap. LIIII-XCII. With a proper background the student would be prepared to learn the computus itself and to deal with the complexities of an Easter Table. Hraban drew together the essential information from the mature work of Beda and pruned away a large part of the historical and theological discussions. He added eight of the nine directions given by Dionysius Exiguus ([65]) with the tables for A.D. 532-621 and by which the student could easily determine whether the current year were annus bissextilis (cf. Dionysius, *Argumentum* VIII) or reckon the number of years from the incarnation to the present (*Arg.* I). There were also explanations of how to determine the current imperial indiction (*Arg.* II), how to use the lunar epacts (*Arg.* III, from DTR LII) and the solar epacts or concurrents (*Arg.* IIII, from DTR LIV), how to determine the current year's position in the nineteen year solar cycle (*Arg.* V) or in the lunar cycle (*Arg.* VI, from DTR LVIII), and how to find the day of the month on which sextadecima luna or Easter Sunday would occur (*Arg.* VIIII).

In most cases these Dionysian *argumenta* provided alternate methods for approaching problems already explained from selections of Bedae *DTR;* but in every case Hraban has added explanatory detail and has often given examples calculated in terms of the year, month, day, or other precisions of the time at which he wrote. From an *Argumentum* XIII, probably of seventh century origin and certainly not from Dionysius, Hraban presented a method of calculating the number of the moon on the first day of January (adapted also by him to the first day of August) and for determining its corresponding weekday; it supplemented the less complex rule given by Hraban (LXXVIIII) from Bedae DTR LVII but which had exceptions in years 8, 11, 19 and which led Beda to

---

(64) These were cited above in note 9.

(65) J.W. JAN, *Historia Cycli Dionysii* (Wittenburg 1718), rpr *PL* LXVII 497-508 ; KRUSCH, *Studien* II (1938) 59-87. Unfortunately Rissel did not indicate these *argumenta*.

invite readers to create a better rule. ([66]) Hraban drew upon parts of
DTR XX to compose his explanations of lunar regulars, including
the warning qualifications that they had originally been set for a
year commencing in September rather than January which was
preferred. ([67]) He also teaches Beda's concept of ferial or solar
regulars. ([68]).

Thus we have found that in many of the explanations drawn
from Beda or Dionysius Hraban added one or more examples of
how to apply the calculation: LV, 5/11; LVI, 21/26; LVIII, 7/11;
LXII, 10/16; LXVIII, 12/14; LXX, 30/36; LXXII, 21/26;
LXXIIII, 55/60. Also not derived from any known source and
perhaps created by Hraban are explanations of how to reckon the
number of years from the beginning of the world to the incarnation
(LXV), and the number of lunar and solar days in any month
(LXXV). He further set in juxtaposition alternative usages,
explanations, and calculations in LVI, 20/22; LXXVII, 19/23;
LXXXII, 1/34; LXXXV, 5/17.

Hraban has followed the example of Beda and explained
eight columns of the table attributed to Dionysius Exiguus.
Writing in A.D. 525, Dionysius had taken over the last nineteen-
year cycle (513-531) of a Cyrillan Easter Table in which the
principles of Alexandrian reckoning had been translated into the
terms of Latin usages. On this basis he had presented an African
(?) bishop Petronius with a table of ninety-five years (532-626). ([69])
Beda must have assumed that his *DTR* would either be bound
together with such a Dionysian table or that the reader would
have one available. Hraban made the same assumption: his
textbook was not self-sufficient but required access to a Dionysian
(or Bedan) table with at least one nineteen-year cycle.

However Hraban offered new tables in cap. LXXXIII, 49/69
and 70/90 which would be especially useful for festivals of the

---

(66) Other attempts to do so may be found in *De argumentis lunae libellus* (*PL*
XC 702-703) and *Liber de computo* (*PL* CXXIX 1284C and 1289A); cf. JONES,
Bedae Opera 354-355. The saltus lunae in year 19 would further disturb the
calculations of luna and feria on Kalends of January or August.

(67) Cap. LXX, 6/49 from Beda, *DTR* XX, 2/12, 26/35, 47/55. Dionysius
had followed Alexandrian practice for a year commencing in September in
conjunction with imperial practice, whereas Victorius used January with the
Julian calendar of Roman usage. Beda maintained Roman usage in this regard;
see JONES, *Bedae Ps* 75.

(68) Cap. LXXIII; Beda, *DTR* XXI. See JONES, *Bedae Opera* 354-358.

(69) The Easter Table prepared by Dionysius may be seen in *PL* LXVII 495-
498, or KRUSCH, *Studien* II (1938) 59-87. The Latin form of an Alexandrian table
used by Dionysius may not have stemmed from Cyril of Alexandria, according
to JONES, *Bedae Opera* 68-70.

Christian year. The same nineteen year cycle was set out but with columns of dates for the moons from which could be reckoned the Christian observances of Lent ([70]) and Rogation Sunday. ([71]) He also explained that the lenten moon and the paschal moon occurred on the same day of the week (LXXXIII, 27/33).

Another characteristic of his table is its inclusion of parallel columns both for the rise of the first moon and for its continuation after midnight which by Roman practice marks the beginning of another day. A possible confusion was clarified by these data for the "first moon of the first month" (after midnight) and for its ascension (before midnight). In order to avoid the occurrence of Easter Sunday on the vernal equinox (thought to be XII Kalendas Aprilis, or 21 March), ([72]) the paschal moon was counted from prima luna rather than from its ascension. The importance of this was two-fold: it preserved the Roman churches' distinction from the Jewish observance of Passover on Nisan XIIII and from the Christian quartodecimans' observation of Easter on that day ([73]); and it avoided the churches' social tension when the date of Easter concurred with pagan rites in the city of Rome. ([74])

Easter tables were provided with coordinated columns of lunar regulars and lunar epacts — data which recur in sequence every nineteen years. These data were applied to locate the

---

(70) Lent begins forty fast days before Easter and was thus called *Quadragesima*. Its terminus occurred on *secundam lunam mensis martii* (LXXXIII 7).

(71) Hraban said that the Rogation term was "in XXI luna mensis maii" (LXXXIII 45). The twenty-first moon of the lunar month following Nisan did fall within the artificial month, May, in A.D. 817, 819, 820, 822, etc.; but the phrase was misleading.

(72) Julius Caesar's reform of the Roman calendar had placed the equinox at VIII Kl. Aprilis. The Alexandrians however observed XII Kl. Aprilis and convinced the Council of Nicaea to agree to this as an early limit for luna paschalis (= Nisan XIIII), perhaps on the basis of an Alexandrian observation of the vernal equinox. However the Julian calendar and the Alexandrian Christians overlooked the procession of the equinoxes which resulted in the gain of about three days in four centuries. From the Council of Nicaea to Hraban's time the true equinox could have advanced just over three days; doubtless he lacked the instruments to observe this difference with a horologe (sun-dial), though he affirmed with Bede that an equinox of 21 March could be thereby confirmed: cap. LIII, 40/47 from *DTR* XXX, 72/77; cf. JONES, *Bedae Opera* 126-127 who exaggerates the observable change by assuming that the Alexandrian/Nicaean affirmation of 21 March referred to the time of Caesar.

(73) Cyril C. RICHARDSON, *The Quartodecimans and the Synoptic Chronology, Harvard Theological Review* XXXIII (1940) 177-190; IDEM, *A New Solution to the Quartodeciman Riddle, Journal of Theological Studies*, New Series XXIV (1973) 74-84; C.W. JONES, *Bedae Opera* 6-10, 18-20.

(74) JONES, *Bedae Opera* 28, quoting Prosper about Roman circuses disturbing the celebration of Easter in A.D. 444.

calendar day and week day on which the paschal moon occurs, from which Easter Sunday may be found. With the epacts one may transfer from lunar to solar cycle; with lunar regulars and concurrents one may determine the day of the week. Week days of the paschal moons recur in a 28-year cycle because of leap year. The calendar day and the week day are coordinated therefore only in a cycle of $19 \times 28 = 532$ years.

Every column of Hraban's tables recurs in 19 years except that for concurrents, and unfortunately confusion exists at this point in all manuscripts. Concurrents indicate the day of the week on the date before which the extra day was added in a leap year. By adding the lunar regular and the concurrent for any year, one can find the weekday on which the Easter full moon appears and thus predict the date of the following Easter Sunday. For example in A.D. 820 the concurrent was 7 and the lunar regular was 2, the sum being 9; when 7 is subtracted for the total days in the week the remainder is 2, indicating the second day or Monday. The full moon of the "first month" in A.D. 820 was given as 2 April; since it was a Monday we learn that 8 April was Easter Sunday.

In the manuscripts of Hraban's table only mss BC give the required concurrent of VII in the fourth year of the cycle, mss FGI shifted the line down one step and give concurrent VI in the fourth year (misleading Rissel and others who have attempted to use the edition which derived from ms F), [75] ms D listed concurrents apart from the table, and there are other arrangements. None of this would cause difficulty for a trained computist in the ninth or any other century who knew that when concurrents were placed on a table in order to locate the weekday, the table became repeatable only 532 years later. He would therefore reckon his concurrents from the *argumenta,* then coordinate his list of concurrents with the other columns of data.

In all these reckonings certain assumptions had to be made which are not self-justifying. Hraban followed consistently the tradition of Beda and Dionysius on determining the date of Easter. The paschal full moon or *luna XIIII* had been equated with the Hebrew observation of the fourteenth moon of Nisan, maintaining continuity of the Church with Israel through the tradition of Christ's passion on Passover, Fulfillment of Passover however was not the death but the resurrection of Christ, said to be on the morning of the third day after Passover and a Sunday. To discover that Sunday however, it was necessary to know in what year Christ had been crucified, and this was a question on which the early Christian communities as well as the

---

(75) F also copied the adjacent column, Luna Prima, one line too high and thus out of coordination.

learned doctors had not agreed. The "diverse opinions" collected by Beda were included by Hraban (LXXXVII), and he added the thoughts of Jerome and Augustine (LXIIII 7-14) ; but finally the Alexandrian calculation promulgated by Dionysius and Beda was accepted by which the Incarnation was placed peculiarly in the second year of a nineteen-year cycle. ([76]) Hraban accepted the vernal equinox on 21 March and also the limits of luna XV-XXI or 22 March to 21 April for the days through which the paschal full moon could range ([77]) — dates which were acceptable to Christians in the city of Rome.

Cap. XCIII-XCIV. In computistical matters Hraban did not invite students to speculate further about these assumptions made in the face of conflicting historical evidence and partly accepted in terms of tradition, the grounds for which no longer pertained. His advice was rather to understand and accept and to apply the reasonable judgments of Beda who passed on the teaching and authority of the Fathers.

Thus Hraban was working in the spirit of *norma rectitudinis* by which Charlemagne and Alcuin had brought order out of chaos in many aspects of life in the empire and the church. ([78]) The emperor had insisted that "properly corrected catholic books" (Admonitio generalis, 72) be used in the schools and had assigned various scholars to prepare such books. Improved texts of the Bible (Alcuin) and the Liturgy (Alcuin), a new Homilary (Paul the Deacon) and Lectionary (Paulinus of Aquileia), as well as an authoritative exposition of the Apostles' Creed (Paulinus of Aquileia) were prepared, and they were promoted by imperial authority. A further step was taken by Hraban who was of the same mind with regard to reckoning and celebrating the birth,

---

(76) LXI 17-20, 36-42 ; Bedae *DTR* XLVII 9-13, 25-30 ; Dionysii *Argumentum* V. The year of incarnation was also described as the fourth indiction (LXII ; Dionysii *Argumentum* I) and the second from last of the lunar cycle (LXXVIII ; Dionysii *Argumentum* VI). These data do not accord with A.D. 1. It seems probable that Dionysius accepted a beginning point which for him was historical rather than computistical, i.e., the date recorded for the Incarnation in the *Fasti Consulares* of the *Chronograph of A.D. 354;* according to JONES, *Bedae Opera* 70.

(77) *Luna XIV* could not precede the vernal equinox nor fall on any part of it (LVIIII 17-19 ; Bedae *DTR* XLV 10-16). On the 34/35 days through which Easter Sunday could range, see LXXXVIII.

(78) On the *norma rectitudinis* as the central idea of Charlemagne's cultural policies, see Josef FLECKENSTEIN, *Die Bildungsreform Karls des Grossen als Verwirklichung der Norma Rectitudinis* (Freiburg in Breisgau 1953), ch. III *et passim.*

death, and resurrection of Jesus Christ. Empire and Church would be served by consistency of observation of its chief feast. The whole reckoning therefore must be seen in its true significance. It is the "Mystical Significance of Easter" (XCIII) in the whole history of mankind (XCV-XCVI) which gives meaning to this life and which looks forward to that last time "in which the blessed will reign with the Lord forever":

> at which time he grants that we gladly receive his grace, he the triune and omnipotent God who created all times and leads always to endless eternity, who is blessed in all ages. Amen.

(XCVI, 59/62)

DATE OF COMPOSITION

The variant readings in an *argumentum* composed by Hraban himself (cap. LXV) indicate that he was organizing the book at first with *annus praesens* DCCCXVIIII. This date was revised however during the summer months of A.D. 820, with details given for July and August. Not only has the annus praesens been stated often in numerals DCCCXX but also in words, in septimum annum Hludowici (cap. LXV 8). It is a leap year (annus bissextilis, e.g. cap. LV 6) and the fourth year of the nineteen-year solar cycle (e.g. cap. LVIII 7-15). The indiction is thirteen, ([80]) and the paschal full moon occurs on the second of April which is Monday, ([81]) so that Easter Sunday would be April eighth or VI Idus Aprilis. The lenten moon was on the twentieth of February, from which could be calculated the beginning of Lent on the twenty-second and Quadragesimal Sunday on the twenty-sixth. ([82]) By comparison with standard Dionysian tables such as those which

---

(80) E.g. cap. LXII 10-11; but here and in other examples there are variant readings which pertain to the previous year, A.D. 819. Cf cap. LXVII 9-11 and cap. LXVIIII 4-5.

(81) Quarto nonas Aprilis (cap. LXXXII 9), secunda die mensis Aprilis (12), secunda feria (29).

(82) Secunda luna martii mensis was said to be the normal quadragesimal term (cap. LXXXIII 6-7, etc), but because of leap-year in A.D. 820 it had to be adjusted from the tabular data: Verba gratia in presenti anno terminus quadragesimalae, qui conscriptus est in undecimo kalendas martias, translatus est in decimo kalendas martias propter bissextum, quem in luna mensis februarii septimo kalendas martias inseruimus (cap. LXXXIII 38-42).

Hraban did not use the phrase, bis sexto kalendas martias, to indicate the extra day inserted in leap-year and which would preserve the normal numerals for noting the several kalends, but rather he renumbered septimo, octo, et seq beyond the usual sixth kalends during leap-year.

Beda and Hraban used it may be determined that only A.D. 820 could have been expressed in these several ways. ([83])

Furthermore Hraban supplied the very day of writing on two occasions : one was inserted into a Bedan explanation of lunar epacts as an example of how it should be applied : Nam uerbi gratia, si hodie cum scribo in undecimo kalendas augusti septima est luna,... (cap. LXVIII 12-13 : Now for instance if, as I write today on the twenty-second of July, it is the seventh moon,....). A few chapters later he applied a Dionysian *argumentum* to calculate from the lunar cycle that the number of the moon on the first day of August in the current year was fourteen and that the day of the week was Wednesday (cap. LXXIIII 44-60). Presumably the materials had been accumulated and often used during previous years of teaching before Hraban set aside some part of each day to write this book, but during those ten days he and his scribes had probably completed seven chapters. Extrapolated at this rate and assuming no long interruptions, the period of writing for Hraban's *Computus* may have been something like 65 to 70 days, from about the beginning of June to the middle of August, A.D. 820.

The choice of that year may have been influenced by the fact that it was the fourth year of a nineteen-year solar cycle in the Dionysian Easter Table whose details correspond exactly to data for A.D. 725, the annus praesens of Bedae DTR. This was both a convenience for the instructor and an advantage for the student, both of whom were expected to resort to Bede's essays for more detailed and advanced explanations of computistical science. In A.D. 820 Hraban wished to make available to Carolingian scholars the calendarial instrument which had been introduced by Dionysius Exiguus (A.D. 532) from Alexandrian antecedents, which had been well-explained by Beda (A.D. 703, 725), and which determines the character of the European calendar to the present time. ([84])

---

(83) Other dates have been given without supporting evidence : A.D. 842-847 by Gottfried Henschen, *Acta Sanctorum :* 4 February (ed J. Bollandus, G. Henschen) III, p. 510 ; or A.D. 828 by Valentin ROSE describing excerpts from Hraban in ms Berlin Phillipps 1833, f. 3, *Verzeichnis der Lateinischen Handschriften der kgl. Bibliothek zu Berlin* XII (Berlin 1893), I. Die Meerman-Handschriften des Sir Thomas Phillipps : no. 138. Pierre DUHEM, *Le système du monde* III (Paris 1916), p. 23 did not cite Hraban's *Computus* but associated him misleadingly with a computistical collection of ca. 810.

(84) The annus incarnationis calculated by Dionysius Exiguus and other conditions convenient to the Roman churches lacked historical necessity, though his criteria were defended successfully by Beda and Hraban. Attempts to revise the annus incarnationis by Florus of Lyon, Heriger of Lobbes and Abbo of Fleury, Marianus Scottus of Fulda and Robert of Hereford foundered on the same conflicting evidence for dates of both the passion and the birth of Jesus which faced Dionysius. Necessity for the Gregorian reform of the calendar in A.D. 1582 was due not to uncertain beginning point or unhistorical criteria but

INTRODUCTION
                      MANUSCRIPTS AND EDITIONS

The sixteen extant manuscripts are listed here with only the
most essential secondary literature; full descriptions of the entire
series will be offered elsewhere. Ninth century manuscripts are
CFGIMP which were written in Fulda, Regensburg, the Bodensee
area and Reichenau, southwestern Germany, and northern Italy.
Others show traces of exemplars which must have been written in
a semi-insular script such as used in Fulda and adjacent scriptoria
during the first half of the ninth century; even the latest ms B from
the seventeenth century was copied from a very early ninth century
Fulda exemplar.

A   AVRANCHES, *Bibliothèque municipale* 114 (s.XII^1 Normandy ?) f.
        98-132. L. DELISLE, *Catalogue général des Bibliothèques Publiques
        de Départements,* Quarto series IV (Paris 1872), p. 483-484; H.
        OMONT, *ibid.,* Octavo series X (Paris 1889), p. 52-53. There is no
        palaeographical support for the assumption that this manuscript
        was written at Mont-Saint-Michel, though it may have been
        brought there during the abbacy of Robert of Thorney (A.D. 1154-
        1186). It shares the exemplar with or was transcribed from ms *V,*
        together with the *AEV* glosses.

B   LONDON, *British Museum,* Additional 10801 (s.XVII), f. 1-59^v.
        *Catalogue of Additions to the Manuscripts in the British Museum of
        the years 1836-1840* (London 1843): Acquisitions of 1837, p. 15.
        Transcribed from a Fulda exemplar, it shares most variants with
        ms *C,* but has some unique readings.

C   OXFORD, *Bodleian Library,* Canonici miscellaneous 353 (A.D.
        827-829, Fulda) f. 2-53^v. W.M. STEVENS, *Fulda scribes at work,
        Bibliothek und Wissenschaft* VIII (1972) 287-317 with nine facsimi-
        les; IDEM, *A ninth-century manuscript from Fulda: Ms Canonici*

---

to the slight over-correction of annus bissextilis which resulted in an increment
to the solar cycle and accumulated to a noticable shift of equinoxes and
solsticies. With the new Gregorian system this was to be corrected arbitrarily by
not observing one leap-year in those years which are multiples of 100, with the
exception of those which are multiples of 400. The location of the additional day
of annus bissextilis was shifted to the day following 28 February, rather than the
day prior to the sixth Kalends (of March). These adaptations and the use of the
nineteen-year cycle in a different way did not mean that the lunar and solar cycles
became perfectly coordinated, but the variation was thereby reduced to only one
day in three thousand years. A summary account was given by John J. BOND,
Handy-Book of Rules and Tables for verifying dates with the Christian Era (London
1866, rpr 1966), p. 6-9; for technical discussion see F.K. GINZEL, *Handbuch der
mathematischen und technischen Chronologie* III (Leipzig 1914), p. 257-266, 417-421.

miscellaneous 353, The Bodleian Library Record IX (1973) 9-16 with three facsimiles; these articles establish provenance of the manuscript in Fulda ante A.D. 836. It will further be demonstrated that it was used by Walahfrid at Fulda during A.D. 827 to Spring 829, in an article, *Manuscript notae in Fulda and Reichenau,* to be published.

D EINSIEDELN, *Stiftsbibliothek,* 319 (A.D. 996 Einsiedeln) p. 157-274. A. BRUCKNER, *Scriptoria medii aevi Helvetica,* V. Einsiedeln (Geneva 1943), p. 183-184; IDEM, *Zur Datierung annalistischer Aufzeichnungen aus Einsiedeln, Carolla Heremitana* XIII (Freiburg i.Br. 1964) 94. The date occurs in an *argumentum* immediately following the computus on ms p. 274.

E EXETER, *Cathedral Chapter Library* 3507 (s.X[2] post A.D. 986, Sherborne?), f. 1v-58. Neil R. KER, *Catalogue of manuscripts containing Anglo-Saxon* (Oxford 1957), no. 116*, IDEM *Medieval Librairies of Great Britain* (2ed London 1964), p. 303. The script is similar to ms Oxford BL Bodley 718 and ms Paris BN lat. 943. Ker found identical *A—S* glosses in another part of this manuscript and ms London BL Cotton Domitian I (Cat., no. 146). Ms *V* was transcribed from the same exemplar, and both have the same *AEV* glosses.

F PARIS, *Bibliothèque Nationale,* Fonds latin 4860 (s.IX[2], Diocese of Konstanz) f. 119v-135v. *Catalogus codicum manuscriptorum Bibliothecae Regiae,* IV (Paris 1744), p.9; T. MOMMSEN, *Monumenta Germaniae Historica, Auctores antiquissimi* IX (Berlin 1892), p. 363-365. The manuscript contains a list of books written by or for the librarian, Reginbert of Reichenau (A.D. 835-842), where Walahfrid had earlier copied the *Computus.* This copy was probably made later during the third quarter of the ninth century at Reichenau for another monastery in the diocese of Konstanz; from there it moved to Mainz about mid-tenth century where it remained until it was acquired by Colbert and used by Etienne Baluze for the *Editio princeps.* Unfortunately the nineteen-year cycles with cap. LXXXIII have been adjusted to a later period and do not accord with Hraban's tables.

G SANKT GALLEN, *Stiftsbibliothek* 878 (ca. A.D. 825, Reichenau) p. 178-240 (*desinit ante cap.* XIII 7 unum momentum...). Bernhard BISCHOFF, *Eine Sammelhandschrift Walahfrid Strabos* (Cod. Sangall. 878), in *Aus der Welt des Buches: Festgabe Georg Leyh* (Beiheft zum Zentralblatt für Bibliothekswesen 75, Leipzig 1950), rpr BISCHOFF, *Mittalterliche Studien* II (Stuttgart 1967), p. 34-51 with additions and four facsimiles; W.M. STEVENS, *Manuscript notae at Fulda and Reichenau,* to be published. Bischoff identified four periods of Walahfrid's hand in this manuscript from ca. A.D. 825

VIII

until the year of his death, 849; this computus is the earliest item he is known to have transcribed at Reichenau and the earliest extant copy. He took this notebook with him to Fulda in 827 where he apparently collated and corrected the computus from ms *C* before departing for Aachen in the Spring of A.D. 829.

H   LONDON, *British Library*, Harley 3092 (s. XI Fulda ?), f. 29-39ᵛ. *Catalogue of the Harleian Manuscripts in the British Museum*, II (London 1812): no. 3092; Paul LEHMANN, *Mitteilungen aus Handschriften II, Sitzungsberichte der Bayerische Akademie der Wissenschaften*, Philol.-hist. Klasse (München 1930), Heft 2, p. 23-24. This was written in the central western Germanic area of Trier or Fulda, according to Lehmann, and was once possessed by Nicholas of Cusa. It has been studied carefully and was annotated very fully on inserted leaves which should be analyzed (but which have not been indicated in this edition). Its variant readings are sometimes shared with ms *D* and are often identical with twelfth century ms *L*. The nineteen-year cycles of *cap. LXXXIII* are lacking.

I   SANKT GALLEN, *Stiftsbibliothek* 902 (s.IX² southwest German or Alemannic area) p. 105-152 (*desinit cap.* XCIV 20 ... nil siderum). A. BRUCKNER, *Scriptoria medii aevi Helvetica*, III. *Die Schreibschulen der Diözese Konstanz:* Sankt Gallen 2 (Geneva 1938), p. 122. Two scribes copied the computus and a third corrected it — all three from different backgrounds. The broken-backed-c indicates Alemannic training for one of them, but it remains uncertain where this codex was written or used. Corrections were so thorough that its probable relation with mss *F* and *G* have been eradicated, and remaining variant readings do not indicate a positive relation with others.

L   LEIDEN, *Bibliothek der Rijksuniversiteit*, B. P. L. 191 BD (s.XII in., middle Rhein area), f. 1-26ᵛ. *Catalogus compendiarus continens omnes codices manuscriptos qui in Bibliotheca Academiae Lugduno-Batavae asservantur*. Deel 14, Inventaris van de Handschriften eerste Afdeeling (Leyden 1932), p. 117; A. CORDOLIANI, *A propos du chapitre premier du De Temporum Ratione de Bède,... II. Les figures du comput manuel du manuscrit BPL 191 BD de la Bibl. univ. de Leyde, Le Moyen Âge* LIV (1948) 217-223; a note in the library from Kurt Weitzmann (December 1949) supports the assumed date of writing by reference to the drawings of f. 4-5ᵛ There are numerous insertions, epitomes, paraphrases, and other variations introduced by the scribe whose exemplar must have been shared with that of the earlier ms *H*. The latter copy lacked the nineteen-year cycle of *cap.* LXXXIII, and the scribe of ms *L* inserted one which does not correspond with Hraban's table.

M MÜNCHEN, *Bayerische Staatsbibliothek,* Codex lat. 14221 (s.IX ex., St Emmeram Regensburg) f. 23-60ᵛ. B. BISCHOFF, *Die südostdeutschen Schreibschulen und Bibliotheken in der Karolingerzeit,* I. Die Bayerischen Diözesen (2ed Wiesbaden 1960), p. 224. This transcription derives from an exemplar similar to Fulda ms *C*.

N MÜNCHEN, *Bayerische Staatsbibliothek,* Codex lat. 14523 (s.X¹, St Emmeram Regensburg) f. 2-48ᵛ. C. SANFTL, *Catalogus veterum codicum manuscriptorum ad S. Emmeram Ratisbonae* ... (A.D. 1809 manuscript available in the Bayerische Staatsbibliothek), p. 174; *Catalogus codicum latinorum Bibliothecae Regiae Monacensis,* II 2 (München 1876): no. 14523. Folios 1-48 form a single codex separate from f. 49-117 (A.D. 854-875) and from f. 118-133 (s.XI med.); it is datable to the first half of the tenth century by comparison with other St Emmeram manuscripts. Textual variants are not identical with other extant copies but many are shared with mss *G* or *I* or *P*.

O MÜNCHEN, *Bayerische Staatsbibliothek,* Codex lat. 17145 (s. XII² from Shäftlarn, Bavaria) f. 41-56ᵛ *(desinit cap,* XLVIII 16 ... cancri). *Catalogus codicum latinorum Bibliothecae Regiae Monacensis,* II 3 (München 1878): no. 17145. Variant readings indicate that the scribe who made intelligent corrections was working from an exemplar like Fulda ms *C*.

P PADUA, *Biblioteca Antoniana* I.27 (s.IX/X, northern Italy) f. 1-44ᵛ. L. GUIDALDI, *II Codice 27 è del Medesimo Scrittoio e Secolo?, I Piu Antichi Codici della Biblioteca Antoniana di Padova (Codici del Sec. IX)* (Padua 1930), p. 21-28 with two facsimiles; B. PAGNIN, *Un presunto calendario bolognese nel codice Antoniano 27, Il Santo* IV 2 (Padua 1931) 316-322; IDEM, *La Provenienza del Codice Antoniano 27 e del Chronicon Regum Langobardorum in esse Contenuto, Miscellanea in Onore di Roberto Cessi* I (Roma 1959), p. 29-41. The scribes were working from an exemplar which gave annus praesens A.D. 820, A.D. 834, A.D. 840, or left blank spaces; other data accord with these several figures. Then most of the dates were corrected to A.D. 879. In addition there is an *argumentum* for A.D. 883, and a Chronicon with the same date. In general appearance nevertheless one would expect that the computus was transcribed early in the tenth century. Place of origin has been suggested as Bologna, Brescia, Leno, or Verona. Numerous glosses appear in the margins, many of which have been published by Patrick McGURK, *Catalogue of Astrological and Mythological Manuscripts of the Latin Middle Ages,* IV. Astrological Manuscripts in Italian Libraries other than Rome (London 1966), p. xv-xvii, 64-72. Variant readings of the text agree in some instances with mss *I* or *F* or *N*, but with none consistently.

R   FIRENZE, *Biblioteca Riccardiana* 885 (s. XIV[1], England), f. 312-
346. *Inventario e stima della Libraria Riccardi: Manoscritti e
Edizione del s.XV* (Firenze 1810): no. 885; L. THORNDIKE, *Notes
upon some medieval astronomical, astrological and mathematical
manuscripts at Florence, Milan, Bologna and Venice, Isis* L (1959)
38-40; S.H. THOMSON, *The Writings of Robert Grosseteste* (Cambridge 1940), p. 94, 96, 116. Thomson thought that the works of
Robert Grosseteste transcribed here should be dated to the second
half of the thirteenth century; Thorndike assumed the first half of
the fourteenth. The computus was transcribed from an exemplar
of A.D. 1199; many lines and sections were omitted; but insertions were made both from the AEV glosses and from portions of
Bedae *DTR* which Hraban had abbreviated or omitted. (On f. 312
the scribe began again to copy his exemplar of the computus but
stopped in the midst of the capitulatio; the repetition has not been
collated for this edition.)

V   LONDON, *British Museum,* Cotton Vitellius A.XII (s.XI ex.,
southern England) f. 10ᵛ-40ᵛ. T. SMITH, *Catalogus librorum manuscriptorum Bibliothecae Cottonianae* (Oxford 1696), p. 82-83; J.
PLANTA, *Catalogue of the manuscripts in the Cottonian Library
deposited in the British Museum* (London 1831), p. 379-380; René
DEROLEZ, *Runica Manuscripta: The English Tradition* (Brugge
1954), p. 222-227, 229-237 et passim; Neil R. KER, *Salisbury
Cathedral Manuscripts and Patrick Young's Catalogue, Wiltshire
Archaeological and Natural History Magazine* LIII (Devizes 1949)
153-156. This scribe used the exemplar of ms E but also copied a
wrapper upon which had been transcribed the *Sententiae* by Abbo
of Fleury, some of which had been composed while he was at
Ramsey Abbey during A.D. 986-988. In this copy of the computus
the name of the author has been erased and that of Gildas
substituted; likewise the name of the recipient has become
Rabanus. It shares the same glosses as AE and some of those in ms
R. Other contents appear to derive from the West Country region
of Glastonbury, Malmesbury, Hereford, and Gloucester, and it has
a calendar which has been attributed to Exeter s.XII. It was seen
at Salisbury between 1535 and 1543 before the 1621 acquisition by
Robert Cotton.

x   Editio princeps: Stephani BALUZII *Miscellaneorum Liber
Primus,* hoc est Collectio Veterum monumentorum quae hactenus
latuerant in variis codicibus ac bibliothecis (Parisiis. Excudebat
Franciscus Muguet Typographus. MDCLXXVIII.), p. 1-92. Etienne Baluze (1630-1718) probably used ms *F* for his edition of
Hrabani *Liber de computo* in a collection in seven volumes appearing from 1678 to 1715 and whose title became *Miscellanea Sacra;*
the entire series was reprinted by G.D. Mansi (Lucca 1764) with
this computus moved to vol. II. Baluze's edition of the computus

supplemented the Colvenerius edition of Hraban's other works (Köln 1626) and was reprinted in the *Patrologiae Cursus Completus, Series Latina Prior* (Paris 1863), CVII, coll. 669-728. Baluze was librarian for Jean Baptiste Colbert from 1667 to 1700 and stated that he used a Colbert manuscript for this edition. His catalogue of the books and manuscripts in that library was drawn up about A.D. 1690 in ms Paris BN n.a.fr. 5692 : cf. no. 6388 on folio 474 ; they became part of the Bibliothèque royale in A.D. 1732, for which a *Catalogus codicum bibliothecae regiae* was published in A.D. 1744. The manuscript in question was designated Colbert 240, Regius 3730a, and subsequently lat. 4860 (= ms F). Many variants in almost every chapter of the computus are shared exclusively with the Baluze edition ; however there are also some readings which could not have derived from ms F — not only the educated guesses of copyists and editors but also examples like the confusion of insular s and r at XXI 11 which could not have ms F as a source. One may speculate that Baluze introduced some readings from a second exemplar no longer available.

Fortunately the single printed version of Hrabani *Liber de computo* which has been available to scholars since 1678 was based upon a relatively reliable ninth century manuscript. Its main fault was the Easter tables in chapter LXXXIII which were not set out for the same nineteen-year cycle as that issued by Hraban ([85]) and which further were corrected at a few points to a user's needs, thus introducing confusion which cannot be attributed to the author.

Fragments or excerpts of Hraban's Computus have thus far been identified in twenty-one further manuscripts which will be described elsewhere. It was named in book lists from medieval libraries not only at Fulda where there were probably five copies remaining in the seventeenth century but also at Sankt Gallen, Murbach, Weissenburg in Alsace, Rolduc near Aachen, St Bertin, St Amand, Exeter Cathedral, Glastonbury, ([86]) Salisbury Cathedral, Erfurt, Michelsburg near Bamberg, and Monte Cassino. From manuscript evidence it is also known to have been in the libraries of or to have been used by scholars from Mont-Saint-Michel, Einsiedeln, Reichenau and two other centres in the

---

(85) F. Rühl, *Chronologie des Mittelalters und der Neuzeit* (Berlin 1897), p.143, and M. Rissel, *Rezeption antiker und patristischer Wissenschaft bei Hrabanus Maurus* (Frankfurt am Main 1976), p. 66-67, speculated that Hraban had modified the nineteen-year cycle. But this notion is derived from the printed edition rather than from the manuscripts which establish that he used the cycle of Dionysius and Beda.

(86) I am grateful to Professor Raymund Kottje of Augsburg for the first eight references, derived from his survey on the works of Hraban mentioned in medieval library catalogues printed prior to 1965. See now his *Hrabanus Maurus — Praeceptor Germaniae?*, *Deutsches Archiv* XXXI (1975), p. 534-545.

Diocese of Konstanz, Speyer, Mainz, Shäftlarn, St Emmeram, Aachen, and Fleury. This list of specifically identifiable places in which Hraban's Computus was known and used will never grow large enough to be fairly representative because school texts like service books were often used and left in the place of use, at altars or in schoolrooms, and they were just so often absent from libraries whose books would be recorded. ([87])

Collations of all variant readings have not yet resulted in sufficient evidence for the construction of a manuscript stemma, ([88]) partly because few of the copies have survived. One important group is *BCMO*, each of which must have been transcribed from an early Fulda exemplar, though not the same one. Four other ninth century manuscripts share many variants with each other but also through different exemplars copied and recopied: *G-I-P-N*. The eleventh-twelfth century pair *HL* must have used a common exemplar which seems to have been derived from tenth century ms *D*; likewise *AEV* share an exemplar which may have been related to ms *D*; and ms *R* appears to derive from the *AEV* group and its glosses. It is extremely fortunate that both the personal copy of Walahfrid Strabo, written mostly in his own hand ca. A.D. 825 at Reichenau (ms *G*), and an early transcript in the author's scriptorium from before A.D. 829 (ms *C*) have survived and display evidence that they were collated at Fulda during Walahfrid's study there, A.D. 827-829.

COMPOTISTICA ET ASTRONOMICA IN THE FULDA SCHOOL

     Arithmetic, Geometry, Astronomy, and Harmonics are the four fields of study which Boethius called <u>quadruvium</u> in the sixth century and which his cousin Cassiodorus named <u>mathematica</u>—thereby distinguishing them not only from three essential literary subjects (later called <u>trivium</u>)[1] but also and especially from the Hellenistic <u>mathematicus</u> and the <u>mathematica</u> <u>nota</u> which might appear to give authenticity to magic and the obfuscations of astrology and prognostications.[2] In the ninth century at Fulda the master Hrában called the four subjects <u>mathematicae</u> <u>disciplinae</u> and understood that mathematics is the language of the sciences —a language created by God but discovered by men.[3]

     It is well known that basic arithmetic in Greco-Roman schools was interlocked with Pythagorean number mysticism in the works for example of Posidonios or Nicomachos of Gerasa, and that some of the most brilliant discoveries in Alexandrian astronomy were made by scholars such as Hipparchos or Ptolemaios whose active pursuits included religious astrology.[4] When their wealthy aristocratic and royal patrons ceased to support schools, libraries, and such advanced studies in astronomy, the context and goals of astrology nevertheless persisted in both Roman and Persian empires. Concern to salvage and carry forward the mathematical disciplines free of such religious encrustations came from within a quite different cultural context, that of the families and schools of a new international society: the Christians. It was a minority culture, effectively described in 1962 by Pierre Riché, <u>Education</u> <u>et</u> <u>culture</u> <u>dans</u> <u>l'Occident</u> <u>barbare</u>, <u>VI</u>e<u>-VIII</u>e <u>siècles</u>,[5] who found that the Cassiodorean tradition of <u>artes</u> <u>et</u> <u>scientiae</u> <u>vel</u> <u>disciplinae</u> was continued by episcopal churches for the training of priests and by monastic communities for the study of sacred scripture. There were clerics and monks who remained faithful to classical forms in their poems and letters and even saints' lives, in the pattern of Gregory and Isidore; but there were also rigorists who felt the need to create a rival Christian culture not based on the cult of the body in Roman baths and brothels and circuses, or on rhetoric for

attainment of high office by an elite, but on personal and
social discipline and on centres of sacred study. Within
this minority culture of a new piety there appears to have
been little of Horace's taste for nuanced Latinity nor an
interest in the "higher" learning offered by various
philosophies; rather, the early Christians seem to have
given their attention to more ordinary aspects of life
focused upon regular worship after the Jewish pattern. This
egalitarian spirit oriented to the cult of God implied an
order which was long in evolving but which was served by
schools in the homes of Christian aristocrats, in episcopal
cathedrals, and eventually in monastic communities.

        Such Christian schooling however did not serve the
common life of Christians by carrying forward the
Hellenistic disciplines in the form of seven liberal arts.
The assumption that artes liberales should have formed the
curriculum of any viable Christian school appears to have
the support of historical landmarks in writings by
Augustinus, Boethius, Cassiodorus, along with lists of
subjects for study in Irish, Anglo-Saxon, and finally
Carolingian manuscripts.[6] Yet the best historians of
ancient and medieval education (Roger, Marrou, Laistner,
Riché) have been consistently disillusioned about how well
this educational programme was carried forward in early
medieval schools, and they will continue to be disappointed;
for Christian schoolmasters who "plundered the Egyptians"
applied their specific gains to new forms and new goals.
Hellenistic schools of Alexandria and Lyons and Carthage did
not suffer damage nor were letters and law debased by the
presence of Origen, Irenaeus, and Augustine of course. But
while Stoic, Academic, neo-Platonic, and even the new
Manichaean teachers were failing to hold students and
patrons for schools in the Roman empire, Christians were
learning their letters and their numbers—carrying away such
valuables while abandoning what they considered to be
pseudo-religious balderdash and ethical pretense. Pierre
Riché has demonstrated the significance of this change by
accepting medieval saint's lives as a primary source for
schooling in grammar during the fifth through eighth and
into the ninth centuries of Visigothic, Italic, Merovingian,
and Carolingian times. Along with letters and biblical
commentaries, these hagiographical texts allowed him to
affirm more securely than ever before the originality of
the barbarian Celts and Anglo-Saxons in generating social
institutions and cultural forms for the new culture of
Christians: "It was in the desert that the West rethought

its culture."

Such history surely sets some teeth on edge. That a
grammaticus would stoop to write the life of a saint or a
biblical commentary in an easily identifiable pattern
immediately suggests to a modern classicist that he no
longer is worthy of professional respect. Such practice of
piety by early medieval schoolmasters and authors has
provided quite sufficient evidence to humanists since the
Enlightenment from Gibbon to the present for ridiculing
vulgar minds of a dark age. However the countervailing
pressure of Leopold von Ranke on the historical profession
to take things just as they are, to accept all sorts of men
and events as equidistant from God, has reversed that
Enlightenment prejudice little by little. Now the early
works De musica need no longer be looked upon as the
repetition of Pythagorean accoustical ratios, repeatedly
transcribed by scribes but of no use for music. Rather one
may recognize the evidence for cantica, the singing which
is known to have been done and to have been studied in
churches, cathedrals, and monasteries.[7] On the other hand
neither grammatica nor cantica require support of
arithmetic ratios that numerology once gave the
Pythagoreans. And indeed many historians have continued to
deny that mathematics had any place in these early Christian
schools. Riché for example is so certain of this that he
could affirm very low limits for the scientific knowledge of
Anglo-Saxons, including "a little arithmetic," by giving the
example of Aldhelm's admission that Difficillima rerum
argumenta et calculi supputatione, quas partes numeri
appellant, lectionis instantia repperi. If "he learned
fractions with a great deal of difficulty," as Riché
proposed, this would be well matched by the current capacity
of most readers of the present essay but would not provide
evidence for an upper level of mathematics in their schools.
In fact Riché does not seem to recognize that his own
evidence for "ecclesiastical arithmetic" or computus along
with astronomy and cosmography was not so limited as he felt
necessary to affirm. Aldhelm was probably referring not to
fractions but to those number groups which one has to keep
in mind when multiplying or dividing with Roman calculi in
one of the early abacus systems.[8]

If the surviving evidence reflects a bit more
arithmetic than expected, computus and cosmographia require
very extensive ranges of knowledge and complex
calculations. They were not subjects for instruction in

Hellenistic schools, though one or another such topic might have been pursued by rare individuals; they were not categories of artes liberales and are never listed as parts of such a curriculum. The manuscript evidence is abundant nevertheless to demonstrate that by the seventh and eighth centuries in Irish and Anglo-Saxon schools these new subjects were studied actively wherever Christians could gather the endowments and support the masters and libraries for schools, that is, in monasteries. By the eighth and ninth centuries within the expanding Carolingian kingdom, the mathematical field was compotistica which gathered to itself and demonstrated the necessity of reckoning with number, of describing all stellar circles and cycles on geometric figures with numerical sequences and formulae, and of accounting for all the phenomena of the KOCMOC (mundus, universe) in their regularities—either by numerical tabulation or by geometric design. In order to pursue the purpose of this reckoning (computus), the "Egyptians" really were plundered and their goods scattered to such an extent that antiquarian search for any one of the classical ars of the quadruvium will surely be frustrated. But compotistica embodies active scholarship of the time in which those "Egyptian" goods were recycled to new purpose. Attention to the new purpose and activities will reveal some of the scholarly interest which preoccupied Carolingian masters in and out of school. Incidentally it may also provide other historians with a viable route to those stolen and hidden goods as well as to a greater appreciation for new riches which could thus be earned by their application, not in the seven Boethian artes liberales or any Augustinian, Isidorean or Aldhelmian expansion of that ideal, but rather in the practice of essentials: grammatica, compotistica, cantica.

There is abundant evidence that the mathematical sciences were diligently pursued during the several centuries of the late Roman empire and the early middle ages. The context was computistical studies[9] in monastic schools, for which the Boethian concept of seven fields of study provided broad support in letters and encyclopedias but no impetus. Charles W. Jones has long emphasized the significance of this orientation to grammatica, compotistica, cantica which thus far has not been acknowledged by the enlightened authors of histories of education.[10] It can perhaps be made clearer in evidence of a master and two of his students at Fulda. Hraban, a master, then abbot of Fulda, and later bishop of Mainz;

Servatus Lupus who became abbot of Ferrières and bishop of
Bourges; and especially Walahfrid Strabo, itinerant tutor
and poet who became abbot of Reichenau.

Walahfrid has received the attention of some very fine
historians who have emphasized his poetry and relations with
Carolingian royalty. In one of the best accounts, Eleanor
Shipley Duckett gave a refreshing biographical reading of
poems and correspondence of Walahfrid. However Miss Duckett
and all others[11] have failed to recognize that he had given
attention to science as well as to humane letters. His
active intellect was more wideranging than later
medievalists supposed, for the evidence strongly indicates
that he concerned himself with arithmetic and astronomy as
well as with grammar and poetry.

Walahfrid's personal notebook is now manuscript no. 878
in the Stiftsbibliothek at Sankt Gallen. It contains none
of the poems which have attracted scholarly attention, nor
the biblical commentaries which brought him abuse.[12] Rather
there are tracts and excerpts from other authors which he
transcribed for study and later use, or because he simply
liked them: for example, a paragraph on how to grow
pears.[13] Two decades ago Bernhard Bischoff demonstrated
that this was Walahfrid's personal manuscript and that
perhaps half of its contents was transcribed by Walahfrid
himself during specific and datable periods of his life.[14]
We can now see his drills on Latin word forms and the
grammatical instructions which one of his teachers provided
from Priscian and Donatus. Metrics and liturgical remarks
indicate concerns which flourished for him. However the
earliest surviving entry is a computus which he learned
about A.D. 825. This was Liber de computo composed by
Hraban in 819 and 820 for students at Fulda.[15] It may have
been brought to Reichenau in 824 by Gottschalk.
Grammatica, compotistica, and cantica are found in
Walahfrid's vademecum from the beginning therefore, and he
added to all three sorts of material in subsequent periods
of his life until he drowned in the Loire river in August
A.D. 849.

It is reasonable to assume that Walahfrid went to
Fulda in A.D. 827 in order to study the sacred scriptures.
One of the tasks he carried out in his two years or less
with Hraban however was a complete re-reading and
correction of his copy of the Liber de computo. The
manuscript which he used for corrections to his own is now

Oxford, BL Canon. Misc. 353 whose primary writing is mature
Fulda script while the remainder was done by scribes who
were being trained under a Fulda master. The surprising
correlation of Walahfrid's manuscript of Hraban's computus
and this early Fulda book allows us to see how the
computus was taught in a monastic school.[16]

If Alcuin and Charlemagne added to the school
curriculum a subject which most of us find strange, it was
nevertheless important for the sciences during many
centuries: students must learn the computus, that is, how
to make and use a calendar. Fortunately the mysteries of
the European calendar do not bother us every day, as they
did thoughtful men in the Middle Ages. Others have made
all the decisions for us about days and weeks, months and
years, so that our calendar appears to proceed like
clockwork, and our problem is to keep up with it! But that
is a strictly modern experience for a very few cultural
groups on the face of this globe, even today. Europe and
its calendar were not always there, and in the eighth and
ninth centuries the variety of attempts to organize
temporal phenomena into calendar sequences was confusing.
There were the sun and its seasons, the moon with phases
and tides, the stars and their formations—some motions of
these were observed to be dependable but others were
obviously so erratic that neither Ptolemy nor Copernicus
could provide dependable accounts of them. There were
regnal years and imperial indictions; there were 8, 11, 16,
18, 19, 76, 84, 95, 100, and 112 year cycles; there were
religious festivals centering on the solstices, equinoxes,
eclipses and comets, or focusing on historical events like
Moses freeing a people, Romulus founding a city, or the
death of a saviour, Jesus Christ. Our calendar is partly
based upon the expectation that in the midst of confusion
the Christian tradition could bring sensible order.

The great effort to do so was that of the English
monk, Bede, who in 703 and again in 725 showed how the 19
and 95 year sequences could be extended into a cycle of
532 years which would coordinate seven-day weeks, lunar
months, and solar years with stellar phenomena, and how to
fit the whole thing to the year of Christ's Incarnation,
so that the date of Easter could always be reckoned and
history could be known in a secure sequence of significant
events.[17]

Older literature on calendar usage before Bede

continued to be used of course; and other explanations were
offered, while Bede's system slowly spread. Letters of
Alcuin and Charlemagne discuss astronomical and calendar
problems in some detail. So much material was generated
about A.D. 809/10 concerning the determination of dates by
different systems that we might postulate a large assembly
of experts at Aachen, meeting under imperial aegis.[18] A
(Breton?) monk, Marchari, wrote to Fulda about such
questions, and the master Hraban replied in A.D. 820 with a
new computus which helped to bring some order into the
temporal confusion.

Hraban's Liber de computo was a handbook which put
Bede's teaching into a form more accessible to students.
He began with a rather poetic definition:

Tempus est mundi instabilis motus rerumque
labentium cursus. (VIIII 5-6)

Time is the motion of the restless world
and the passage of decaying things.

and with the query: whether anything in this world can be
understood without mathematical calculations. The student
is immediately introduced to the concept of number and how
to use various types of numerals; he must understand the
terminology for fourteen divisions of time and their
proportional relationships. By question and answer he
learns to observe the stars and their patterns, the movements of sun and moon and their imperfect correlations.
By the middle of the book he was deeply embroiled in
technical details of how to make and use a continuous
calendar and its rationale in Latin Christian history.

At the end of the computus in the Fulda manuscript and
also copied by Walahfrid[19] are answers to problems which a
master had posed for his students when they began their
study of the calendar itself. The first problem was this:
On which feria (weekday) did the Kalends (first day) of
each month occur? Bede had explained the system, and so
did Hraban who also applied it to the example of August
when he was writing. Taking the year 820 as his example,
a student has worked out in the Fulda manuscript the feria
for the Kalends of each month from March 820 to February
821; thus the problem was solved and a system learned which
could be applied to any year.[20]

The second problem was this: In the Roman system, one
must know how many Nones, Ides, and Kalends there are in
each month. Bede and Hraban had provided explanations of
how the Romans came to their system and what the terms
meant, but not how to use them. But in both Walahfrid's
manuscript and the Fulda manuscript we see that Fulda
students were drilled in the number of Nones and of Ides
and of Kalends, and in the application of the terms by
groups of months, according to a corresponding number of
Nones and Ides. The first Kalends of February for example
correspond with our first of February, but a Roman reckoned
backwards to the second (pridie Kal. Feb. = 31 Jan.), third
(III Kal. Feb. = 30 Jan.), and so on to the nineteenth
Kalendae (XIX Kal. Feb. = 14 Jan.). Following Kalends in
March, May, July, October were six Nones but four Nones in
other months; then came eight Ides in each month.
Alternate months of 30 and 31 days would have 17 and 19
Kalends respectively, with the exception of March whose 16
Kalends (which fall in the month of February) increase to
17 every fourth year when the sixth Kalends is repeated
(annus bissextilis).

Students were also taught to divide months into weeks.
How do you correlate 12 months and 52 weeks in the Roman
system? On which days of the month will a weekday recur?
Begin with the Kalends of January: whatever feria occurs
on the Kalends of January will recur also on the sixth
Ides and on the eighteenth Kalends, the eleventh Kalends,
and the fourth Kalends (of February). In order to continue
reckoning the sequence of weeks, the student must recall
what he had drilled on in the preceding problem: the
number of Nones, Ides, and Kalends for each month. Answers
to these problems would have been simple, almost
instinctive for a Roman. But Fulda students learned to
think in Roman terms only with effort,[21] and they needed
the aid of memory verses which are so abundant in medieval
manuscripts.[22]

If such drills are elementary, they are also necessary
in order to go further. Perhaps Walahfrid took the teaching
more seriously than most students, for he applied it in an
uncommon way. At Fulda he copied into his notebook a
complete calendar (ms p. 324-327), but in a form rarely
found. Rather than using one page for each month, he
grouped the months exactly as here described, according to
whether they had the same number of Nones, Ides, and
Kalends: January, August, and December together on one

page; March, May, July and October on a second page; April,
June, September, and November on a third; and a column for
February--a space of less than two folios. While calendars
of this type had been used in the Roman Empire, none is
known for five centuries before Walahfrid created his at
Fulda[23] on the basis of Hraban's teaching.

     Walahfrid also transcribed Bede's De natura rerum and
De temporibus. Continuing this work both at Fulda and at
Aachen, he collected various computistical argumenta or
reckonings of indictions, bissextus, saltus lunae, lunar
and solar epacts, and other technical information essential
for the computus. He grouped these together according to
whether they used the annus mundi (first year of creation)
as a starting point, or the annus Domini (year of Jesus'
birth). With his paragraphs ordered in this way it was
obvious that the annus mundi was different for the various
chroniclers and that the result would differ for those who
worked from Hebrew, Greek, or Latin Bibles. And as to the
annus Domini: when was it, after all? The earliest
Christian calendars seem to reckon from the Passion of
Christ, not from his birth; and the Gospels do not supply
data about either the Passion or the Birth which are
reconcilable with each other or with the few bits of
information about the contemporary Hebrew calendar which
can be verified. What happens then to Bede's sequence of
years? Is the Incarnation not datable? These problems
did not wait on the twelfth century to be debated.
Walahfrid raised them in the first half of the ninth
century, and so did others. They are problems demanding
arithmetic and astronomy, a good knowledge of history, and
a practical logic.

     Walahfrid's correspondence indicates that his first
introduction to arithmetic was under the instruction of
Tatto at Reichenau who used Boethius' Arithmetic. Boethius
had studied in Alexandria toward the end of the fifth
century and before A.D. 500 had published textbooks in
Arithmetic, Geometry, Astronomy and Harmonics. His
Arithmetica and Musica were recorded in the libraries of
Reichenau, Fulda, Aachen, Tours, and numerous other centers
of study. If we have seen Walahfrid at his reckoning, we
also know of the study of arithmetic from the correspondence of another great literary figure, Servatus Lupus, who
was a scholar and teacher at Fulda for at least seven years.
At Fulda Lupus compared Ciceronian texts with those he
could borrow from other libraries and continued such

textual studies throughout his life, thus earning the
approval of modern scholars as a philologist.[24] But he
likewise had an interest in mathematical subjects and wished
to supplement Boethian arithmetic with the one prepared in
the fifth century under the name Calculus by Victorius of
Aquitaine. While at Fulda Lupus wrote to Einhard--an older
layman from Fulda who had served Charlemagne and Louis the
Pious as an engineer-architect, retiring after his wife's
death to Seligenstadt as lay abbot. In this letter[25] Lupus
requested a transcription of the muster of capital letters
prepared by the royal scribe Bercaudus (litterae unciales
some call them, noted Lupus, but now capitalis quadrata[26]),
as well as Einhard's assistance with the Calculus Victorii
Aquitani. Two codices survive which contain these precise
contents, along with four lesser items[27]: ms Bern Burgerbibliothek 250, f. 1v-11v; and ms Basel Stadt- und
Universitätsbibliothek O.II.3. They were both written ca.
A.D. 836 in the same scriptorium, possibly Seligenstadt but
probably Fulda. On the basis of the system of marginal
corrections it appears that Servatus Lupus must have taken
ms Bern 250 as his personal copy; it is not surprising
therefore that it later formed a part of the Fleury library
where it could have received the addition of an arithmetic
table of Roman numerals entered right to left and headed
Gerbertus Ratio Numerorum Abacique Figuras (f. 1 recto),
along with a sequence of symbols for weights. (It is
interesting to note that the numeral nine is represented
here as IX rather than VIIII.) Folio 12 was originally a
blank cover, but it received the incipit and first table of
Abbo's Ephemerida or Compotus vulgaris and a further large
gathering was added to receive the remainder of one of the
earliest copies of that computus at the end of the tenth
century.[28]

     It is curious that Victorius of Aquitaine has been
passed over in virtual silence by cultural historians of
this century, despite the numerous manuscripts of his
Calculus from the eighth century onwards.[29] But Lupus' turn
from Boethian arithmetic to that of Victorius is worth
attention especially because it illustrates the Christian
movement away from mystical speculations with number towards
numerical description of observed phenomena. Pythagorean
number mysticism and studies of cosmic harmonies provided
the concepts for many Hellenistic manuals of arithmetic.
One by Nicomachus of Gerasa (fl. A.D. 100) was translated
from Greek into Latin within a century by Apuleius of
Madaura (s. II) as Introductio arithmeticae; though neither

the Greek nor the Latin version has survived, it was known
to Martianus Capella, Cassiodorus, and Isidore.[30] During or
immediately after his studies in Alexandria,[31] Boethius
published his __Arithmetica__ (A.D. 500) which drew heavily upon
the Pythagorean Nicomachus, and this was the version
available to Walahfrid at Reichenau and Lupus at Fulda as
well as in many other ninth century schools. In general it
must have been fitting for Christian students who were
interested in theology and the study of sacred scriptures to
read in any version:

> All that has by nature with systematic method been
> arranged in the universe seems both in part and as
> a whole to have been determined and ordered in
> accordance with number, by the forethought and the
> mind of him that created all things; for the
> pattern was fixed like a preliminary sketch, by
> the domination of number preexistent in the mind
> of the world-creating God, number conceptual only
> and immaterial in every way, but at the same time
> the true and eternal essence, so that with
> reference to it, as to an artistic plan, should be
> created all these things, time, motion, the
> heavens, the stars, all sorts of revolutions. It
> must needs be, then, that scientific number, being
> set over such things as these, should be
> harmoniously constituted in accordance with
> itself, not by any other but by itself.

Introductio arithmeticae I vi 1-2

Augustine, Boethius, Cassiodorus, Isidore, Aldhelm, Bede,
Alcuin, Hraban, and many others expressed the same sentiment
and supported it from Genesis I 14, Sapientia II 21, and
other quotations from Biblical writers. But Pythagorean
cosmological notions could also follow from such a
beginning and indeed were conveyed somewhat by Nicomachus.
Religio-philosophical principles deriving from the monad
(sameness) and dyad (otherness), together constituting the
opposing elements of the cosmos, could not be avoided. The
numerical system was derived step by step from these
elements. Rational integers are classified as perfect,
abundant, and defective or diminutive numbers which are also
ordered by four distinctions of evenness and oddness, by the
prime and relatively prime, and by figuration. The
proportions and ratios to be studied were presented as
displaying the harmony of the universe.[32] However Boethius
pared away much of the number symbolism that supported such

principles and retained arithmetic operations. These assumed
the form of positions of pebbles in sand (a kind of abacus)
and number relations were explained pictorially in terms of
adding or subtracting units or according to positions. Thus
plane and solid numbers required geometric analogies. It is
seldom noticed that Euclid's arithmetic books VII, VIII, and
IX of the Elements shared the same Pythagorean sources and
concepts but developed them into geometrical proofs rather
than analogies.[33] It should be remarked also that Martianus
included numerous passages from Euclid along with selections
of Apuleius' translation of Nicomachus. This mixture of
Nicomachan and Euclidean uses of Pythagorean concepts in the
presentation of "Arithmetic" by Martianus Capella however
was not available at Fulda and probably not at Reichenau
until at least the second half of the ninth century. Lupus'
wish to study another book of arithmetic with Einhard
indicated a need for unadorned number and perhaps more
practical calculation techniques. That at least is what he
found in the Calculus Victorii Aquitani: addition,
subtraction, multiplication, division, and fractions for
twenty pages, to which others added a further table of
numerals and a computus with tables of observed stellar
motions and systems of recurring dates.

Euclidis Elementa VI-XI were known at Fulda, Reichenau,
and several other Carolingian schools through selection and
epitome in Boethius De arithmetica. In addition the Fulda
library catalogues listed Boethius De geometria, as also one
of the Reichenau book lists of A.D. 836-842 had recorded
libri Boetii geometricorum, along with works of grammar,
computus, geography, metrics.[34] If we attempt to determine
how much of Euclid was available at Fulda during the time of
Hraban, Walahfrid, and Lupus, we are faced with several
historiographical problems: first, the inability of scholars
to find Boethius' geometria during the past century or more;
second, strong doubts that it contained a significant part
of Euclid in usable form even if known; thirdly, the presence
of non-Euclidean agrimensores texts which put algebraic,
geometric, and trigonometric concepts to work for business,
construction, and surveying purposes and thus are cited as
evidence for the decline of geometry; fourth, the presence of
Ars geometriae et arithmetricae which is assumed to derive
from the agrimensores tradition rather than the Euclidean and
thus to be further evidence of decline. It is unfortunate
that a firm consensus has been reached among historians of
medieval science[35] which has discouraged attention to the
texts and to the contents in which their manuscripts were

produced.

In Hellenistic scholarship geometry was not simply
Euclidean but had a dual nature. The axiomatic approach was
systematized in the Elements of Euclid, in which earlier
postulates and propositions were collected without the more
cumbersome mechanics of proof and arithmetic theories
developed by earlier mathematicians. Alternately,
professional theory and practice drew from many of the same
works, some dating from Babylonian times, and are exemplified
by Heron of Alexandria. Neither master's work was completely
intact in Greek original or Latin translation for the use of
Carolingian schoolmasters. The more proximate tradition
which did survive in part was that of Roman handbooks on
applied mathematics which carry forward algebraic and
arithmetic practices and which, as Otto Neugebauer has
pointed out, represent a substantial increase rather than
decline in study of the physical sciences.[36] Many such
handbooks and tracts which mixed geometria and gromatica
were known in the Fulda school during the ninth century
through ms Vat. Pal. lat. 1564; perhaps transcribed in the
scriptorium of Louis the Pious (A.D. 814-840), this large
collection of agrimensores is well illustrated with working
diagrams and was used in Fulda by mid ninth century.[37]

It was fortunate for engineers like Einhard, as well as
surveyors who were assigned the duty of delineating
boundaries between the kingdom of Lothaire, Charles the
Bald, and Louis the German,[38] that instruction from such a
Corpus agrimensorum Romanorum could be found in the
Carolingian schools. As A. P. Juschkewitsch and A. C.
Crombie have expressed it: the needs of building, trade,
transport, commerce, financial institutions of noble and
ecclesiastical as well as multiple royal households,
instruments of credit, legal acts and records and so forth
provided impetus for reflection concerning ratios and
proportions, equations of first and second degree, similar
configurations in discrete circumstances including the
ancient theorem now bearing the name of Pythagoras,
calculation of surface areas and of volumes, and so on.[39]

Carolingian scholars responded to these needs not only
by searching out Roman tracts but also by preparing several
works on geometry itself during the ninth century. The
manuscripts contain works, separately or in combination,
with the following incipits:

Geometria est disciplina magnitudinis immobilis et
  formarum ... (TKr 584)

  ms München CLM 560 (Reichenau s.IX[1]) f. 122-149
    Neapoli BN V.A.13 (France s.IX med) f. 9-14?
    Paris BN lat. 13030 (Corbie, 850-875/80)
    f. 59^V-83
    Bamberg Staatsbibl. H.J.IV.22 (No. France,
    s.IX/X) f. 1-16

Tu qui vis perfectus [profectus] esse geometricus ...
  (TKr 1591)
  ms Bamberg H.J.IV.22 (s.IX/X) f. 6^V: this
    incipit indicates a discussion which often
    forms a part of the above work but is also
    found separately in tenth century codices
    and later.

Geometriae disciplina primum ab Egiptus reperta
  dicitur ... (TKr 585)
  ms Neapoli BN V.A.13 (ca. A.D. 850-875/80)
    f. 15-15^V
    London BL Harley 3017 (Fleury s.IX[2]) f. 178^V-
    180
    Bern Burgerbibl. Bongar. 87 (Luxeuil s.IX[2])
    f. 8^V-18

Geometrice artis peritiam qui ad integrum ... (TKr 584,
  656)
  ms München CLM 13084 (Freising, A.D. 854-875)
    f. 48-69
        CLM 13047 (s.IX-X) f. 48^V-69^V

These sometimes are entitled simply Geometria, or
Arithmetica, or Boetii de geometria and their contents
vary.[40]

    One modern scholar may dismiss or ignore these books,
assuming that the contents lack Euclidean precision,[41] while
another may propose that a significant attempt was being
made to teach theoretical geometry from such parts of Euclid
as could be extracted from the agrimensores.[42] They are well
worth greater attention and closer analysis than they have
received. However there was also a new sort of textbook
available during the ninth century, Ars geometriae et
arithmeticae, with incipits

Quia vero, mi Patrici geometrarum exercitatissime,
    Euclidis de artis geometricae figuris obscure
    prolata,... (TKr 1234: Boethius, Geometry,
                      simplification of Euclid; ms of s.XIV)

Principium mensure punctum vocatur cum medium tenet
    figurae. (TKr 1126: Euclid, Elementa in 3 books;
        ms of s.XII.)

Punctum est, cuius pars nulla est, linea vero praeter
    latitudinem longitudo ... (cf TKr 1152: Euclid,
                              Elements, tr Adelhard
                              of Bath, versio II:
                              ...cui pars non est...)

While the best modern citations lead one to twelfth and
thirteenth centuries and especially to Adalhard of Bath, the
twenty-six extant manuscripts of the complete text show that
it was widely available from s.IX to XI, including probably
Reichenau, Murbach, and Aachen.[43] It is found in

        ms München CLM 560 (Reichenau s.IX[1] or IX med)
            f. 134^V-145^V
        Neapoli BN V.A.13 (No France s.IX med) f. 8-13

But how much of Euclid is actually to be found in the Ars
geometriae et arithmeticae? In order to locate sources for
an eleventh century work designated "Geometrie II," Menso
Folkerts compared four groups of manuscripts containing
excerpts from the Elements in Latin; in the attempt to
establish a source M = Euclidis elementorum a Boethio, ut
videtur, translatorum textus genuinus, he published an
edition of a text Mc which is our Ars geometriae et
arithmeticae (or Geometria I), with equivalent Euclid on
facing pages.[44] He has thereby identified the presence of
a quite unexpected amount of the Latin Euclid in this work,
as may be summarized here:

Euclid Book I.   Definitions 1 to 23 complete.

                  Postulates 1 to 5 complete.

                  Axioms 1 to 4 complete.

                  Propositions 1 to 4, 6 to 18, 21, 23,
                  26 to 28, 31 to 37, 39 to 48.

42

Euclid Book II.   Definitions 1, 2 complete.

                        Propositions 1, 3 to 6, 9 to 12, 14.

Euclid Book III.   Definitions 1 to 6, 8 to 11.

                        Propositions 3, 7, 9, 10, 12 to 14, 16, 18, 19, 22, 24, 27, 30 to 33.

Euclid Book IV.   Definitions 1, 2 complete.

                        Propositions 1 to 4, 6, 8, 12, 11.

Euclid Book V.   Definitions 1 to 18 complete.

The large portions of book I which are present suggest that either the entire book or at least a fulsome epitome had been available as a source for this text of Ars geometriae et arithmeticae. Complete texts for Definitions, Postulates, Axioms, and Propositions 1-3 with diagrams indicate that a student could be taught the method for verification of theorems. Much less material was incorporated from books II-V, and its overall value and coherence remains to be assessed. However it may be proposed initially that there is an abundance of material from which a student could gain an understanding of Euclidean geometry.

It should be emphasized furthermore that these contents of Ars geometriae et arithmeticae are not oriented towards gromatical purposes and that most of it cannot be found in the Corpus agrimensorum Romanorum.[45] One is compelled to assume that there was available some form of Euclid which was probably condensed but was not mixed. Rather than project purely anonymous and totally lost source, it would seem reasonable to take at face value the manuscript rubrics which name Euclid and Boethius and to assume that in the ninth century at least Boethius De geometria had not disappeared.

The ancient assertions that Boethius had translated Euclid into Latin[46] had been confirmed furthermore by the discovery of Latin fragments of Euclid's books XI-XIII from the late fifth or early sixth century in ms Verons Biblioteca

capitole XL (38). Three bifolia survive from having been
reused at Luxeuil s.VIII in for copying the Moralia of
Gregory and then brought to Verona in the ninth century;
worse damage came in the nineteenth century from attempts to
decipher the palimpsest by use of a destructive reagent.[47]
Survival of the Latin unmixed Euclid into the ninth century
has also been confirmed by discovery of portions of books I
and II in München Universitätsbibl. Fol. 757 (s.IX in) f. 1
and 2. These fragments[48] do not overlap with each other
and can be expected to stimulate further discoveries.
Geometria was a subject of greater interest in Carolingian
schools than has hitherto been acknowledged, and like their
Hellenistic forebears ninth century masters pursued its dual
nature in theory and practice.

     While he was studying at Fulda during A.D. 827–829
Walahfrid also added to his notebook Bede's De natura rerum
(ms St Gallen Stiftsbibl. 878, p. 242–262).[49] Both Bede and
Hraban explained the nature of the cosmos in terms of a
spherical shape for the heavens and a spherical shape for
the earth. They conveyed little of the antique concept of
an orbis quadratus in which the lands around mare nostrum
and all of Europe and Asia were assumed to make up one
quarter of the globe, while another quarter lay across the
Atlantic to the west, a third beyond the known parts of
Africa to the south, and a fourth diagonally southwest. It
was often proposed that each zone was inhabited (e.g. by
Krates of Mallos), and there were supporting accounts of
Ethiopians to the south. But there were no equivalent
reports from the western quarters and certainly none from
the one in which Antipodae were supposed to live. Whether
there were in fact people (antipodae) living opposite them
on the globe therefore was often denied (e.g. Plutarch,
Lactantius) for lack of evidence,[50] as had been affirmed by
Augustine, Isidore, and Bede--all of whom also taught that
the earth is a sphere.[51] Hraban did not enter into that
antique issue, apparently assuming the matter settled long
since.

     Another description of the globe had been more common
which projected the three known continents, Asia-Libya-
Europa, over the whole sphere. This was assumed by Sallust
and Lucan whose works survive only in Carolingian
manuscripts which are illustrated with drawings of the
orbis terrae. It was also a model for periplus literature
in Dionysios Periegetes, Orosius, Priscian, and Isidore.
This globe was usually depicted in the form of a tripartite
rota terrarum whose earliest representations survive as a

symbol of imperial authority on coins dated to the reign of
Nero and as a teaching device in the earliest extant
manuscript of Isidori De natura rerum: ms El Escorial, Real
Monasterio R.II.18, f. 24^v which dates from the latter part
of the seventh century. Isidore's illustrated schoolbook
was presented to his former student Sisebuto, by then king
of the Visigoths in central Spain, in A.D. 612; but within a
year he had added a chapter De partibus terrae illustrated
with this tripartite rota terrarum. When the work was used
in West Saxon, South Umbrian, and perhaps Irish schools it
acquired further additions which resulted in a "long
version." When Hraban was a student ca. A.D. 787 to 796/801
there were in Fulda at least one and perhaps two copies of
the long version De natura rerum available which had been
brought from England, and two more were transcribed during
this time in the Anglo-Saxon script of Fulda.[52] One of each
is complete enough to display the rota terrarum on its last
leaf depicting the globe: ms Basel F.III.15f (Anglo-Saxon
s.VIII^2) f. 13; and Basel F.II.15a (Fulda s.VIII ex) f. 16^v.

Isidore's De natura rerum outlined the knowledge
necessary for a man or woman to function intelligently in
the world, and it was widely used by schoolmasters. But it
was only a beginning for anyone taking up a specific topic
of cosmology. When Walahfrid transcribed Bedae De natura
rerum at Fulda during 827-829 however he would find the
specific term globus terrarum and later he would record
Bede's full scale argument for the earth's sphericity in
De temporibus ratione XXXII-XXXIV: 1) the seasons are the
same for those living on the same latitude; 2) the hours of
the day are simultaneous for inhabitants along a meridian
of longitude; and 3) one sees stars disappear or reappear on
the horizon when one travels North to South, or vice versa;
therefore

> Causa autem inaequalitatis eorundem dierum terrae
> rotunditas est; neque enim frustra et in scripturae
> divinae et in communium litterarum paginis orbis
> terrae vocatur. Est enim re vera orbis idem in
> medio totius mundi positus, non in latitudinis
> solum giro quasi instar scuti rotundus sed instar
> potius pilae undique versum aequali rotunditate
> persimilis; ...
>
> Bedae DTR XXXII 2-6

The cause of the unequal length of the days is the
roundness of the earth; for it is not in vain that

both in Sacred Scriptures and in ordinary literature
it is called the orb of the earth. It is certainly
a fact that the orb is placed in the centre of the
whole universe, not only in latitude as if round
like a shield, but rather with the rotundity of a
ball no matter which way it is turned; ...

Implicit in these statements was a third Hellenistic
way of thinking about the earth and its parts which could be
learned from Isidori DNR X. De quinque circulis mundi and
from Bedae DNR IX. De circulis mundi, each of whom described
a model of the earth divided into five zones of latitude
parallel to the equator. Thus were conceived two temperate
zones north and south of the centre torrid zone, with arctic
and antarctic at the extremes. Euro-Asia and the northern
coast of Africa were in the northern temperate zone, and it
was normally assumed that the corresponding climate of the
southern temperate zone made it habitable, whether or not
anyone actually lived there. This concept of the earth was
essentially astronomical, for the imagined lines of latitude
were projected from those of the celestial globe in the
Eudoxan tradition of tropics (of Cancer and of Capricorn).[53]

These several Hellenistic concepts of the earth, its
shape, and its parts were taken up by Hraban who repeatedly
referred to the orbis terrarum and its rotunditas
alternately as globus terrae, e.g. in ch. XLIX 3-14 of his
computus. He did not quote the succinct arguments of Bedae
DTR XXXII-XXXIV on these matters, but he conveyed the same
ideas and data in his chs. XXXVI-LI and especially ch. LII
61-71 concerning the variation of daylight hours as one
moves north, which was also a reliable measure of latitude
for the globe (Bedae De temporibus VII 4-11).[54]

Added to both Fulda copies of Isidori De natura rerum
are notes on the position of signs of the zodiac and a table
by which they could be coordinated with months of the solar
year. Hraban had given little space in his computus to the
Zodiac--just enough to teach this same information (chs.
XXXIX, XLIX, LI). But the signs of the Zodiac served a
double purpose, both of which were astronomical: one was to
provide a system of orientation to an otherwise chaotic
array of stars. In order to see anything in particular in
the night sky one must always delineate a section, pick out
the brighter stars therein, and then impose upon them
connecting lines or figures of some sort. For this purpose
the Hellenistic texts of Aratus or Hyginus served very well

with their popular descriptions of star-groups within the
circular band of the sky through which passed the zig-zag of
the moon as well as the more regular path of the sun and the
very erratic movements of the planets. It is within that
band that the twelve signs of the zodiac lie, and the writers
also described other star-groups outside the zodiac. Of
course they wax eloquent about such phenomena and relate the
sweetest or the goriest stories about the imaginary beasts,
men, and women in the celestial sphere, often with
illustration.

Following Hraban's _computus_ in the Fulda manuscript
Oxford BL Canon. Misc. 353 on f. 53^v-55 is yet another
elaborate description of the heavens according to groups of
stars which then could have been visualized as storybook
beasts and heros and thus be easily remembered. It begins
Duo _sunt_ _extremi_ _vertices_ _mundi_, _quos_ _appellant_ _polos_...[55]
One could gain a basic celestial orientation from it but no
astrological prognostications whatsoever. One must plunder
the Chaldaeans as well as the Egyptians (Hrabani _De_ _computo_
XXVIII-XXXII), but reject their superstition. Of course
there was still worship of Woden and Thor in the region of
Fulda, despite Boniface's success in establishing Christian
missions and monasteries and dioceses. Hraban related his
astonishment at the ignorance and superstition expressed by
many on the occasion of a lunar eclipse. He was preparing
a sermon when he suddenly heard people let out a great roar:
some said they saw horns growing out of a beast or heard the
grunts of heavenly pigs, while others threw spears to drive
away the portents.[56] By contrast with such pagan frenzy,
the Christian monks and laymen of the Fulda monastic school
were offered rational explanations for such phenomena and
had better opportunities to grow in both faith and reason
under Hraban's tutelage.

If Walahfrid and Lupus followed Hraban's teaching very
closely they would have learned that the star-groups are of
such distance from the earth that observation of their
motion across the heavens required about two hours (_De_
computo XXXIX 28-30). Pointing out the necessity of direct
observation and of instruction by someone expert in these
things (ch. XL), Hraban then provided Bede's method for
determining the positions of moon and sun in the Zodiac on
any day (ch. XLI, XLII), and noted that the extra ten-and-
a-half hours in any month are to be neglected in this
calculation but accounted for in reckoning phenomena for the
twelve month period. To Bede's description of

characteristics by which planets may be recognized, Hraban
added a few lines on the positions of the seven planets at
the time of writing:

> Modo autem, id est anno dominicae incarnationis
> DCCCXX mense iulio nona die mensis, est sol in
> XXIII parte cancri, luna in nona parte tauri,
> stella saturni in signo arietis, iovis in librae,
> martis in piscium, veneris quoque stella et
> mercurii, quia iuxta solem in luce diurna modo
> sunt, non apparet in quo signo morentur.

                          De computo XLVIII 14-19

Right now, that is, 9 July A.D. 820, the sun is   *
in the 23rd part of cancer, the moon in the 9th
part of taurus, saturn in the sign of aries,
jupiter in libra, mars in pisces, but it is not
clear in which sign are venus and mercury because
they are near the sun during daylight.

On several successive occasions during the summer of A.D.
820 Hraban provided similar explanations for his students
in terms of the year, month, weekday, or number of the moon
at the time of writing. In chapter XLVIIII he told how, if
you know only the year from the Incarnation, you can observe
the heavens for the position of the sun in the zodiac, refer
to a 19 year calendar cycle, and then determine the day of
the month and day of the week upon which the reckoning is
made.

It would appear from these accounts that two kinds of
instruments are assumed to be at hand though not mentioned.
One is a simple sighting tube (tubu) on a scale, and the
other is a celestial star-chart or globe also with a
zodiacal circle which is graded 30 parts for each sign.
With the former it might be possible to go further than
Hraban did with reference to Venus and Mercury and to plot
their courses about the sun in regular paths, rather than
in erratic paths about the earth. In fact this was done in
a diagram in ms Karlsruhe Landesbibliothek Reichenau 167,
p. 35 (35ᵛ); it is a hand with Irish symptoms in a
Reichenau codex dated A.D. 836-848, almost co-terminus with
the abbacy of Walahfrid.[57]

An instrument something like that assumed for
determining the part I-XXX of a zodiacal sign in which sun,
moon, or another planet was located at a time of

observation was certainly available at Salzburg and
Regensburg in two manuscripts, Wien NB 387 (A.D. 808-816)
f. 165^V and its duplicate in München CLM 210. They
are clearly replicas of an instrument then extant by which
one may follow the 24 hour cycles of the sun and through
twelve signs of the zodiac along with multiple phases of the
moon. Apparently one revolves the separate plates on which
the series of cyclical data are depicted (held together by a
central "horse," typical of astrolabes), sets a pin
vertically in one of the inner circle of 365 small holes,
and observes the phenomena in order to determine the data,
which was lacking. A good illustration of the replica may
be seen in Donald Bullough, The Age of Charlemagne (London:
Elek, 1965) 189, but there has been no scholarly analysis of
the illustrations that they deserve, so far as I know.

*   A planisphere was also drawn into those manuscripts,
and a reproduction of it from CLM 210, f. 113^V, may also be
found in Bullough 129 (plate 50). It depicts the figures
which one learns from the literature, both astronomical and
popular, along the zodiacal band but also a range of other
figures for orientation to different parts of the night sky.
These star-groups are numbered according to the Ptolemaic
star-chart, according to Marcel Destombes (1975), and the
celestial map may be held up above one's head in order to
relate to the stars roughly according to the Ptolemaic
projection from the south celestial pole. It was not this
projection however which was known in Fulda. Rather the
scholars to whom Hraban referred his students above would
most likely have resorted to the planisphere in ms Basel
A.N.IV.18, f. 1^V which seems to have been prepared in Fulda
s.IX in. Here are most of the same constellations but they
are facing in different directions. It is as if one were
seeing the celestial globe from outside of it, an
orientation developed most successfully by Hipparchus. The
Salzburg and the Fulda planispheres are the earliest and
best representatives of a long tradition of astronomical
studies in manuscripts which also included double views of
the celestial globe.[58] These early ninth century
planispheres, whether Ptolemaic or Hipparchan in
orientation, are also heavily marked by perfect indented
circles and segments of circles. The available photograph
of the Salzburg planispheres has recently been studied by
John D. North (University of Groningen) who tentatively
suggested that this and another planisphere diagram from
Verona "show that at least from the time of the ninth century
the necessary skills for making such a plate [for an

anaphoric clock] were once again available in Europe."[59]   No
one has yet submitted the Fulda planisphere to a comparable
test; and neither has there ever been an edition or
discussion so far as I know of the heavily illustrated
astronomical schoolbook also found in ms Basel A.N.IV.18
which was written in Fulda during the second quarter of the
ninth century.[60]

   While intellectual and cultural historians have sought
diligently for the Hellenistic tradition of artes liberales
in Carolingian schools and have sometimes been pleasantly
surprised by the literary activity discussed there, they
have usually not found evidence of significant work in the
subjects of the quadrivium. Nevertheless there were social
and economic and political activities in the age of
Charlemagne and many generations of his successors, as well
as personal and intellectual demands for habits of thought
which led to accurate and repeatable results in surveying,
cartography, assaying ores and precious metals, telling of
time at different latitudes at different times of year. It
was the study of the evidence for telling time which brought
Charles W. Jones to reflect that "Continual reference to the
seven liberal arts by Romanesque writers has led to a common
assumption that the curricula of monastic and episcopal
schools were planned according to that concept. Such
evidence as I have encountered points to a different
conclusion. ... We have no evidence that even [grammar] was
intended as a liberal rather than a vocational subject...."
The actual books in use indicated that their users were using
a new curriculum. "A good computistical text contained
selections from arithmetic, physics, metrics, music,
history, astronomy, theology, and the like."[61] Walahfrid's
personal notebook is an example. The whole array of
materials for various scientific and mathematical studies at
Reichenau and Fulda here outlined has only been found and
assembled in consequence of a search for computistical
texts. Computus was primary but then attracted other
scientific texts which were required to answer increasingly
complex questions. It was the interplay of general social
demands with the other needs of Christian monastic schools
for grammatica, compotistica, and cantica which provided the
surviving evidence of manuscript collections of the
argumenta and algorithms for numerical calculations and
solutions of algebraic, geometrical, and trigonometrical
problems especially in engineering and astronomy. It was
within this lively context, not in a vacuum, that when
finally available the great value of a complete Euclid or a
complete Ptolemy would meet with recognition.

50

NOTES

1 Boethii De institutione arithmetica I 1, ed. F. Friedlein
(Leipzig: Teubner, 1867) 7-10; apparently Boethius used
the spelling quadruvium, though it is reported as
quadrivium in the Vita Isidori written by Braulio of
Saragosa, according to du Cange, Glossarium. Du Cange
reports the term trivium from the same lost source,
though it seems not to recur before the ninth century.
See Jacques Boussard, "Les influence anglais sur l'ecole
carolingienne du VIIIe et IXe siècles," Settimana di
Studio XIX (Spoleto 1972) 449-450.

2 A pagan mathematical philosophy was strongly opposed
e.g. by Ambrosii De Abraham II 11, 10, ed Patrologia
Latina XIV 494; but the difficulty passed by the sixth
and seventh centuries. See M. L. W. Laistner, "The
Western Church and Astrology during the early Middle
Ages," Harvard Theological Review XXXIV (1941) 251-275;
Jacques Fontaine, "Isidore de Séville et l'astrologie,"
Revue des Etudes Latines XXXI (1954) 271-300; idem,
Isidore de Séville et la culture classique dans
l'Espagne wisigothique I (Paris: Etudes Augustiniennes,
1959) 348-350.

Cassiodorii Institutiones de artibus ac disciplinis
liberalium artium I.xxii, II Praefatio and iiii, ed
R. A. B. Mynors (Oxford: The Clarendon Press, 1937).

3 Hrabani De computo I De numerorum potentia emphasized
notitiam, quam constat omnium disciplinorum esse
magistram, and repeated the pietistic defences for his
emphasis which were provided by Isidori Origines III 4,
1 and 4, 4 and especially by Alcuin's letter to
Charlemagne, Ep. 144, ed E. Dümmler, Monumenta Germaniae
Historica, Epistolae IV, p. 239. Hraban also
differentiated sharply between the terms astronomia
which dealt only with natural phenomena and astrologia
which mixes natural and superstitious: De clericorum
institutione II 25 (ed Alois Knöpfler, München 1900);
the latter however is to be handled with discernment,
as heathen philosophy can also contain some good (II 26,
cf. III 18). This is the tradition of Augustine,
especially his De civitate Dei.

4 In addition to the citations of note 2 above, see the
remarks of George Sarton on "Ptolemy and his time" in

Ancient Science and Modern Civilization (Lincoln:
University of Nebraska Press, 1954; rpr New York:
Harper, 1959) 59-64, 70-71.

5   English translation: Education and Culture in the
    Barbarian West, sixth through eight centuries, tr John
    J. Contreni (Columbia, S. Car.: University of South
    Carolina Press, 1976.

6   Henri Marrou, Saint Augustin et la fin de la culture
    antique (2ed Paris: E. de Boccard, ca. 1949) 211-235,
    provided tables of the Hellenistic (Boethian) artes
    listed variously by ten Greek and Latin writers; Varro
    and Martianus Capella added architectura and medicina.
    But a system of physica with seven parts (including
    mechanica, medicina, both astronomia and astrologia)
    also appeared in Cassiodorus' Variae III 52 and in
    Isidore's Differentiae II 39, 150-152. The latter work
    and the same author's De natura rerum were carried to
    Ireland or Saxon England by mid-seventh century whence
    a new tradition developed, especially encouraged by
    Aldhelm. It spread to Carolingian schools through
    expositions attached to Isidori Origines I-III and
    Alcuini De rhetorica et virtutibus. Further uses of
    this tradition in manuscripts and texts has been
    indicated by B. Bischoff, "Eine verschollene Einteilung
    der Wissenschaften," Archives d'Histoire Doctrinale et
    Littéraire du Moyen Age XXV (1958) 5-20, rpr idem,
    Mittelalterliche Studien I (Stuttgart 1966) 273-288.

7   Cf for example the customary acknowledgement that
    "Boethius and Cassiodorus furnished the Occident with
    its foundation of musical thought," Paul Henry Lang,
    Music in Western Civilization (New York: W. W. Norton,
    1941) 35. Yet they are said to have emphasized ratio
    which "makes the theorist superior to the mere musician"
    (p. 59-61), and this explains the absence of vocal and
    instrumental music for many centuries. Lack of
    surviving texts in the expected forms however has not
    prevented Richard Crocker, Paul Evans et alii from
    revealing the cantica which was of course practiced and
    developed for both voice and instrument; see "The Early
    Frankish Sequence: a new musical form" in the collection,
    "Toward a Medieval Aesthetic," by Charles W. Jones,
    Richard L. Crocker, and Walter Horn, in Viator VI (1975)
    309-390, esp. p. 341-349; The Early Medieval Sequence
    (Berkeley/Los Angeles: University of California Press,

52

1978) and "Alphabet notations for early medieval music,"
p. 79-104 below.

8   The reckoning which Aldhelm had learned with difficulty
    at Malmesbury was not sufficient for his new master at
    Canterbury in A.D. 670-672 (presumably Hadrian), and like
    most students he "regarded all my past labour of study
    (as being) of little value." See now the translation by
    Michael Herren in Aldhelm, the prose works (Cambridge:
    D. S. Brewer, 1979) 152-153, Letter I. In reckoning
    however calculi usually meant those pebbles or stone
    roundels (tesserae, contorniates, aspices) which were
    moved on a board for calculations. Thus Aldhelm
    probably meant not fractions but those number groups
    which must be kept in mind as one does multiplication
    and division; it was this calculatio which; he thought he
    had learned so well earlier but which now he kept in mind
    only "sustained by heavenly grace." The phrase calculi
    supputatio, quas partes numeri appellant, indicates
    nummerorum partes colligere (Gregory of Tours, Historia
    Francorum X 21) necessary for the reckoning board (abacus
    in the first form) as found in the schools of Gregory,
    Aldhelm, Bede, Hraban, et alii but presumably too much a
    part of oral learning to require exposition in school
    manuals. See Maria Rissel, Rezeption antiker und
    patristischer Wissenschaft bei Hrabanus Maurus (Bern:
    Herbert Lang, 1976) 47; Karl W. Menninger, Number words
    and number symbols (Eng. tr. Cambridge, Mass.: M.I.T.
    Press, 1969) 305-306, 315-331; Elizabeth Alföldi-Rosenbaum,
    "The finger calculus in antiquity and in the middle ages,"
    Frühmittelalterliche Studien V (1971) 1-9.

9   See the survey of computistical literature from the
    second century until the time of Bede by C. W. Jones,
    Bedae Opera de temporibus (Cambridge, Mass.: Mediaeval
    Academy of America, 1943) 6-122; his bibliographical notes
    to the chapters of Bedae De temporibus ratione, as well as
    his Bedae Pseudepigrapha: Scientific Writings Falsely
    Attributed to Bede (Ithaca: Cornell University Press,
    1939), discuss much of the work done in later centuries
    also. This has been supplemented by my introduction to
    Rabani De Computo (Corpus Christianorum continuatio
    mediaevalis XLIV; Turnholt: Brepols, 1979) 165-187; All
    Latin computistical works thus far identified prior to
    A.D. 1582 which have been printed and many which are still
    in manuscript will be listed in my Catalogue of
    computistical tracts, in preparation.

10 Preface to Bedae venerabilis Opera, pars I Opera
    didascalica (Corpus Christianorum Series Latina CXXIII
    A, 1975) v-vii, xii-xv; and "Carolingian Aesthetics: Why
    modular verse?" Viator VI (1975) 309-340.

11 E. S. Duckett, "Walahfrid Strabo of Reichenau,"
    Carolingian Portraits: a study in the ninth century (Ann
    Arbor, Michigan, 1962) 121-160. In addition to the great
    surveys of medieval literature by Adolf Ebert (vol. II,
    1880), Max Manitius (1911), and J. De Ghellinck (1939),
    as well as the two large volumes of essays in Der Kultur
    der Abtei Reichenau, ed. Konrad Beyerle (Munich 1925),
    the following studies should be noted: Konrad Plath,
    "Zur Entstehungsgeschichte der Visio Wettini des
    Walahfrid," Neues Archiv XVII (1892) 261-279; Ludwig
    Traube, "Zu Walahfrid Strabos De imagine Tetrici," ibid.
    XVIII (1893) 664-665; Emil Madeja, "Aus Walahfrid Strabos
    Lehrjahren," Studien und Mitteilungen O.S.B. XL (1920)
    251-256; Friedrich von Bezold, "Kaiserin Judith und ihr
    Dichter Walahfrid Strabo," Historische Zeitschrift CXXX
    (1924) 377-439; and Otto Herding, "Zum Problem des
    karolingischen 'Humanismus' mit besonderer Rucksicht auf
    Walahfrid Strabo," Studium Generale I (1948) 389-397.

12 Beryl Smalley, Study of the Bible in the Middle Ages
    (2 ed. Oxford 1952) 56-50; Karl Langosch, "Walahfrid
    Strabo," in Die deutsche Literatur des Mittelalters,
    Verfasserlexikon, ed. Wolfgang Stammler and Karl
    Langosch, IV (Berlin 1953) 734-769, esp. coll. 739-745;
    see also V (1955) 1111-1112.

13 "De pomis" on ms p. 368-369 was excerpted from a Fulda
    copy of Palladius; later Walahfrid used it in a poem
    addressed to his sponsor and friend, Grimald, abbot of
    S. Gallen, as one of the twenty-seven short poems
    gathered into his "Hortulus," ed. MGH. Poetae latini V
    434 seq.

14 Bernhard Bischoff, "Eine Sammelhandschrift Walahfrid
    Strabos (Cod. Sangall, 878)," Aus der Welt des Buches:
    Festgabe Georg Leyh (Leipzig 1950) 30-48; reprinted with
    additions in Bischoff's Mittelalterliche Studien:
    Ausgewählte Aufsätze zur Schriftkunde und Literatur-
    geschichte II (Stuttgart 1967) 34-51 and four
    reproductions. See also Gustav Scherrer, Verzeichnis
    der Handschriften der Stiftsbibliothek von St. Gallen
    (Halle 1875) 307-309, and Anton Bruckner, Scriptoria

54

medii aevi Helvetica I (Geneva 1935) 93-94 and Tafel
XXII.

It appears to me that the script on ms p. 344-347
(called Hand M) is actually two different hands; leaving
rubrics aside, one hand copies ms p. 344 lines 4-17 and
19-32, while another copies p. 345-347. Furthermore the
script of p. 344 and the preceding script of p. 322-323
and 340-344 line 2 (called Hand L) show distinct Fulda
characteristics, as does a note in the lower margin of
p. 341. With the exception of one later addition, all
the material in ms p. 322-351 was found at Fulda when
Walahfrid was there as a student, A.D. 827-829.

15 *Rabani Mogontiacensis Episcopi De Computo*, edidit Wesley
M. Stevens (Corpus Christianorum continuatio mediaevalis
XLIV, 1979) 199-321; edidit Etienne Baluze (Paris 1678),
rpr *Patrologia Latina* CVII 669-728. Baluze had transcribed Hraban's computus from ms Paris BN Lat. 4860
(s.IX[2]) f. 119[v]-135[v].

16 The Bodleian ms Canon. Misc. 353 was copied in the Fulda
scriptorium during the decade before mid A.D. 836, as I
have shown through palaeographical analysis of the
scripts: "Fulda Scribes at Work," *Bibliothek und
Wissenschaft* IX (Wiesbaden 1972) 284-316; see also my
general description of the manuscript and its contents:
"A Ninth Century Manuscript from Fulda," *Bodleian Library
Record* IX (Oxford 1973) 9-16; both articles include
facsimiles.

My discovery that Walahfrid used this Fulda manuscript to
correct his own copy depends upon a correlation of
corrections in his copy of Hraban's computus with
marginal notae in ms Canon. Misc. 353. Positively, the
correlation is thoroughgoing; negatively, there is no
other rationale for the presence of these notae, so far
as I have been able to conceive. Part of the evidence
was presented to a meeting of the History Colloquium of
the universities of Manitoba and Winnipeg, 11 December
1970, and is now being prepared for publication.

17 Charles W. Jones, "Bede's Place in Medieval Schools,"
*Famulus Christi: Essays in commemoration of the
thirteenth centenary of the birth of the venerable Bede*,
ed Gerald Bonner (London: SPCK, 1976) 261-285, esp.
p. 266-268.

18 These developments in Carolingian school instruction have
    been outlined in Jones, "An early medieval licensing
    examination," History of Education Quarterly III (1963)
    19-29; cf Stevens, "Introduction," Rabani De Computo
    (1979) 173-175. The putative Aachen dialogue on
    computus from A.D. 809 was found in ms Paris BN lat. 2796
    (A.D. 814?) f. 98-101, ed. E. Dümmler, MGH. Epistolae IV,
    p. 565-576, rpr Jones op. cit. with English translation,
    p. 24-28.

19 These computistical notes appear in substantially the
    same form in both ms St Gallen 878, p. 327, and ms Canon.
    Misc. 353, f. 53^v; see my "Walahfrid Strabo--a student at
    Fulda" (1972) 15-16. The usefulness of such material is
    attested by the presence of similar notes
    in innumerable other manuscripts: see Jones, Bedae
    Pseudepigrapha (1939) 74-75.

20 Bedae Liber de temporum ratione XXI 5-7; Hrabani Liber de
    computo LXXIII and LXXIIII; discussed by Jones, Bedae
    Opera de temporibus (1943) 157-158.

21 Modern scholars, like those of the ninth century, resort
    to their own schoolbooks and reference works in order to
    keep such matters straight. A good summary may be found
    for example in Allen and Greenough's New Latin Grammar
    (Boston 1931) 428-429.

22 Walahfrid's four computistical poems are found together
    with Hraban's computus in ms Paris BN lat. 4860 (s.IX^2)
    f. 156-157 whose contents derive from Reichenau and
    whose writing is from the area of Constance, and in CLM
    14523 (S. Emmeram s.X^1) f. 1-1^v; they were edited by
    E. Dümmler, MGH. Poetae latini II, no. lxxxix, who
    expressed uncertainty about their authenticity, but
    unnecessarily according to B. Bischoff, "Eine Sammel-
    handschrift Walahfrid Strabos" (1950) 46.

23 The earliest Fulda calendar dates from about A.D. 800,
    and Walahfrid's version is the earliest evidence of it.
    Dr. Winifried Böhne is preparing an edition of the
    eight Fulda calendars which survive from ninth to
    eleventh centuries (according to his letter of 21
    October 1975, his edition should be ready by 1977).
    Walahfrid's Fulda calendar was published by E. Munding,
    Die Kalendarien von St. Gallen I (Beuron 1948; Texte und
    Arbeiten XXXVI) 6, 19-20, 36, without understanding that

56

it was a Fulda calendar or that it had been copied by
Walahfrid. Munding worked from a transcription which
had been made for Bischoff during World War II.

24 C. H. Beeson, Lupus of Ferrières as Scribe and Text
    Critic (Cambridge, Mass.: Mediaeval Academy of America,
    1930): facsimile edition of Cicero, De oratore, in ms
    London BL Harley 2736, which was corrected by Lupus.
    Other manuscripts used or corrected by him are listed by
    Robert J. Gariépy Jr, "Lupus of Ferrières: Carolingian
    scribe and text critic," Mediaeval Studies XXX
    (Toronto 1968) 90-105.

25 Ep. 5, ed. E. Dümmler, MGH. Epp. VI (1902), p. 17;
    concerning this correspondence see Emmanuel von
    Severus, Lupus von Ferrières (Münster in Westphalen 1940)
    30, 47-48, 171-180.

* 26 B. Bischoff, "Die alten Namen der lateinischen
    Schriftarten" (1934), revised in Mittelalterlichen
    Studien I (Stuttgart 1966) 1-5, esp. p. 4. The ms Bern
    250 is described briefly in the Katalog Karl der Grosse
    (Aachen 1965) 222-223, and plate 36 provides a
    facsimile of folio 11^V displaying the Bercaudus muster
    of capital letters.

27 Folio 10^V De Geometrica Nunc Loquitur, with lengths of
    measure as indicated by finger reckoning and a
    rectangle whose sides and diagonals are given in pedes
    and digiti. Folio 11^V Luna; Quot Horis Lucet Luna...,
    table of 29 + 1 and an explanation of how to use it:
    Traditur autem secundum quosdam argumentum, qui horam
    punctus determinant.../... pro duobus quadrante pro uno.
    A verse of twelve lines may have been added later in the
    ninth century: Prima cleonei tolerata aerumna leonis/
    ... Cerberus extremi suprema est meta laboris, ed. A.
    Riese, Anthologia Latina (Leipzig: Teubner, 1894), no.
    641.

28 The whole contents of ms Bern 250 correspond with item
    22 in the Fleury catalogue of A.D. 1552, according to
    Denis Grémont, in A. Vidier, L'historiographie à
    S. Benoît-sur-Loire (Paris: A. & J. Picard, 1965) 42
    n.112. Folios 12-28 contain most of Abbo's computistical
    writings that appear in ms Berlin Staatsbibliothek
    Phillipps 1833 [Cat. 138], nearly a duplicate and also
    s.X ex Fleury; but attempts to identify the author's hand

among the several scripts of these two manuscripts have
not been convincing.

29 Victorius was not mentioned in the works for example of
Haskins, Laistner, Stahl, or Riché, and is only cited as
a computist by D. E. Smith and other historians of
mathematics. Nevertheless his Calculus was published
and discussed by G. Friedlein, "Der Calculus des
Victorius," Zeitschrift für Mathematik und Physik VII
(1871) 42-79. See A. Van de Vijver, "Les oeuvres
inédites d'Abbon de Fleury," Revue Bénédictine LI (1935)
137-140; C. W. Jones, "The lost Sirmond manuscript of
Bede's computus," English Historical Review LII (1937)
218-219; idem Bedae Pseudepigrapha (1939) 53 for other
editions, including PL XC 677-680 which is inadequate.

30 Nicomachus of Gerasa: Introduction to Arithmetic, tr
M. L. D'Ooge (New York: Macmillan, 1926), ch. IX
"Translators and commentators of Nicomachus." See also
William H. Stahl, Martianus Capella and the seven
liberal arts, vol. I (New York: Columbia University
Press, 1971) 149-170.

31 Boethius probably studied not only in Athens but also in
Alexandria under a student of Proclus, according to R.
Bonnaud, "L'education scientifiques de Boèce,"
Speculum IV (1929) 98-206, esp. p. 148; P. Courcelle,
"Boèce et l'école d'Alexandre," Mélanges d'Archéologie
et d'Histoire LI (1934) 185-223.

32 Cf F. E. Robbins and L. C. Karpinski in Nicomachus of
Gerasa, p. 124; V. F. Hopper, Medieval number symbolism
(New York 1938) 97-98, who failed to distinguish between
the purposes of the different works from which he
selected his quotations, thus restricting the usefulness
of the book.

33 Noticed by Wm. H. Stahl, Martianus Capella I (1971)
160-161. The relationship between Pythagorean proofs
and Euclidean proofs was called to my attention by Mary
Catherine Bodden in June 1978; cf her dissertation
(Toronto 1979) on the arithmetical portions of "St
Dunstan's Classbook," ms Oxford BL Auct. F.4.32, from
Glastonbury s.IX[1].

34 The Fulda library was developed for the most part during
the ninth but with significant growth also in the tenth

58

and eleventh centuries; thereafter little was added, as
the inventories of A.D. 1550 and 1561 show. Geometria
Boetii was found in a group of twenty texts in Ordo XLVI:
Arithmetica, geometria, astronomia et computus, musica,
an organization stemming from Johannes Knöttels ca. A.D.
1500. See Karl Christ, Die Bibliothek des Klosters
Fulda im 16. Jahrhundert: Die handschriften
Verzeichnisse (Leipzig: Otto Harrasowitz, 1933) 34-40;
for entries in the complete inventories of A.D. 1550 and
1561, see p. 159, 268, 272, 275. However the item
Boetius in geometriam (ms Vat. Pal. lat. 1928, no. 503)
was also accompanied by the incipit: Geometria est
disciplina magnitudinis immobilis et formarum, quae est
descriptio contemplativa..., ed PL LXII 1572: Boetii
in geometriam).

35 Concerning "études géometriques de l'Occident barbare,"
Paul Tannery asserted "Ceci n'est point un chapitre de
l'histoire de la science; c'est une étude sur
l'ignorance, à une époque qui précède immediatement en
Occident de la mathématique arabe," in "La géometrie au
XI siècle" (1897), rpr idem, Mémoires scientifiques, ed.
J. L. Heiberg, vol. V. Sciences exactes au Moyen Age
(Paris: E. Prévot, 1922) 79. M. L. W. Laistner,
Thought and Letters in Western Europe (1957) 173:
"Geometry as such was scarcely known or understood."
Cf. M. Clagett, "The medieval translations from the
Arabic of the Elements of Euclid," Isis XLIV (1953)
16-17, who emphasizes that scholars "have been unable
to show that a complete translation was made from the
Greek, either by Boethius or anybody else." His note 1
cites most of the known sources for fragments of Euclid
in Latin before the twelfth century, but also gives
conservatively restricted reports about their contents.

36 Neugebauer, The Exact Sciences in Antiquity (2ed
Providence, R.I.: Brown University Press, 1957) 146.
For the mass of materials which have accumulated under
the name of Heron of Alexandria (ca. A.D. 62) see
Moritz Cantor, Die römischen Agrimensoren und ihre
Stellung in der Geschichte der Feldmesskunst (Leipzig:
Teubner, 1875; rpr Wiesbaden 1968), who analyzed the
definitions, explanations, and formulae according to
exactitude of results. In summary Cantor found
accurate conclusions for the elements of triangles and
quadrangles, surface area and volume of a cone, a
truncated cone, cylinder, cube, and parallelopiped; and

relatively accurate conclusions for the section of a
circle and a sphere, for the volume of a section of a
sphere, and for the areas of regular multisided plane
figures (involving near approximations of π). He also
found approximate conclusions to various degrees of
accuracy for the area of a triangle or a trapezoid
(lacking use of square roots), and for the area of other
figures which were treated as if they had parallels or
right angles, thus distorting the results. Finally he
identified confused materials, mistaken terms of
reference (e.g. area reckoned in terms of mass) or
reference to bodies whose shape is not known, and errors
in transcription (e.g. regarding pyramids). His summary
of this exactitude is found on p. 38-49.

37 B. Bischoff, "Die Hofbibliothek unter Ludwig dem
    Frommen," in Medieval Learning and Literature, Essays
    presented to Richard William Hunt, ed J. J. G. Alexander
    and M. T. Gibson (Oxford at the Clarendon Press, 1976)
    3-22, esp. p. 19-20; Paul Lehmann, Johannes Sichardus
    (München: C. H. Beck, 1912) 99, 115-118, and "Fulda und
    die antike Literatur," Aus Fuldas Geistesleben, ed
    Josef Theele (Fulda 1928) 9-23, esp. p. 14 and 21.
    Similar manuscripts became available at Corbie by mid-
    ninth century: Neapoli Bibl. Nazionale V.A.13 in which
    the hand of Hadoard (custos librorum) has been
    identified; and Wolfenbüttel Herzog-August-Bibl. Gud.
    105, a Corbie copy of either the Palatine manuscript or
    a common exemplar and of later provenance from the
    monastery of St Bertin at St Omer. See Bischoff,
    "Hodoard und die Klassikerhandschriften aus Corbie"
    (1961), rpr Mittelalterliche Studien I (1966) 58-60;
    Lehmann op. cit. (1928) 21; B. L. Ullman, "Geometry in
    the Mediaeval Quadrivium," Studi di bibliografia e di
    storia in onore di Tammaro de Marinis IV (Verona 1964)
    263-285, esp. p. 266-267 and 273-276.

38 The basilica of Steinbach (Hesse) was built by Einhard
    (Eginhardus) apparently ca. A.D. 820; as with the Köln
    Cathedral (period VII, begun by bishop Hildebald, 787-
    818), the Steinbach church was built with pillars, not
    columns: Wilhelm Weyres, "Der karolingische Dom zu
    Köln," in Karl der Grosse, III. Karolingische Kunst,
    ed W. Braunfels and H. Schnitzler (Düsseldorf:
    L. Schwann, 1965) 412-413. A reliquary in the form of
    an "Arc de Triomphe" from the Maastricht cathedral ca.
    A.D. 820-830 is inscribed Einhardus peccator:

60

B. de Montesquiou-Fezenac, "L'Arc de Triomphe d'Einhardus," Cahiers archéologiques IV (1949) 79-103. Einhard's own life and work has been little studied, but the surprising unity of style of artists working at Aachen after ca. A.D. 781 may have been due to the discrete influence of Einhard, according to Carol Heitz, "Nouvelles interprétations de l'art carolingian," Revue de l'art (1968) no. 1-2, p. 104-113, esp. p. 104-113.

Mathematical dimensions and ratios have been discussed by L. Hugot, "Die Pfalz Karls des Grossen in Aachen," in Karl der Grosse III (1965) 534-572; W. Horn and E. Born, The Plan of St Gall (Los Angeles and Berkeley: University of California Press, 1979) I, sections 1.14-17, and III, Appendix III. "The Significance of the Plan of St Gall to the History of Measurement," by A. H. Dupree.

39 The comments of Juschkewitsch and Crombie are found in Scientific Change: Historical Studies in the Intellectual, Social, and Technological Invention, from Antiquity to the Present, ed. A. C. Crombie (London: Heinemann, 1963) 294 and 319.

40 The editions are inadequate: see the first two in PL LXII 1352-1358, 1358-1364; others may be found through TKr references: A Catalogue of Incipits of Mediaeval Scientific Writings in Latin, ed L. Thorndike and P. Kibre (rev. ed. London: Mediaeval Academy of America, 1963). Other works of the ninth century of more gromatic nature have not been included here.

41 John E. Murdoch, "Euclides Graeco-Latines," Harvard Studies in Classical Philology LXXI (1966) 249-302.

42 B. L. Ullman, "Geometry in the Mediaeval Quadrivium," Studi di Bibliografia e di Storia in honore di Tammaro de Marinis IV (Verona 1964) 263-285.

43 M. Folkerts, "Boethius" Geometric II. Ein mathematisches Lehrbuch des Mittelalters (Wiesbaden: F. Steiner, 1970) 33 n. 59; some of the occurences of "Geometria I" are noted in his manuscript descriptions, but he is primarily interested in a later composition which had "Geometria I" as a source. For Reichenau see the Catalogue of A.D. 820/821, in Mittelalterliche Bibliothekakatakoge Deutschlands und der Schweiz, ed

Paul Lehmann (München 1918) I 250, no. 16-17; "De
opusculis Boetii. De arithmetica libri II de
geometria libri III...." CLM 560 is dated in the first
half or mid-ninth century by B. Bischoff, Die südostdeutschen Schreibschalen und Bibliotheken in der
Karolingerzeit (2ed Wiesbaden: O. Harrassowitz, 1960)
262 n.3. For Murbach see W. Milde, "Der Murbacher
Bibliothekakatalog des 9. Jahrhunderts," Empharion,
Beihaft 1967, cit. Folkerts 34 n.70-71; it is dated ca.
A.D. 840.

44 Folkerts (1970) 173-207: Appendix I.

45 Folkerts' text shows that additional excerpta Euclidis
    are present in the Corpus agrimensorum for which further
    portions of Boethius' Geometria may perhaps have been
    the source. The same is true of independent bits of the
    Elements which were found in Martianus Capella,
    Cassiodorus, and Isidore of Seville--none of whom
    intended to provide a teaching manual of the systematic
    quality of Ars geometriae et arithmeticae. Analysis of
    these texts has been undertaken by my student, Michael
    R. Scott, who presented some of the results to the
    Canadian Society for the History and Philosophy of
    Mathematics, meeting at the University of Western
    Ontario in June 1978.

46 Epistola Theoderici regis ad Boethium circa 24 annos
    natum c.a. 506 scripta, in Cassiodorii Variae I 45; and
    Cassiodorii Senatoris Institutiones II 6, 2, ed.
    R. A. B. Mynors (Oxford at the Clarendon Press, 1963)
    152.

47 Euclidis latine facti fragmenta Veronensis, ed. M.
    Geymonat (Milano: Istituto Editoriale Cisalpino, 1964).

48 Identified by B. Bischoff, published by M. Geymonat,
    "Nuovi frammenti della geometria 'Boeziana' in un
    codice del IX secolo?" Scriptorium XXI (1967) 3-16.

49 Edited by Charles W. Jones, Bedae venerabilis opera,
    pars I. Opera didascalica (Corpus Christianorum, Series
    Latina CXXIII A (Turnholt, Belgium: Brepols, 1975):
    Preface p. v-xvi; Introduction to DNR p. 173-188; text
    p. 189-234.

50 H. J. Mette, Sphairopoiia (München: C. H. Beck, 1936)

62

66-78, on Krates (ca. 170 B.C.); Plutarch, De facie in orbe lunae VII; Germaine Aujac, "La sphéropée, ou la mécanique ou service de la découverte du monde," Revue d'Histoire des Sciences XXIII (1970) 93-107.

51 Sphericity of the earth was the prevalent view at all times since Plato, Eudoxus, and Aristotle, and the Carolingians were no exceptions. See Charles W. Jones, "The Flat Earth," Thought IX (1934) 296-307.

Modern speculations that Boniface held the earth to be flat and therefore accused Virgil of Salzburg of heresy for believing it to be round is without evidence, but it also ignores the literature concerning the orbis quadratus from which the issue of antipodae arose. See my discussion of Boniface in Speculum LI (1976) 752-755, and "The figure of the earth in Isidore's De natura rerum," to appear in Isis LXXI (1980).

52 J. Fontaine, Isidore de Séville, Traité de la nature (Bordeaux: Féret, 1960) 31-37, described the three manuscripts under the sigla A and Z (England), F and W (Fulda).

53 G. Aujac, "L'image du globe terrestre dans la Grèce ancienne," Revue d'Histoire des Sciences XXVII (1974) 196-205.

54 Maria Rissel, Rezeption antiker und patristischer Wissenschaft bei Hrabanus Maurus (Bern: H. Lang, 1976) 54-56 discussed at some length the question of whether Hraban abandoned the spherical concept of the earth; but she ignored his De computo XXXVI-LI in which he supplied the same basic data and arguments.

55 This is a Carolingian tract whose earliest manuscript seems to be Wien NB 387 and its twin CLM 210 (Salzburg A.D. 810-818) f. 74^v-76. It is found together with Hraban's computus in seven copies; for editions see TKr 473 and 1689.

56 Karl Helm, "Hrabanus Maurus und die Volkskunde," Beiträge zur Geschichte der deutschen Sprache und Literatur LXXI (Halle 1949) 466-470.

57 The accompanying text is similar to but varies from the explanation of these phenomena transmitted by Martianus,

Chalcidius, and Macrobius which may have been available
at Reichenau by this time, whereas only Chalcidius was
known at Fulda so early. See Charles W. Jones, "A note
on concepts of the inferior planets in the middle ages,"
Isis XXIV (1936) 397-399; the concept was not derived
from Heracleides, according to Bruce Eastwood
(University of Kentucky) who is analyzing the textual
and manuscript tradition.

58 The most useful attempt to differentiate the two types
    of planisphere is by Patrick McGurk, "Germanici Caesaris
    Aratea cum scholiis, a new illustrated witness from
    Wales," National Library of Wales Journal XVIII, 2
    (Winter 1973) 197-216, with illustrations from ms NLW
    735C (Limoges s.XII). According to McGurk (1979) the
    double views of the heavens are not north and south as
    all have previously supposed, but right and left
    hemispheres--a type of division which may be the
    continued legacy of the orbis quadratus.

59 J. D. North, "Monasticism and the first mechanical
    clocks," in The Study of Time II, ed J. T. Fraser and
    N. Lawrence (New York: Springer, 1975) 381-398, esp.
    p. 386-388. His second planisphere in this probe was
    ms Berlin Staatsbibl. Phillipps 1830 [cat. 129],
    f. 11-12.

60 I hope to take up both tasks as soon as possible.

61 Jones, Saints' Lives and Chronicles in Early England
    (Ithaca: Cornell University Press, 1947) 200-201 n.11;
    "This examination has led me to believe that a
    systematic study of manuscript textbooks would correct
    more than one current generalization about early
    medieval education" (p. 201).

X

WALAHFRID STRABO – A STUDENT AT FULDA

Walahfrid Strabo, the mid-ninth century abbot of Reichenau, has received the attention of some very fine historians who have emphasized his poetry and relations with Carolingian royalty. In one of the best accounts, Eleanor Shipley Duckett gave a refreshing biographical reading of poems and correspondence of Walahfrid. However Miss Duckett and all others[1] have failed to recognize that he had given attention to science as well as to humane letters. His active intellect was more wide-ranging than later medievalists supposed, for the evidence strongly indicates that he concerned himself with arithmetic and astronomy as well as with grammar and poetry.

Born about 808/9, Walahfrid had been the child of a poor family in Allemania (Swabia) when he was brought to the island of Reichenau. He became one of those pueri oblati, who renounced property rights and devoted themselves to the riches of eternal life, even as children, on the analogy that no offering was too small! He liked monastic life and proved to be an apt pupil, despite the "twisted eye" which led others to call him Strabo.

We are only beginning to understand the practical and psychological significance that writing had for the young and old in Carolingian schools.[2] No doubt Walahfrid began writing with an iron griffel on a wax tablet, and at least by the year 825 he was eager enough to have acquired his own vellum copy-book. The earliest entries were made at Reichenau when Walahfrid was about sixteen years of age and already a close friend of Gottschalk when they both were students of Wetti (d. November 824). In 827 Walahfrid went north to the now famous monastery and school at Fulda in order to study with Hraban, and there he added to his notebook. But within two years both Gottschalk and Walahfrid were in trouble: Gottschalk wanted to be free of his vows and appealed over the abbot's head to a council of bishops, and Walahfrid felt Hraban's antagonism for his loyalty to a rebel. In the spring of 829 Walahfrid joined the royal family of Louis the Pious at Aachen as tutor for Judith's son, Charles, known later as emperor Charles the Bald. He continued to make entries in the notebook during his time at court and also later as abbot of Reichenau. The final notations were made within weeks of his death in the summer of 849.

Walahfrid's personal notebook is now manuscript no. 878 in the Stiftsbibliothek at Sankt Gallen, Switzerland. It contains none of the poems which have attracted scholary attention, nor the biblical commentaries which brought him abuse.[3] Rather there are tracts and excerpts from other authors which he transcribed for study and later use, or because he simply liked them: for example, a paragraph on how to grow pears.[4] Two decades ago Bernhard Bischoff demonstrated that this was Walahfrid's personal manuscript and that perhaps half of its contents was transcribed by Walahfrid himself during specific and datable periods of his life. Three previous attempts to identify Walahfrid's script had not been successful; there are many mediaeval autographs, and it is a temptation to add to the number prematurely. In June 1967 however I was able to spend a week at Sankt Gallen, and not only was I impressed with the validity but also expect to publish further evidence to substantiate Bischoff's analysis.[5]

The earliest entry in this manuscript as it now stands (ms p. 178-240) is Hraban's Computus which Walahfrid was trying to learn about 825 at Reichenau — perhaps transcribing a copy which had been brought there from Fulda by Gottschalk in 824. As a student he drilled on Latin word forms (ms p. 174-176) and copied out grammatical instructions from versions of Priscian and Donatus (ms p. 5-69) which allow us to see the student struggling with language before he became an accomplished poet. Both sides of the seven liberal arts are represented from the beginning therefore; later he came back to these materials and added to them, perhaps using them at Aachen with the young Charles, or in the school at Reichenau; and neither the trivium nor the quadrivium was neglected during the subsequent periods of Walahfrid's life. But when all the grammar and metrics which we expect to find in his notebook have been mentioned, most of its contents have still not been touched. In fact, it is natural history and science which seem to dominate his attention and in particular the computus, that is, material on calendar usage.

A surprising correlation of two manuscripts will allow us a view of Walahfrid, as he was just beginning to grasp the elements of computistical science. One of the tasks carried out by Walahfrid at Fulda in 827-829 was a complete re-reading and correction on every page of his copy of Hraban's *Computus*. I have identified the manuscript which he used to correct his own: it is now Canonici Miscellaneous 353 in the Bodleian Library at Oxford, and the primary writing is mature Fulda script while the remainder is by scribes who were being trained

by a Fulda master. Both Walahfrid's manuscript and this Fulda manuscript contain not only Hraban's *Computus* but also additional notes
by which we may see how the computus was taught and learned at
Fulda.[6] I shall discuss them together as sources of Walahfrid's experience in the classroom.

Hraban's handbook[7] began with the concept of number, how to
use numerals and how to apply fourteen divisions of time. By question
and answer the student learned to observe the stars and their patterns,
the movements of sun and moon and their imperfect correlations. By
the middle of the book he was deeply into the technical details of how
to make and use a continuous calendar based upon a nineteen year
cycle, as well as the rationale for such a calendar in Latin Christian
history.

Additional notes which appear at the end of the computus in these
two manuscripts[8] are answers to problems which a master had posed
for his students when they began their study of the calendar itself. The
first problem was this: On which *feria* (weekday) did the *Kalends* (first
day) of each month occur? Bede had explained the system, and so did
Hraban who also applied it to the example of August when he was
writing. Taking the year 820 as his example, a student has worked
out in the Fulda manuscript the feria for the Kalends of each month
from March 820 to February 821; thus the problem was solved and a
system learned which could be applied to any year.[9]

The second problem was this: In the Roman system, one must
know how many Nones, Ides, and Kalends there are in each month.
Bede and Hraban had provided explanations of how the Romans came
to their system and what the terms meant, but not how to use them.
But in both Walahfrid's manuscript and the Fulda manuscript we see
that Fulda students were drilled in the number of Nones and of Ides
and of Kalends, and in the application of the terms by groups of months,
according to a corresponding number of Nones and Ides.

Students were also taught to divide months into weeks. How do
you correlate 12 months and 52 weeks in the Roman system? On which
days of the month will a weekday recur? Begin with the Kalends of
January: whatever feria occurs on the Kalends of January will recur
on the sixth Ides and on the 18th Kalends, the 11th Kalends, and the
fourth Kalends (of February). In order to continue reckoning the
sequence of weeks, the student must recall what he had drilled on in
the preceding problem: the number of Nones, Ides, and Kalends for

each month. Answers to these problems would have been simple, almost instinctive for a Roman. But Fulda students learned to think in Roman terms only with effort,[10] and they needed the aid of memory verses which are so abundant in medieval manuscripts.[11]

If such drills are elementary, they are also necessary in order to go further. Perhaps Walahfrid took the teaching more seriously than most student, for he applied it in an uncommon way. At Fulda he copied into his notebook a complete calendar, (ms. p. 324-327), but in a form rarely found. Rather than using one page for each month, he grouped the months exactly as here described, according to whether they had the same number of Nones, Ides, and Kalends: January, August, and December together on one page; March, May, July and October on a second page; April, June, September, and November on a third; and a column for February — a space of less than two folios. While calendars of this type had been used in the Roman Empire, none is known for five centuries before Walahfrid created his at Fulda.[12] The teaching was that of Hraban; application by a quick-witted student may be seen in this short practical calendar.

Not all of Walahfrid's time at Fulda was spent in either grammatical or computistical studies. On one page of his notebook (ms p. 277) he took a compass and drew a circular maze, that is, a game, which no doubt distracted attention from his assignments. Also all of us who have studied in Germany for more than a year will have empathy with Walahfrid when I mention his attention to medicine while at Fulda. He must have sought medical aid where he could find it and copied out a letter falsely ascribed to the Greek physician Hypocrates (ms p. 328-331). It provided a medical regimen for an entire year, and it also told him what to do for headaches, constipation, vertigo and ulcers. To this he added further medical prescriptions with German translations.[13] He must have been trying to overcome the real and very disturbing communications gap between a local physician and a transient foreign student by translating key words and phrases into his own language. Many of us know the situation well.

In both the manuscript of Walahfrid and the Fulda manuscript which Walahfrid used, there is evidence that attention to the calendar led to further learning. In the Fulda manuscript (f. 53v-55) the student proceeded to copy an elaborate description of the heavens according to groups of stars which once could have been visualized as storybook figures of beasts and heros and thus be easily remembered. Hraban had given little space to the Zodiac, so this longer text of Hellenistic origin was added. It begins, *Duo sunt extremi vertices mundi, quos appellant polos* . . . , and clearly assumes the globular shape of the whole cosmos and of the earth. So did Bede's *De natura rerum* which was Walahfrid's next addition to his notebook while at Fulda (ms p. 242-262). The same teaching had been stated or assumed in several parts of Hraban's *Computus,* but no doubt good students raised questions. At Boniface's monastic foundation at Fulda, the ninth century abbots and masters taught both Christian doctrine whose orthodoxy never failed and a cosmology which was never publicly challenged: the world was a globe, the universe a sphere, and all was in God's own hand.[14]

His notebook shows that Walahfrid continued to collect materials on these subjects later at Aachen and at Reichenau. He added Bede's *De temporibus* and took up the question basic to any calendar — when does it begin? To Bede's and Hraban's account of the *annus mundi,* the year of creation, he added three more accounts and noted the conflicts and the confusion which would result if anyone tried to use them. He copied a variety of computistical paragraphs by which one might find the *indiction,* the *bissextus,* the *saltus lunae,* the *lunar epacts,* and other technical information for reckoning dates. And he grouped these paragraphs together according to whether they used the *annus mundi* or the *annus Domini.* Apparently Walahfrid recognized that there is an uncertainty about the year of Incarnation, as well as about the year of creation — a problem which he could not resolve; nor can we. In his notes we can see a scholar moving from elementary questions, through technical data, into the most difficult problems of computistical science — and the arithmetic, astronomy, and history which it required.[15]

I have emphasized some aspects of schooling which were important to Walahfrid at Reichenau, at Fulda, and throughout his life, but which have rarely claimed the interest of modern historians and biographers. I hope that attention to a few computistical notes has led us into the heart of Carolingian education, into the mind of a student studying.

NOTES

[1] E. S. Duckett, "Walahfrid Strabo of Reichenau", *Carolingian Portraits: a study in the ninth century* (Ann Arbor, Michigan 1962) 121-160. In addition to the great surveys of medieval literature by Adolf Ebert (vol. II, 1880), Max Manitius (1911), and J. De Ghellinck (1939), as well as the two large volumes of essays in *Der Kultur der Abtei Reichenau*, ed. Konrad Beyerle (Munich 1925), the following studies should be noted: Konrad Plath, "Zur Entstehungsgeschichte der Visio Wettini des Walahfrid," *Neues Archiv* XVII (1892) 261-279; Ludwig Traube, "Zu Walahfrid Strabos De imagine Tetrici," *ibid.* XVIII (1893) 664-665; Emil Madeja, "Aus Walahfrid Strabos Lehrjahren," *Studien und Mitteilungen O.S.B.* XL (1920) 251-256; Friedrich von Bezold, "Kaiserin Judith und ihr Dichter Walahfrid Strabo," *Historische Zeitschrift* CXXX (1924) 377-439; and Otto Herding, "Zum Problem des karolingischen 'Humanismus' mit besonderer Rucksicht auf Walahfrid Strabo," *Studium Generale* I (1948) 389-397.

I wish to thank Professor Hugh Mackinnon of Waterloo University for his useful comments on my paper at the meeting of the Canadian Historical Association at Memorial University on 31 May 1971. Portions of this material were presented also at the annual Medieval Institute, Western Michigan University, 21 March 1970.

[2] See the suggestions of Jean Leclercq, *The Love of Learning and the Desire for God* (Eng. tr. New York 1961) 128, 175 passim.

[3] Beryl Smalley, *Study of the Bible in the Middle Ages* (2 ed. Oxford 1952) 56-60; Karl Langosch, "Walahfrid Strabo," in *Die deutsche Literatur des Mittelalters, Verfasserlexikon,* ed. Wolfgang Stammler and Karl Langosch, IV (Berlin 1953) 734-769, esp. coll. 739-745; see also V (1955) 1111-1112.

[4] "De pomis" on ms p. 368-369 was excerpted from a Fulda copy of Palladius; later Walahfrid used it in a poem addressed to his sponsor and friend, Grimald, abbot of S. Gallen, as one of the twenty-seven short poems gathered into his "Hortulus," ed. MGH. Poetae latini V 434 seq.; see Bischoff (note 5) 47.

[5] Bernhard Bischoff, "Eine Sammelhandschrift Walahfrid Strabos (Cod. Sangall. 878)," in *Aus der Welt des Buches Festgabe Georg Leyh* (Leipzig 1950) 30-48; reprinted with additions in Bischoff's *Mittelalterliche Studien: Ausgewählte Aufsätze zur Schriftkunde und Literaturgeschichte* II (Stuttgart 1967) 34-51 and four reproductions. See also Gustav Scherrer, *Verzeichnis der Handschriften der Stiftsbibliothek von St. Gallen* (Halle 1875) 307-309, and Anton Bruckner, *Scriptoria medii aevi Helvetica* I (Geneva 1935) 93-94 and Tafel XXII.

If I may supplement the brilliant analysis by Bischoff, whose work so often leads others to new discoveries: the script on ms p. 344-347 (called Hand M) is actually two different hands; leaving rubrics aside, one hand copied ms p. 344 lines 4-17 and 19-32, while another copied p. 345-347. Furthermore the script of p. 344 and the preceding script of p. 322-323 and 340-344 line 2 (called Hand L) show distinct Fulda characteristics, as does a note in the lower margin of p. 341. With the exception of one later addition, all the material in ms p. 322-351 was found at Fulda when Walahfrid was there as a student, including especially the computistical argumenta discussed below.

I wish to express my thanks to Professor Bischoff for discussing this manuscript with me in August 1967, as well as to Professor Johannes Duft, Librarian of the Stiftsbibliothek at Sankt Gallen, for making it available to me.

[6] The Bodleian ms Canon. Misc. 353 was copied in the Fulda scriptorium during the decade before mid A.D. 836, as I have shown through palaeographical analysis of the scripts: "Fulda Scribes at Work," *Bibliothek und Wissenschaft* IX (Wiesbaden 1972) 284-316; see also my general description of the manuscript and its contents: "A Ninth Century Manuscript from Fulda," *Bodleian Library Record* IX (Oxford 1973) 9-16.

My discovery that Walahfrid used this Fulda manuscript to correct his own copy depends upon a correlation of corrections in his copy of Hraban's computus with marginal notae in ms Canon. Misc. 353. Positively, the correlation is thoroughgoing; negatively, there is no other rationale for the presence of these notae, so far as I have been able to conceive. Part of the evidence was presented to a meeting of the History Colloquium of the universities of Manitoba and Winnipeg, 11 December 1970, and is now being prepared for publication.

7 Hraban's Liber de computo, ed. *Patrologia Latina* CVII. 669-728; a critical edition of the text is being prepared from sixteen manuscripts, of which two discussed here are the earliest known.

8 These computistical notes appear in substantially the same form in both ms St. Gallen 878 p. 327 and ms Canon. Misc. 353 f. 53^v. The usefulness of such material is attested by the presence of similar notes in innumerable other manuscripts.

9 *Bedae Liber de temporum ratione* XXI 5-7 (ed. Jones); *Hrabani Liber de computo* LXXIII (ed. PL CVII 709B) and LXXIIII (ibid. 710C). See the discussion by Charles W. Jones, *Bedae Opera de temporibus* (Cambridge, Mass., 1943) 157-158.

10 Modern scholars, like those of the ninth century, resort to their own schoolbooks and reference works in order to keep such matters straight. A good summary may be found for example in Allen and Greenough's *New Latin Grammar* (Boston 1931) 428-429.

11 Walahfrid's four computistical poems are found together with Hraban's computus in ms Paris BN lat. 4860 (s.IX^2) f. 156-157 whose contents derive from Reichenau and whose writing is from the area of Constance, and in CLM 14523 (S. Emmeram s.X^1) f. 1-1^v; they were edited by E. Dümmler, *MGH. Poetae latini* II, no. 1xxxix, who expressed uncertainty about their authenticity, but cf Bischoff 46.

12 This calendar was published by E. Munding, *Die Kalendarien von St. Gallen* I (Beuron 1948; *Texte und Arbeiten* XXXVI) 6, 19-20, 36, without understanding that it was a Fulda calendar or that it had been copied by Walahfrid. Munding worked from a transcription which had been made for Bischoff during World War II.

13 The medical regimen here attributed to Hippocrates, *Ep. ad Antiochum et Antonium*, is very common in medieval Latin manuscripts. Walahfrid was probably at Fulda when he copied it and could have discovered it in *Bedae Liber de temporum ratione* XXX 13-39; Hraban had used parts of this chapter in his *Liber de computo* LIII *De solstitiis et aequinoctiis* but had omitted the ps-Hippocratean material; see Jones 365-366. Also at Fulda Walahfrid transcribed the medical prescription on his ms p. 331-334; on the interlinear German words see E. Steinmeyer, *Die althochdeutschen Glossen* IV (Berlin 1898) 455. Further medical notes were added after Walahfrid had left Fulda (ms p. 368-377 and 392-393).

14 This work begins to appear in manuscripts during the early ninth century and was attributed to either Aratus or Hyginus or not at all; ed. PL XC 368-369.

The historical evidence is overwhelming that educated men of the Middle Ages conceived of the earth as a sphere, though some purveyors of "enlightment" wish it were not so — e.g. such fabulists as Cosmas Indicopleustes, J. L. E. Dreyer, and John Herman Randall. See the discussion by C. W. Jones "The Flat Earth," *Thought* IX (1934) 296-307.

The only significant exception may be the sixth century bishop, Isidore of Seville, whose language usage suggests that he imagined the world not as a perfect sphere but rather as somewhat flattened at the poles and bulging at the equator — which it is!

15 At Fulda were recorded *Bedae Liber de natura rerum* (ms p. 242-262) and further computistical argumenta (ms p. 344-347). At a later period he added *Bedae Liber de temporibus* (ms p. 262-276 with table of contents on p. 243), various argumenta (ms p. 176-177, 284-302), and excerpts of chronicles (ms p. 240-241, 277-283, 302-305).

In addition to the interest they have as evidence for Walahfrid's studies, some of these computistical materials have never been found in other manuscripts and should be published.   *

On ms St. Gallen 878 see also the essays of Alfred Cordoliani, "Les manuscrits de comput ecclésiastique de l'abbaye Saint-Gall du VIIIe au XIIe siècle,"

Revue d'Histoire de l'Eglise Suisse XLIX (1955) 179-181, and "L'evolution du comput ecclésiatique à Saint-Gall du VIIIe au XIe siècle," ibid. 288-323, esp. p. 293. Their usefulness is seriously restricted by M. Cordoliani's acceptance of dates and provenance for this and other manuscripts which were incorrect and by his attempt to cite folios rather than pages (pagination is standard in Swiss collections); see Johannes Duft, "Berichtigungen zu A. Cordoliani," ibid. L (1956) 388-393, esp. p. 391.

COMPUTUS-HANDSCHRIFTEN WALAHFRID STRABOS

Zwei Handschriften ermöglichen es uns, etwas zu dem Unterricht und den Erfahrungen zweier junger Männer in den berühmtesten monastischen Schulen des neunten Jahrhunderts zu sagen. Es sind die Schulen der Klöster Reichenau am Bodensee und Fulda, fast 350 Kilometer nördlich in Hessen. Die Handschriften sind Oxford Bodleian Library, mss. canonici Miscellaneous 353 und Sankt Gallen, Stiftsbibliothek, HS. 878 in der Schweiz. Die beiden Handschriften betreffen den jungen Walahfrid Strabo und seine Kameraden, von denen ich einen später provisorisch benennen will.

Walahfrid Strabo ist für uns zum erstenmal als achtjähriger auf der Insel Reichenau faßbar. Nach dem Studium dort bis 827 und Stationen in Fulda (827-829) und Aachen (829-838) kehrte er auf die Reichenau als etwa dreißigjähriger zurück. Nach politischen Schwierigkeiten wurde er 842-849 Abt. Die Korrekturen der Handschriften zeugen eindrucksvoll von seinem lebhaften Studium wissenschaftlicher Texte und von seiner unablässigen Besorgnis um Genauigkeit. Alle Handschriften sind damals schon ein- oder zweimal korrigiert worden, und danach noch miteinander verglichen worden[1].

Im neunten Jahrhundert wurde die Leitung des Abschreibens von Texten gewöhnlich einem erfahrenen Schreibmeister anvertraut, einem Meister, der die alten oder fremden Schreibtypen interpretieren konnte. Er besaß die Fähigkeit, Kontraktionen und Abbreviaturen sowie stenographische Zeichen zu entziffern und die ungewöhnliche lateinische Sprache richtig zu lesen. Er mußte die Novizen beim Erlernen der Grundlagen betreuen und Nachlässigkeit und Unachtsamkeit zuvorkommen. So war das Skriptorium Arbeitsstätte und Unterrichtszimmer zugleich, wo die Studenten nachlässig werden konnten, und Verbesserungen der Texte ziemlich übereilt eingefügt wurden. Manchmal wurde der

---

1   Nota bene: Ich möchte mich hiermit bei Gebhard und Ingrid Schmidl (Dachau), Manfred Heiderich (Winnipeg), Kathi Reimann (Winnipeg), Helmut Bender (Passau), und besonders Linwood DeLong (Winnipeg) bedanken, die mir bei meinen begrenzten Sprachfähigkeiten mit Rat und Hilfe beistanden.

Meister zu anderen Verpflichtungen gerufen. Dadurch wird sich eine große Menge von fehlerhaft erhaltenen Handschriften wohl erklären lassen. Dennoch sind auch viele Handschriften erhalten, die von Studenten sorgfältig abgeschrieben und korrigiert worden sind, das heißt, von Studenten, die gut gelernt hatten[2].

Die Analyse der Canonici Hs 353 (Oxforder Handschrift) ergab, daß sie nicht nur im Laufe des Jahrzehnts vor 836 nach Christus zu Fulda abgeschrieben wurde, sondern auch, daß die Arbeit von Schreibern mit drei verschiedenen Graden von Kompetenz durchgeführt wurde[3]: Es gab drei fachkundige Fuldaer Schreiber, zwei erfahrene Fachmänner, die nicht zuerst in Fulda ausgebildet worden sind und den Fuldastil erst lernen mußten, sowie schließlich sechs Anfänger, die damals von dem Schreibmeister unterrichtet wurden. Man kann an der Arbeit der Anfänger und derjenigen, die nicht aus Fulda stammten, feststellen, daß die Schreibformen ihrer *r*'s und *a*'s sich entwickelten, und daß die Studenten es lernen mußten, drei Typen von *a* und zwei Typen von *d* und zwei Typen von *r* nach fuldaischen Bräuchen abzuwechseln. Sie versuchten die Ligaturen, oft aber auf mangelhafte Art und Weise, besonders die schöne halbkreisförmige *ct*-Ligatur, die die erfahrenen Fulda-Hände sogar in einer Dienstkopie eines *computus* ausgezeichnet gemalt haben[4].

Die Analyse der Handschrift St. Gallen 878 von Bernhard Bischoff ergab siebzehn Schreiber auf verschiedenen Entwicklungsstufen[5]. Am wichtigsten war die Hand des Walahfrid Strabo. Sie kann man durch vier Abschnitte seines Lebens verfolgen: von 824 oder 825 nach Christus, als er fünfzehn Jahre alt war, bis zum Jahre seines Todes 849. Die Abschnitte I und II entstanden auf der Reichenau, wo Walahfrid bei Erlebald, Wetti, Haito, Tatto, und wohl Grimalt Grammatik und Komputistik erlernte[6]. Abschnitt III gehört in die Fuldaer Zeit, wo er eine ungewöhnliche Art eines kompakten Fuldaer Kalenders kopierte und dazu Bedas *liber de temporibus* und sein Buch *De natura rerum*. Während seiner Zeit in

---

2　Wattenbach, *Schriftwesen*, pp. 317-344, besonders pp. 325-335; Lindsay, "Collectanea varia", pp. 10-15; Bischoff, *Paläographie*, 1979, p. 61; jetzt neubearbeitet und vermehrt: *Latin Palaeography. Antiquity and the Middle Ages*, 1990, p. 43.

3　Ms Oxford BL Canonici misc. 353 ist "sec. forsan IX exeuntis" von Cox, *Catalogi codicum manuscriptorum Bibliothecae Bodleianae Pars tertia* datiert. Aber durch paleographische Analyse hat Stevens, "Fulda scribes at work", pp. 287-317, bewiesen, daß die Canonici Handschrift in dem Fuldaer Skriptorium "ante 836 med." geschrieben worden sein muß.

4　Computus heißt Rechnen aller Arten, besonders aber Zeitrechnung durch Arithmetik und durch die Zyklen des Mondes und der Sonne, die das christliche Osterfest datierten und erklärten.

5　Bischoff, "Eine Sammelhandschrift Walahfrid Strabos", in: *Mittelalterliche Studien* II, pp. 34-51. Darin sind die Abbildungen IIa und IIb leider vertauscht; Walahfrids Hand WI ist mit der Aufnahme IIb dargestellt, Hand WII mit der Aufnahme IIa.

6　Walahfrids Lehrer auf der Reichenau sind oft erwähnt worden: z. B. von Plath, Madeja, Duckett, und besonders Langosch in: *Die deutsche Literatur des Mittelalters, Verfasserlexikon*, Band IV, pp. 734-769 und Band V, pp. 111 -1112.

Aachen (829 - 838) als Erzieher Karls, des Sohnes von Judith und Ludwig dem Frommen, hat Walahfrid in Abschnitt IV verschiedene Einzelheiten in sein handschriftliches *Vademecum* eingetragen, einschließlich einer Reihe voneinander abweichender Daten für die Geburt Jesu Christi. Seine Rückkehr ins Kloster Reichenau als Abt gab ihm die Möglichkeit, die Abschrift einiger früher unvollständig kopierter Texte, über Grammatik und Metrik fortzusetzen. Weitere Studien sollen eine noch ausführlichere Geschichte dieses langen vierten Abschnitts von Walahfrids Leben beschreiben[7]. Die Paläographie bietet die Möglichkeit, Geschichte wiederzugewinnen, die die Chroniken und Annalen uns nicht liefern.

Walahfrids Kopie des *Liber de computo* von Hrabanus Maurus ist das früheste Exemplar, das noch existiert[8]. Wir sprechen hier oft über Fulda und Hrabanus Maurus. Im Jahre 743 war Fulda als Kloster durch Bonifatius und den ersten Abt Sturmi aus Bayern gegründet worden. Einer der berühmtesten Äbte war Hraban, ein Fuldaer Mönch, der bei Alcuin in Aachen und Tours studierte; als Abt leitete er das Kloster Fulda von 822 bis 842; er ist 856 als Bischof von Mainz gestorben. Hraban hat viele Bücher geschrieben, unter anderem Bibelkommentare, eine illustrierte Enzyklopädie, und 820 ein Buch über den *computus*. Nur fünf Jahre nach der Abfassung des computistischen Textes Hrabans wurde dieser auf der Reichenau von Walahfrid selbst und auch von einer zweiten Hand abgeschrieben. Im neunten Jahrhundert gab es einen aktiven Austausch von Männern zwischen Fulda und der Reichenau. Der junge Godescalc kam im Sommer 824 mit Gefährten aus Fulda nach Niederzell auf die Reichenau. Entweder er oder ein anderer Fuldaer brachte ein frühes Exemplar von Hrabans *Liber de computo* mit. Walahfrid fing bald an, diesen Text sowie Texte über Grammatik zu untersuchen und zu korrigieren[9]. Das läßt sich wie folgt rekonstruieren:

---

7 Zwischen Frühling 829 und seinem Tod im August 849 verbrachte Walahfrid den Sommer 829 in Weißenburg; vom Herbst 829 bis 833 war er bei Judith und Karl am Kaiserhof in Aachen; 833 bis Frühling 834 war er mit dem jungen Karl in Prüm, als man Judith nach Tortona schickte; Ostern 834 bis Sommer 838 war er wieder bei Judith und Karl in Aachen; Sommer 838 war er auf der Reichenau, Frühling 839 bis 840 kurz als Abt; 840 wurde er durch Ludwig den Deutschen abgesetzt, und bis 842 muß er wohl zwischen Speyer, Weißenburg, Prüm, oder Fulda herumgewandert sein; 842 bis 849 war er Abt auf der Reichenau, manchmal aber auf diplomatischen Reisen zwischen Karl dem Kahlen und Ludwig dem Deutschen, um zu vermitteln.

8 *Rabani Mogontiacensis Episcopi de computo*, ed. Stevens, 1979.

9 Es muß während des neunten Jahrhunderts einen aktiven Austausch der Mönche zwischen Fulda und der Reichenau gegeben haben. Die Fuldaer Äbte haben oft von einer Reihe von Bischöfen aus Rom Privilegien erbeten, die die Beherrschung ihrer Gesellschaft und Besitze gegen die Ansprüche der Würzburger Bischöfe ermöglichten. Fulda hatte auch einige Geschenke von Grundstücken in Italien bekommen, und deren Aufseher könnten durch ihre Reisen einen guten Empfang auf der Reichenau gefunden haben.

Zuerst schrieb er ein fehlendes Wort an den Rand; vor das Wort setzte er zwei Punkte sowie unser Semikolon, und diese zwei Punkte wiederholte er im Text, wo das Wort hingehörte. War eine ganze Zeile verloren gegangen, so gebrauchte er dafür eine Reihe von vier Zeichen oder *notae*. Die fehlenden Wörter wurden mit einem *Nota*-Zeichen davor an den oberen oder unteren Rand geschrieben, die *nota* interlinear in den Text eingesetzt. Dieser Schreiber verfolgte offenbar sehr erfolgreich den Unterricht seines Reichenauer Lehrers.

St. Gallen 878
Walahfrid
Ms

WII, A. D. 825 - 827

p. 206 (H)

anno ˙ʄ.
unum diem solari adicimus   eamque
lunae mensis februarii tribuamus
(*Computus* LIIII 34 - 35)

ʄ. similiter lunari unam diem
adiciamus   (WII, Rd unten)

p. 16 A (WII)
lectus. lecta. lectum. ✗ Genetivo
lecti. lectae. lecti
(Donatus minor *De participio*)

✗ Participia venientia a verbo
passivo tem(pore...) generis
masculini. feminini. et neutri
numeri singularis. figurae simpli-
cis casus nominativi et vocativi
(... ...)untur sic. nominativo
lectus lecta lectum.
            (WII, Rd unten)

p. 28 (WII)
feminini. ʄ. vel omnis
(Grammatik)

ʄ. vel communi duum vel trium
generum. hic ufens. ufentis. haec
mens. mentis. hic et haec et hoc
prudens. prudentis. In omnes
latina masculini vel feminini.
            (WII, Rd unten)

p. 39 (WII)
eramus. ritis. runt. ☉ Infinitivo
    (runt *corr ex* rint)
(Grammatik)

☉ Et ultima modo, cum auditus
fuero .ris .rit. pluralis cum
auditi fuerimus .ritis .rint
            (WII, Rd oben)

p. 191 (H)
D. Quot minuta. ✳ M. XXVIII.
(*Computus* XXXIII 32 - 33)

✳ M. VII. milia CC. D. Quot
momenta   (WIII, Rd oben)

p. 218 (WI)
post sex ✳ redeant.
(Computus LXXI 15 - 16)

✳ annos futurae sint, quae primo
probis sextum anno sunt eadem et
ante annos XI transierunt et post
sex.   (WIII, Rd unten)

Abbildung 1

Der junge Godescalc ist sicher vor dem Tod Wettis am 3. November 824 in Niederzell auf der Reichenau angekommen; er war von Wettis grotesken bildhaften Visionen des Lebens nach dem Tode geprägt, die wohl beabsichtigt waren, jeden Menschen aus seinen Sünden aufzuschrecken.

Walahfrids nächste Studienzeit dauerte von 827 bis 829 in Fulda, wo 820 der *computus* Hrabans entstanden war. Er schlug das Werk hier nochmals auf, um zwei fehlende Zeilen einzusetzen, und kennzeichnete diese Einschiebsel mit einer Art Sternchen ✕̈ , ein *X* mit zwei Punkten dazu. Die Korrekturen sind erkennbar, nicht nur anhand der Entwicklung von Walahfrids Schrift zwischen Zeitabschnitt I bis III, sondern auch deutlicher durch den Gebrauch einer anderen Feder und Tintenfarbe. Zwei Beispiele werden das verdeutlichen:

1.   Die Handschriftseiten 178 bis 194 wurden von einer Hand geschrieben, die von Bischoff als Reichenauer *Hand H* bezeichnet wurde. Auf Seite 191 im Computus-kapitel XXXIV hatte die Hand H zuerst das Wort *alterutra* geschrieben; doch verstand der Schreiber die Bedeutung nicht. Er nahm an, daß die letzten drei Buchstaben -*tra* eine Suspension für *terra* darstellten. Also änderte er das Wort *alterutra* zu den beiden Wörtern *alterum terra*; der Text wurde dadurch unverständlich. Später setzte er erneut das richtige Wort ein, mit breiterer Feder und mit schwarzer Tinte. Meine eigene Kollation der Handschriften hat gezeigt, daß eine *q nota* am linken Rand der Fuldaer Handschrift steht.

2.   Walahfrid selbst stellte die Abschrift der Seiten 194 bis 240 der Reichenauer Handschrift in der heutigen Bibliothek von St. Gallen her. Auf Seite 210 in Computuskapitel LX liest man einen Satz mit zahlreichen Korrekturen von zwei einfachen Wörtern, die lauten sollten:

antiquissimi romanorum.

Irgendwie müssen die Gedanken des fünfzehnjährigen Jungen abgeschweift sein, so daß er dem Sinn des Textes nicht mehr folgte. So verkannte er das insulare *r* im Fuldaer Exemplar und schrieb mit seiner gewöhnlichen, leicht rotbraunen Reichenauer Tinte:

antiquissimuŕomanorum.

Der Fehler war über dem Buchstaben mit einem winzigen Minuskel *r* notiert. Sowohl das interlineare *r* wie auch ein oberer Strich des fehlerhaften *s* wurden ausradiert. Mit der gleichen Feder und der gleichen Tinte zog er einen neuen Strich, der den Buchstaben *s* in *r* veränderte. Noch ein Strich änderte den vorherigen Buchstaben -*u* in ein -*us*. Das Ergebnis waren die zusammenpassenden Wörter *antiquissimus romanorum*, doch der Satz ließ sich noch nicht gut lesen. Diese Änderungen sind mit Sicherheit auf der Reichenau erfolgt.

Etwas später wurden die fehlerhaften Buchstaben -us ausradiert, um mit einer breiteren Feder und schwarzer Tinte den Buchstaben -i einzusetzen, damit der Text *antiquissimi romanorum* sich schließlich in der Handschrift Walahfrids richtig lesen ließ. Meine Kollation der Texte ergab noch eine *q nota* am Rand der Fuldaer Handschrift.

Fulda HS	Walahfrid HS	Hrabanus De computo
f. 19ᵛ	p. 191 [H]	XXXIV 18 - 22 Quibus aeque qualitatibus disparibus quidem per se sed alterutra ad invicem...aer humidus et calidus, ignis [...] calidus et siccus.
f. 33ᵛ	p. 210 [WI]	LX 14 ... antiquissimusomanorum antiquissimu omanorum antiquissimuʼromanorum antiquissimi romanorum
f. 39ᵛ 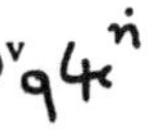	p. 219 [WI]	LXXIII 2 ...ad epactas solis

Abbildung 2

Der Duktus der späteren Korrekturen zusammen mit der schwarzen Tinte zeigt sich sehr oft nicht nur in der zweiten Reihe von Walahfrids Korrekturen in der Reichenauer Handschrift, sondern auch in zahlreichen Korrekturen des Computustextes in der Fuldaer Handschrift. Walahfrid ist 827 nach Fulda gefahren, um dort zwei Jahre lang zu studieren. Könnte die zweite Stufe von Korrekturen seines *vademecum* in Fulda gemacht worden sein?

Die Oxforder Handschrift Canonici miscellaneous 353 ist merkwürdig, weil sie den Text von Hrabans *computus* enthält, wie dieser im Skriptorium des Autors geschrieben wurde; die Handschrift ist die zweitälteste Kopie. Der Fuldaer Schreibmeister schrieb zunächst zwei Zeilen auf dem ersten Folio jedes Quaternio selbst. Die erste Lage übertrug er einem erwachsenen Fuldaer Schreiber, um ihn weiter schreiben zu lassen. Den zweiten Quaternio gab er an eine Gruppe von Novizen, die er den Fuldastil schreiben lehrte. Die dritte Lage vertraute er wieder einem Schreiber an, der eine reife, noch besser entwickelte Schrift zeigt und so weiter[10].

---

10   Der Lagenaufbau dieser Fuldaer Handschrift und die Charakteristika jeder Hand sind von Stevens in "Fulda scribes at work" und "A ninth century manuscript from Fulda" ausführlich beschrieben worden.

Verschiedene Zeichen oder *signes de renvoi* wurden für kleine Korrekturen gebraucht. Gewöhnliche Textfehler von Buchstaben oder von grammatischen Formen aber waren auf fol. 3 bis 40 und fol. 54. notiert. Die Schreiber markierten diese Fehler am Rand mit einem Griffel ohne Tinte und mit einer fast unsichtbaren, schräg verkürzten Linie ∕ sowie mit zwei oder drei gekreuzten Linien ✕ ✕ . Dabei wurden in den meisten Texten die angemessenen Korrekturen in der Nähe dieser Zeichen eingetragen. Sobald man sie einmal entdeckt hat, sind solche Linien mit dem bloßen Augen unschwer zu erkennen, doch können sie auf Mikrofilmen oder mit anderen Arten der Photographie kaum sichtbar gemacht werden[11].

Noch wichtigere Korrekturen wurden durch eine Reihe besonderer Zeichen dargestellt, die etwas verschieden sind. Zwei von ihnen finden sich in Walahfrids Handschrift; beide sind interessant, besonders

      das Antissima   ⊃   mit einem Punkt auf Folio 7, und

      das Sternchen   ✕   mit vier Punkten auf Folio 36.

Jetzt aber lenke ich zuerst Ihre Aufmerksamkeit auf Folio 13v, wo eine Art von tironischen *notae* zeigt, daß ein zweimal geschriebenes Wort expungiert werden sollte; das Wort wurde nicht ausradiert, sondern mit Tinte umkreist. Eine zweite tironische Note ist genauer zu erfassen. Sie steht am linken Rand von fol. 39v neben den Wörtern *ad epactas solis*: ꝗ und bedeutet *quaere epacta* (Epakte ist ein Ausdruck der Kalenderwissenschaft)[12]. Mit derselben Feder und der leicht rotbraunen Tinte sind zwei Abbreviaturen am Rand von fol. 26 und 27 eingetragen: *ſt* und *ſnt* mit Strichen darüber; sie bedeuten *interrogandum est*. Man soll hier etwas nachfragen! Gemeint sind Fragen über die Charakteristiken der Planeten, die man identifizieren kann, und über die Position von Sternen in bestimmten Konstellationen außerhalb des zodiakalen Bandes, die man erkennen konnte.

Wollte ein Schreiber eine Handschrift korrigieren, mußte er auch die Interpunktion verbessern. An den Rändern der beiden Folien 9 und 11 steht ein hohes *ả* mit einem Strich

---

11   Bischoff, "Über Einritzungen in Handschriften des frühen Mittelalters", in: *Mittelalterliche Studien* I, pp. 88-92, ohne diese Fuldaer Handschrift zu erwähnen. Die Einritzungen in Handschriften brauchen noch weitere Analysen.

12   Für das Entziffern von Tironischen Noten danke ich Professor Dr. Bernhard Bischoff (gestorben 1991) in seinem Brief vom 27. Dezember 1971. Siehe auch Kopp, *Paleografica critica* II, pp. 8, 125, 368-370; Bischoff, *Paläographie*, pp. 103-105 (engl. Ausgabe: pp. 80-82). Die Tironischen Noten lernte man in der Schule: Steffens, *Lateinische Paläographie*, pp. xxxi-xxxiii.

darüber; es bedeutet *distinctum est.* So gibt es viele Arten von *notae* an den Rändern von diesem *computus* in der Fuldaer Handschrift, die bestimmte Funktionen erfüllen[13].

Dieselbe Feder und die leicht rotbraune Tinte wurden gebraucht, um noch eine andere *nota* zu machen, die am Rand dieser Handschrift achtzehnmal wiederholt wird: ein verlängertes *q* , das ich schon zweimal erwähnt habe. In allen achtzehn Fällen erweist es sich, daß die *notae* von der Hand desselben Schreibers eingetragen sind. Verschiedene Erklärungen werden für den Gebrauch der *q notae* am Rande eines Textes vorgeschlagen. Einem Autor zufolge muß die *q nota* als Gewohnheit der Karolingerzeit und späterer Zeiten immer als *quaere* aufgelöst werden. Das heißt: jeder soll merken, daß etwas zu korrigieren ist, oder der Leser soll über die Worte nachdenken. Man weiß aber, daß Servatus Lupus in dieser Zeit *r* oder *rq require* am Rande gebraucht hat, um unzuverlässige Wörter und Phrasen in Texten zu notieren. Nach anderen Autoren konnten die *q notae* später die Quellen eines Traktats bestimmen[14]. Es dürfte eine Vereinfachung der stenographischen *nota* ᓚ sein, um r*equire* oder *requirendum est* anzuzeigen. Aber an den Rändern der Fuldaer Handschrift weisen die langen und eleganten *q notae* keine wichtigen Auslassungen auf. Es zeigt sich, daß keine Fehler irgendeiner Art in den nebenstehenden Texten aufzufinden sind, die Texte sind normalerweise perfekt. Die *q notae* beziehen sich auch nie auf eine Quelle von Hrabans Werk. Möglich ist, daß die Passagen etwas anzeigten, was ein Lehrer bemerkenswert fand. Ein Leser könnte bestimmte Passagen betonen wollen; oder ein Schreiber wollte ein oder zwei als wichtige Exzerpte aufnehmen, um sie in seinem eigenen Traktat zu verwenden. Welches Interesse aber, welchen Zweck können diese Passagen gehabt haben? Die Fälle von *q notae* in der Fuldaer Handschrift sind zu verstreut, sie sind zu willkürlich, um eine logisch befriedigende Erklärung für sie zu finden.

Nachdem die überraschende Beziehung der mehrfach korrigierten Wörter *antiquissimi romanorum* in Walahfrids Handschrift und einer *q nota* am Rand der Fuldaer Handschrift bemerkt ist, muß man die Sache weiter verfolgen. Wo die *q nota* in der Fuldaer Handschrift stand, habe ich achtzehnmal den Text ausgeschrieben. Dazu habe ich den Text paralleler Zitate aus Walahfrids Handschrift gestellt. Es gibt zwei Fälle, die man nicht überprüfen

---

13 Paul Lehmann hat einige Notizen zu einem Auftrage Ludwig Traubes, "Untersuchungen zur Überlieferungsgeschichte römischer Schriftsteller", 1891, hinzugefügt: Traube, *Vorlesungen und Untersuchungen*, Band III, p. 15, in der er die Codex ms Vat. Lat. 474 (s. IX), f.95 notiert: Hucusque ab abbate et praeceptore lupo requisitum et distinctum est. Rafti, "L'interpunzione nel libro mano-scritto", pp. 239-298. Diese *ft* und *d* notae sind nicht bemerkt worden von Rand, *A Survey of the Script of Tours* I, pp. 24-25; nicht von Lowe, "The oldest omission signs", pp. 349-380.

14 Pellegrin, "Les manuscrits de Loup de Ferrières", pp. 20-21; Wieland, "Medieval MSS: an additional expansion for the abbreviation 'Q'", pp. 19-26; Mundó, "L'escriptura i la codicologia", pp. 99-103; Mostert, "Marginalia in Handschriften: een onderschatte bron", pp. 102-107.

kann, weil ein Quaternio der Handschrift Walahfrids verloren ist. Man kann jedoch die Textvarianten in den sechzehn Fällen vergleichen, die in den beiden Handschriften vorkommen. In neun von diesen hat entweder der fünfzehnjährige Walahfrid oder die Hand H auf der Reichenau einen Fehler gemacht, später wurde eine Korrektur in Fulda von Walahfrid angefertigt, die den Sinn des Textes geändert oder erklärt hat. Es gibt auch sieben Fälle, worin die Fuldaer Handschrift fehlerhafte Varianten enthielt, welche aber der junge Walahfrid nicht akzeptierte. Dreimal wurde die Fuldaer Handschrift korrigiert; viermal erfolgte keine Korrektur, weder in der einen noch in der anderen Handschrift. Offenbar hatte Walahfrid die Absicht, einen genau geschriebenen Text in seinem persönlichen Vademecum zu haben, und er hat ihn durch die Kollation seiner und der Fuldaer Kopie mit Hilfe eines anderen Schreibers in Fulda erhalten.

Das heißt: die *q nota* wurde in der Fuldaer Handschrift so gebraucht, wie sie auch am Ende des Quaternions in einigen Handschriften aus St. Martins in Tours benutzt wurde[15]. Sie sollte nicht *quaere* sondern *requisitum est* bedeuten, "kollationiert nach dem Exemplar".

Wenn es stimmt, daß in der Fuldaer Handschrift die *q nota requisitum est* bedeutet, hat das für uns zwei wichtige Konsequenzen. Erstens sind die beiden frühesten noch vorhandenen Handschriften von Hrabans *Liber de computo* oft und sorgfältig korrigiert, aber auch miteinander kollationiert worden. Einige Korrekturen wurden wahrscheinlich gemacht, nachdem man einen Schreibmeister in Fulda gefragt hat, der wohl der Autor Hraban selbst sein könnte. Zweitens erfolgte die Kollation bestimmt während Walahfrids kurzem Aufenthalt in Fulda von 827 bis Frühling 829. Dadurch ist ein neuer *terminus post quem non* für das Schreiben der Oxforder Handschrift (Canonici miscellaneous 353) bestimmt. Sie muß demnach zwischen 820 und 829 geschrieben worden sein. Sie wird eine der seltenen Fuldaer Handschriften sein, die innerhalb eines so engen Zeitabschnittes datiert werden kann. Die Hoffnung ist berechtigt, daß die verstreuten Bücher eines so wichtigen Skriptoriums eines Tages in ihrer chronologischen Abfolge beschrieben werden können.

Ich möchte Ihnen nicht nur eine datierte Handschrift und den Nachweis eines sorgfältig korrigierten Textes hinterlassen, sondern auch noch etwas über den jungen Walahfrid Strabo sagen. Er wird oft als Poet, Liturgiker und Bibelkommentator erwähnt. Eine solche literarische Einengung ist jedoch eine einseitige Ansicht, die nicht mit seinem persönlichen Vademecum und mit den vielen naturwissenschaftlichen Themen im Einklang steht. In seiner

---

15 Die ersten Belege aus Tours, in denen die *q* nota so gebraucht wird, sind ms Paris BN N. a. lat. 1575 [St. Martin 50], (s. VIII l), und ms Épinal B. mun. 149(68) (A. D. 744/745): *Codices latini antiquiores*, V: 682 und VI: 762; Lowe hat gesagt, *the signatures are often accompanied by entries in Notae Tironianae, attesting the collation of the quire, the longest of which reads: requisitum est ab Arice abbate* (Épinal f.70, Faksimile).

Handschrift finden sich zwar einige Zeilen des Dichters Horaz und auch andere Verse, aber nur zwei Seiten von liturgischem Interesse, und nichts über die Bibel. Demgegenüber stehen viele sehr lange Texte über grammatische und komputistische Fragen, in etwa gleichem Verhältnis. Die oben genannten Korrekturen, die für die maßgebende Orthographie oder richtige Interpunktion sorgten, weitere für Verbformen und Kasusbildung und einige für das Hinzuschreiben ausgefallener Zeilen, werfen gelegentlich auch verschiedene Fragen von wissenschaftlicher Bedeutung auf. Man fragt: Was sind Epakten? Wie gebrauchen wir die Indiktionen? Wie und wann wird der *saltus lunae* gerechnet? Wie stehen die Himmelskörper zueinander, besonders die Planeten und die Sternbilder innerhalb des Zodiakus, aber auch die hellsten Sterne und die Konstellationen außerhalb des Zodiakus? Die ersten Fragen waren natürlich nötig, um das Bedaische System der Zeitrechnung zu verstehen; die anderen Fragen zu den Sternen waren eigentlich nicht nötig, obwohl alle Komputisten sie stellten. Die Komputisten forschten immer nach den Dingen des Kosmos, obwohl das keine praktische Bedeutung hatte, und auch keine Hilfe bei der Bestimmung des Ostersonntags leistete. Wir sollten daher nicht überrascht sein, wenn sich ergibt, daß die beiden Handschriften das Studium des Kosmos durch weitere Texte über den Himmel, die Sterne, und die Konstellationen unterstützten.

Während seines ganzen Lebens hatte Walahfrid ein lebhaftes Interesse am *computus*[16]. 825 hatte er Hrabans Traktat abgeschrieben; zwischen 827 und 829 kopierte er auch Bedas *Liber de natura rerum* und Bedas *Liber de temporibus* mit Hilfe zweier weiterer Schreiber in Fulda. Weitere komputistische *argumenta* wurden zuerst in Fulda, danach in Weißenburg, und später wohl zwischen 830 und 838 am Aachener Hof von Ludwig und Judith aufgenommen. Ein Problem ist in diesen Texten verborgen, das Walahfrid gefunden und angezeigt hat. Als er um 825 auf der Reichenau Hrabans Kapitel LXV *De anno mundi* in sein Vademecum [p. 206: WI] kopierte, hatte er bemerkt, daß es zwei Methoden gab, die Jahre vom Anfang der Schöpfung bis zur Geburt Christi zu berechnen:

---

16   Walahfrids Ausbildung in den Naturwissenschaften, die aus seinen Briefen hervorgeht, ist nur von Rudolf Wolf erklärt: *Nicht nur wurde er in Grammatik und Dialektik unterrichtet, sondern Strabo erzählt, daß er im Sommer 822 unter Leitung von Tatto das Studium der Arithmetik nach Boethius begonnen habe; dann habe er das Rechnen mit den Fingern und den Gebrauch des Abacus gelernt, nachher die Zeiteinteilungen der Hebräer, Griechen und Römer, sowie die Berechnung des Kalenders. Im Jahre 825, beim Schlusse der vorbereitenden Studien, hörte er bei Tatto auch Astronomie. Derselbe erklärte den Grundriß des Boethius, die Schriften Bedas über Sonnen-, Mond- und Planetenlauf, lehrte die Sternbilder, den Tierkreis, die Ursachen der Finsternisse, den Gebrauch des Astrolabs und Horoskops, der Sonnenuhr und des Tubus kennen* (Geschichte der Astronomie, p. 76). Boethius und Beda haben über den Astrolab nicht diskutiert, doch konnte Wolf noch etwas gut erklären: *Ein Tubu: Eine bloße Röhre ohne Gläser, die entweder überhaupt als Surrogat der Diopter oder speziell zur Orientierung nach dem Polarstern benutzt wurde.* Siehe noch dazu: Stevens, "Walahfrid Strabo", pp. 9 - 16.

1) *secundum Hebraicam veritatem* anni III milia DCCCCLVI

2) *secundum Septuaginta interpretes* anni V milia CXCVIIII.

Die beiden Methoden waren von Hraban aus den Werken Bedas genommen. Doch hat Hraban in Kapitel XCVI nach hebräischer Wahrheit die Jahre bis zur Inkarnation mit 3952 berechnet, nicht mit 3956.

Die Summe 3956 ist in Hrabans Kapital LXV falsch gerechnet. Darin hat Hraban seine eigene Rechnung gemacht, aber sie beruht auf einer unrichtigen Jahreszahl des *Secunda aetas*, die aus einer Kopie von Bedas *Liber de temporibus* XVI. 6 stammt, wahrscheinlich aus der Handschrift Karlsruhe Reichenau 167, f. 21 - 22, wo man die Jahre CCXCVI statt CCXCII lesen muß. Die Quelle Hrabans für sein Kapitel LXVI aber muß Beda, *De temporum ratione*, Kap. LXVI. 1 - 268 sein, das ihm den Verlauf der Zeit vom Anfang der Schöpfung bis zur Geburt Christi anni *III milia DCCCCLII* in allen Handschriften richtig gegeben hat. Walahfrid hat den Unterschied zwischen den Resultaten in den beiden Kapiteln festgestellt[17].

In Aachen hat Walahfrid einige Angaben gefunden, die die Verschiedenheit der Berechnungen noch mehr hervortreten ließen. Es gab einen *Liber Albini magistri*, den der berühmte Lehrer Alcuin aus York nach Aachen mitgebracht hatte[18]. Walahfrid hat das Buch Alcuins gefunden, und in der letzten Zeile auf Seite 277 seines Vademecums die Rubrik *Excerptum de libro Albini magistri* gemalt. Auf den nächsten Seiten 278 - 283 findet sich eine kurze Chronik mit der frühen Hand WIV abgeschrieben: *Adam cum* [esset] *CXX annorum genuit Seth, Seth autem habens annos CV genuit....* Es steht aber in Frage, was der *Liber Albini magistri* enthalten hatte. In drei Handschriften findet man viele Materialen über das Zeitrechnen, die aus dem Buch Alcuins genommen worden sein könnten:

1. Vat. Pal. lat. 1448 (s. IX in), f. 1 - 44 im Monasterium St. Maximin (Trier) geschrieben, f. 45 - 59 und 60 - 69 im Monasterium St. Alban (Mainz), f. 70 - 79 wohl in Mainz[19].

---

17 Viele andere magistri, die Hrabans *computus* benutzt haben, haben das Problem des *annus mundi* bemerkt. In Kapitel LXV waren die Ziffern oftmals korrigiert. Alle Varianten sind in der Annotation der neuen Edition zu finden, *Rabani de computo*, hrsg. von Stevens, 1979.

18 Jones, *Bedae Opera*, p. 110 hat gesagt, daß mehrere Stücke des *Libellus Annalis* in erster Linie insular sind und wohl aus dem Irland des siebten Jahrhunderts kommen könnten. Er vermutete, daß mehrere Materialien aus einer Kopie von Beda *De temporum ratione*, die einige Zusätze über *annus praesens DCCLVIII* enthalten haben, bevor sie Alkuin von York aus auf das Festland brachte.

19 Codex ms Vat. Pal. lat. 1448 (s. IX in), f. 71v: Ostertafel anni 779-797; f. 72: Argumenta paschalis DXXLXXX; f. 73: Argumenta über Finsternisse; f. 73v-74: Calculatio quomodo ... [annus

2. Vat. Pal. lat 1449 (s. IX 1) war für die Verwendung der Schule St. Nazarius (Lorsch) geschrieben; auf f. 2v findet sich der monumentale Titel dargestellt:

*IN CHRISTI NOMINE
INCIPIT LIBELLUS
ANNALIS BEDAE
PRESBITERI
FELICITER*

und auf f. 72 sind die zwölf Verse an Karl von Alcuin, die als Widmung für sein *Liber Albini magistri* dienen sollen:

*Ut praecepta mihi dederas, dulcissime domine,
Sic celeri currens calamo dictare libellum
Annalem, veterum simul argumenta sophorum,/...*

*Qua Christum laeti cernatis perpete visu,
Posco; tuum memorans Flaccum, sine fine valeto*[20].

3. Vat. Pal. lat. 1447 (s. IX 1), f. 1 - 32, die wohl in Mainz geschrieben wurde[21], enthält manche derselben Texte; auf f. 19 ist auch ein *LIBELLUS ANNALIS* zitiert. In jedem Codex finden sich einige Daten 777, 780, 779 - 797, die als *anni presentes* in den argumenta benutzt worden sind. Die *Calculatio Albini magistri*, die Alcuin aus Northumbrien auf das Festland mitgebracht haben muß, ist jedesmal darin zu finden. Die erwartete Chronik und die fünf argumenta, die von Walahfrid ausgewählt und abgeschrieben waren, werden dennoch darin vergebens gesucht.

Doch gab es in der Aachener Zeit Walahfrids noch ein anderes Kompendium von Zeitrechnung, in dem er seine Chronik finden konnte. Diese Chronik, "Adam cum esset ...", ist aus Northumbrien nach Aachen gebracht worden. Der ursprüngliche Text stammt aus dem fünften Jahrhundert, bevor er die northumbrischen Einträge bis zum Jahr 793 bekommen hat; danach sind alle Einträge fränkisch. So wie Walahfrid diese Chronik gefunden hat, reichen die Jahre bis *anno Karoli XLII* oder *annus domini DCCCVIIII*, ein Jahr das auch als *annus mundi IIII mille DCCLXI* ausgedrückt werden kann, welches nur

---

praesens DCCLXXVI]; f. 74-74v: Verse Alcuins: Haec dilecte comes propriis...; f. 74v-75v: Geographia; f. 76-79: Chronik: Adam pater generis humani dei manibus ex terra creatus vixit annis nongentis triginta...; Lindsay und Lehmann, "The early Mayence Scriptorium", pp. 22-26.

20 Am Hof wurde Alcuin als Horatius oder Flaccus oder Albinus oft genannt. Die Verse sind von Dümmler herausgegeben worden in: *MGH. Poetae Latini* I, pp. 294-295. Siehe auch Bischoff, "Die Hofbibliothek Karls des Großen", p. 45; idem, *Lorsch im Spiegel seiner Handschriften*, pp. 37-43, besonders p. 43, und p. 81 Anm. 5; *Initia carminum* Nr. 16886. Dieselben Verse Alcuins sind in ms Vat. Pal. lat. 1448, f. 72 später in s. IX/X zugefügt worden.

21 Hanselmann, "Der Codex Vat. Pal. lat. 289. Ein Beitrag zum Mainzer Skriptorium im 9. Jahrhundert", pp. 78-87, dazu eine *Liste der m. E. mit Sicherheit für Mainz in Anspruch zu nehmenden Handschriften des 9. Jahrhunderts.*

dann möglich wäre, wenn man eine Zählung 3952 für das Jahr der Inkarnation ansetzte, nicht 3956. Auf diese Weise konnte Walahfrid zu dem Schluß kommen, daß die Jahre Karls und die Jahre Ludwigs, Jahre des Vaters und des Sohnes, nicht nach demselben Maßstab gerechnet wurden. So konnte er entdecken, daß man mehr als eine Methode hatte, das laufende Jahr aufgrund der Geburt Jesu festzustellen, wenn das letztere Datum nicht festgelegt war. Walahfrid fügte eine Zeile aus dem Text der Chronik ein, die auf der Seite 278 (WIV) ausgelassen worden war, und er kennzeichnete sie mit dem *antissima* Ɔ

   In der Chronik wird das Jahr 809 hervorgehoben, ein Jahr das man nur richtig verstehen kann, wenn man die Rechenmethode gebraucht, die Dionysius Exiguus um 525 zusammengefaßt, und die Beda um 725 erweitert hat[22]. Es ist die Jahreszahl eines Ereignisses von Bedeutsamkeit in der Geschichte der Naturwissenschaft: eine Konferenz über Kalenderprobleme[23]. Die Anweisung dafür kam von Karl dem Großen; das Verzeichnis der gestellten Fragen und gegebenen Antworten existiert noch. Es wurde von dem Konferenzpräsidium unter Adalard von Corbie herausgegeben. Unter den Resultaten sind zwei sehr große Sammlungen von Arithmetik, Computus, und Astronomie; in beiden Sammlungen befinden sich einige Kleinigkeiten zu Theologie und Geschichte, die sich auf das Zeitrechnen durch das Jahr und durch viele Jahre beziehen: komputistische Traktate, astronomische Exzerpte aus Plinius und interessante Abbildungen von wissenschaftlichen Instrumenten, die im Gebrauch waren. Die beiden Sammlungen sind ein Drei-Bücher-*computus* und ein Sieben-Bücher-*computus* genannt worden[24].

   Aus dem Sieben-Bücher-*computus* hat Walahfrid ein Horologium genommen, das die Stunden des Tageslichtes zu jeder Stunde aller Monate angab (pp. 301 - 302, frühe Hand WIV). Auf dem 51. nördlichen Breitengrad funktioniert es ordentlich für Aachen

---

22   Jones, *Bedae Opera*, pp. 6-139, besonders pp. 68-75: Dionysius Exiguus, und pp. 105-139: Beda.

23   *MGH. Epistolae* IV, pp. 565-567; Jones, "An early medieval licensing examination", pp. 19-29, mit Übersetzung ins Englische; der Titel paßt nicht zum Inhalt. Fünf Kopien existieren noch. Bemerkungen über diese und andere komputistische Encyklopädien sind bei Cordoliani, "Une encyclopédie carolingienne de comput: les Sententiae in laude compoti", pp. 237-243, und Bischoff, "Die Bibliothek im Dienste der Schule", pp. 228-229, zu finden.

24   *Compilatio computistica et astronomica A. D. DCCCVIIII;* der Drei-Bücher-Computus ist in zwei Handschriften gefunden: München Staatsbibl. CLM 210 (A. D. 816-818), Wien, ÖNB Lat. 387 (A. D. 816-820); der Sieben-Bücher-Computus ist in sechs Handschriften gefunden: Madrid BN 3307 [L. 95] (A. D. 820-840), f. 5-76v, Monza Bibl. Communale f - 9/176 (A. D. 869), f. 7-92v, Paris BN Lat. 12117 (A. D. 1031-1060), f. 168-183, N. a. lat. 456 (s. IX/X), f. 10-37, Vat. Lat. 645 (s. IX 2), f. 12-92v, Reg. lat. 309 (s. IX 2), f. 4v-120. Mehrere Teile der Sammlungen werden auch in anderen Handschriften ausgeschrieben. Arno Borst, "Die Enzyklopädie von 809", wird die beiden Haupthandschriften hier besprechen. Siehe auch die unveröffentlichte Arbeit von King, "An investigation...".

oder Köln, Fulda oder auch Winnipeg, aber nicht für die Reichenau oder Salzburg auf dem 47,5. Grad. Daraus hat er auch die Daten von sechs Finsternissen notiert, die von 786 bis 812 in der Gegend von Aachen und St. Amand zu sehen waren (p. 351, frühe Hand WIV). Es war auch der Sieben-Bücher-*Computus*, aus dem Walahfrid acht Seiten von komputistischen *argumenta* genommen hatte, wovon die drei ersten *argumenta* drei Systeme erklären, um das *annus mundi* zu finden (pp. 284 - 292, frühe Hand WIV)[25]. Jetzt hatte Walahfrid vier Daten für die Inkarnation Jesu: *annus mundi* 3952, 3956, 5199, und 5500. Er fügte noch ein *argumentum* hinzu, das die Indiktionen des Kaiserhofes berechnete, und er korrigierte den Text; nach dem Bedaischen *annus mundi* 3952 der Inkarnation würde es mit dem in Aachen damals gültigen *annus mundi* 4761 im Einklang stehen, das heißt *annus Domini* 809.

Verzeihen Sie mir bitte alle diese Ziffern und Summen. Doch sind sie die Termini, die die karolingischen Schulmeister und Studenten gebrauchen mußten, um zu verstehen, was ein Datum im Verhältnis zu anderen Daten wirklich bedeutete.

Jetzt darf ich die Ergebnisse der paläographischen Analyse zusammenfassen. Aus dem Vergleich der Inhalte und der Schriften der beiden Handschriften habe ich versucht, zwei junge Schreiber bei der Arbeit in ihren jeweiligen Skriptorien auf der Reichenau und in Fulda zu beschreiben, wie jeder schrieb, etwas notierte, und Korrekturen machte. Nach zwei oder drei Jahren arbeiteten die Jungen auch in Fulda zusammen. Dort pflegte der eine Schreiber aus einem Fuldaischen Exemplar des *computus* seines Abts laut vorzulesen, während der zweite denselben Text in seiner eigenen Reichenauer Kopie korrigierte. Der zweite Junge hieß Walahfrid Strabo; wer ist der erste?

Wohl ist der *computus* Hrabani (820) durch Godescalc aus Fulda nach der Reichenau um 824 mitgebracht worden[26]. Dort hat er die melodischen Verse *Ut quid iubes pusiole/...* nicht nur über die Freiheit von den Sünden geschrieben sondern auch um Freiheit von der Verbannung gesucht, das heißt, er klagte über die zwei Jahre, die er auf der Reichenau verbracht hatte, als eine Verbannung auf einer Insel im Meere:

---

25 Noch fünf argumenta in St. Gallen 878, pp. 344-347 (frühe Hand WIV), hat Walahfrid aus dem Sieben-Bücher-Computus genommen.

26 Der Name Cotescalc [Godescalc] war unter Einträgen lebender Mönche um 826-830 im Reichenauer Verbrüderungsbuch, ms Zürich, Zentralbibliothek Rh. hist. 27, erwähnt; Freise, "Zur Datierung", pp. 537-39. Für diese angeführte Stelle danke ich Herrn Privatdozent Dr. Freise, Universität Münster.

> *Exul ego   diuscule*
> *hoc in mari   sum, domine*
> *annos nempe   duos fere*
> *nosti fore,   sed iam iamque*
>   *miserere*
> *Hoc rogo humillime...,*

genau wie Walahfrid damals über die schwere Schulzeit nach dem Tod Wettis unter dem neu eingesetzten Lehrmeister Tatto geklagt hatte[27]. Godescalc hat seinen Freund als *pusiolus, filiolus, miserulus, puerulus, pusillulus,* und *consodalis* angeredet, ebenso als *tiro* und *cliens.* In dieser Zeit hat man sich nie mit solchen Ausdrücken an Christus im Gebet gewendet, bestimmt nicht Godescalc. Statt dessen hört man den jungen Godescalc sich selbst wie auch seinen noch jüngeren Freund Walahfrid als *frater, fraterculus* anreden.

Bald sind die beiden *oblati pueri* 827 - 29 in Fulda wieder zusammen. Schließlich ist Godescalc in seinen Versuch gegen den Abt Hraban erfolgreich gewesen, sich aus dem monastischen Leben zurückzuziehen[28]. Danach wissen wir wenig über Begegnungen zwischen Godescalc und Walahfrid, obwohl es sehr wahrscheinlich ist, daß Walahfrid während seiner Zeit am Hof versucht hat, dem oft beschuldigten Godescalc zu helfen. Professor Bischoff hat über die Freundschaft gesagt: *Immerhin muß daran erinnert werden, daß der eine Reichenauer Mitschüler und Freund, der Gottschalk zeitlebens der Treue hielt, der Dichter Walahfrid war*[29].

Diese Reihe von Lebenserfahrungen reicht nicht als Beweis aus, daß die Eintragungen *fn, d,* und *q notae* in der Fuldaer Handschrift sicher von Godescalc geschrieben worden sind. Doch ist die Möglichkeit attraktiv. Godescalc war auf der Reichenau gewöhnlich mit Walahfrid zusammen in der Mensa und im Schlafsaal, in der Schule, in der Bibliothek, und im Skriptorium, sowie bestimmt in Fulda. Er war bei Walahfrids zweitem Reichenauer Schreibabschnitt und dem drittem Fuldaer Schreibabschnitt immer anwesend. Der erste Schreiber unserer Analyse dürfte Godescalc heißen, der auch ein junger Student und Poet zur selben Zeit in denselben Klöstern war.

---

27   Godescalc hat "de magistro meo Uuectino" gesprochen, Lambot, *Oeuvres,* p. 170; Bischoff, "Gottschalks Lied für den Reichenauer Freund", *Mittelalterliche Studien* II, pp. 32-34.

28   So hat Erzbischof Otgar auf der Synode von Mainz 829 gesagt, ohne daß der *puer oblatus* sein Erbe mitnehmen konnte, welches die Eltern dem hl. Bonifatius geschenkt hatten; von Hefele, *Conciliengeschichte,* pp. 75, 139. Später hat Hinkmar von Reims das Schlußwort. Zu dieser traurigen Geschichte siehe Vielhaber, *Gottschalk der Sachse*; Freise, "Zur Datierung", pp. 537-538; und die von Freise ausführlichen "Studien zum Einzugsbereich der Klostergemeinschaft von Fulda", pp. 1003-1269, besonders pp. 1021-1029 über Godescalc. Ein Zitat aus den Versen Alcuins bei Godescalc ist von Lambot, *Oeuvres,* 1945, p. 426, notiert worden.

29   Bischoff, "Gottschalks Lied", p. 33.

Walahfrids Kamerad notierte Probleme in dem Fuldaer Exemplar von Hrabans *computus* mit einer Reihe von *q notae*, als die Interpretation des Textes in Frage gestellt wurde. Wenn doch eine Korrektur des Textes für nötig gehalten wurde, hat Walahfrid sie allein in seiner eigenen Handschrift gemacht. Die Freunde arbeiteten damals an einem komputistischen Text, der nur einer von vielen Texten in einem *vademecum* ist, das fortlaufend durch weitere Texte ergänzt wurde.

Zeitrechnen ist ein Thema, mit dem sich der Poet Walahfrid sein ganzes Leben lang beschäftigte, wie man aus seiner Kollation sehen kann: das heißt, *requisitum est.*

BIBLIOGRAPHIE

Bischoff, B., "Die Bibliothek im Dienste der Schule", in: *Settimane di Studio del Centro Italiano di Studi sull' alto Medioevo* XIX, Spoleto 1972, pp. 385-415; nachgedruckt in: *Mittelalterliche Studien* III (1981), pp. 213-233.

----, "Eine Sammelhandschrift Walahfrid Strabos (Cod. Sangall. 878)", in: *Aus der Welt des Buches. Festschrift Georg Leyh*. Beiheft zum *Zentralblatt für Bibliothekswesen* LXXV, Leipzig 1950, pp. 30-48; nachgedruckt und ergänzt in: *Mittelalterliche Studien* II (1967), pp. 34-51, mit vier Abbildungen.

----, "Gottschalks Lied für den Reichenauer Freund", in: *Medium Aevum Vivum. Festschrift für Walther Bulst*, Heidelberg 1960, pp. 61-68; nachgedruckt in: *Mittelalterliche Studien* II (1967), pp. 26-34.

----, "Die Hofbibliothek Karls des Großen", in: *Karl der Große* II, Düsseldorf 1965, pp. 42-62, mit sechs Abbildungen; nachgedruckt in: *Mittelalterliche Studien* III (1981), pp. 149-169, Tafel V bis X.

----, *Lorsch im Spiegel seiner Handschriften*, München 1974.

----, *Mittelalterliche Studien* I, (Stuttgart 1966); II (1967); III (1981).

----, *Paläographie des römischen Altertums und des abendländischen Mittelalters*, Berlin 1979; neubearbeitet, vermehrt, und übersetzt von Dáibhí Ó Cróinín und David Ganz: *Latin Palaeography. Antiquity and the Middle Ages*, Cambridge 1990, (enthält 23 Tafeln mit Handschriften-Facsimilia).

----, "Über Einritzungen in Handschriften des frühen Mittelalters", in: *Zentralblatt für Bibliothekswesen* 54 (1937), pp. 173-177; nachgedruckt und erweitert in: *Mittelalterliche Studien* I (1966), pp. 88-92.

Codices latini antiquiores, hrsg. E. A. Lowe, V: Oxford 1950 und VI: 1953.

Cordoliani, A., "Une encyclopédie carolingienne de comput: les Sententiae in laude compoti", in: *Bibliothèque de l'Ecole des Chartes* 104 (1943), pp. 237-243.

Cox, H. O., *Catalogi codicum manuscriptorum Bibliothecae Bodleianae*, Pars tertia, Oxford 1854.

Die deutsche Literatur des Mittelalters, Verfasserlexikon, ed. W. Stammler und K. Langosch, Berlin IV (1953), V (1955).

Duckett, E. S., "Walahfrid Strabo of Reichenau", in: *Carolingian Portraits*, Ann Arbor 1962, pp. 121-160.

Dümmler, E., "Gedichte Alcuins an Karl den Großen", in: *Zeitschrift für Deutsches Altertum* XXI (1877)

Freise, E., "Studien zum Einzugsbereich der Klostergemeinschaft von Fulda", in: *Die Klostergemeinschaft von Fulda im früheren Mittelalter*, hrsg. K. Schmid et alii, Band 2.3, München 1980, pp. 1003-1269.

----, "Zur Datierung und Einordnung fuldischer Namengruppen und Gedenkeinträge", in: *ibid*. Band 2.2 (1978), pp. 526-570.

Hanselmann, J. H., "Der Codex Vat. Pal. lat. 289. Ein Beitrag zum Mainzer Skriptorium im 9. Jahrhundert", in: *Scriptorium* 41 (1987), pp. 78-87.

380

Hefele, C. J. von, *Conciliengeschichte*, Freiburg 1855-1874; von H. Leclercq übersetzt und erweitert: *Histoire des conciles* IV/1, Paris 1911.

Initia carminum Latinorum saeculo undecimo antiquiorum, bearbeitet von D. Schaller und E. Könsgen mit J. Tagliabue, Göttingen 1977.

Jones, C. W., *Bedae Opera de temporibus*, Cambridge, Mass. 1943.

----, "An early medieval licensing examination", in: *History of Education Quarterly* 3 (1963), pp. 19-29.

Karl der Große, Lebenswerk und Nachleben, hrsg. von W. Braunfels et alii, Band II: *Das geistige Leben*, hrsg. von B. Bischoff, Düsseldorf 1965.

King, V. H., *An investigation of some astronomical excerpts from Pliny's Natural History found in manuscripts of the earlier middle ages*, Oxford 1969, nicht herausgegeben.

Kopp, U. F., *Paleografia critica*, Mannheim 1817-1829, vier Bände; nachgedruckt als *Lexicon Tironianum*, hrsg. von B. Bischoff, Osnabrück 1965.

Lambot, C., *Oeuvres théologiques et grammaticales de Godescalc d'Orbais*, Louvain 1945.

Lindsay, W. M., "Collectanea varia", in: *Palaeographia Latina* II, St. Andrews 1923, pp. 5-55.

Lindsay, W. M. und Lehmann, P., "The early Mayence Scriptorium", in: *Palaeographia Latina* IV, St. Andrews 1925, pp. 15-39.

Litterae medii aevi. Festschrift für Johannes Autenrieth, hrsg. M. Borgolte und H. Spilling, Sigmaringen 1988.

Lowe, E. A., "The oldest omission signs in Latin manuscripts: their origin and significance", in: *Miscellanea Giovanni Mercati* VI, Città del Vaticano 1946, pp. 36-79 mit zehn Abbildungen; nachgedruckt in: *Palaeographical Papers, 1907-1965*, hrsg. von L. Bieler, Oxford 1972, pp. 351-380, Tafeln 61-70.

----, *Palaeographica Studia*, ed. L. Bieler, Oxford 1974.

----, vide *Codices latini antiquiores*.

Madeja, E., "Aus Walahfrid Strabos Lehrjahren", in: *Studien und Mitteilungen O.S.B.* XL, München 1920, pp. 251-256.

Monumenta Germaniae Historica, Antiquitates I. *Poetae Latini aevi Carolini* , hrsg. von E. Dümmler, Berlin 1881.

----, *Epistolae* IV, *Epistolae karolini aevi* II, hrsg. von E. Dümmler, Berlin 1895.

Mostert, M., "Marginalia in Handschriften: een onderschatte bron", in: *Millenium* II (1988), pp. 102-107.

Mundó, A. M. "L'escriptura i la codicologia", in: *Lambard* I (1977-1981), pp. 99-103.

Pellegrin, E., "Les manuscrits de Loup de Ferrières", in: *Bibliothèque de l'Ecole des Chartes* CXV (1957), pp. 5-31.

Plath, K., "Zur Entstehungsgeschichte der Visio Wettini Walahfrids", in: *Neues Archiv* 17 (1892), pp. 261-279.

Rabani Mogontiacensis Episcopi de computo, ed. W. M. Stevens, in: *Corpus Christianorum continuatio medievalis* XLIV, Turnhout, Belgien 1979, pp. 163-332.

Rafti, P., "L'interpunzione nel libro manoscritto: mezzo secolo di studi", in: *Scrittura e civiltà* XXI (1988), pp. 239-298.

Rand, E. K., *Studies in the script of Tours* I: *Survey of the Script of Tours*, Text and Facsimiles, Cambridge, Mass. 1929.

Steffens, F., *Lateinische Paläographie*, 2. Aufl., Berlin/Leipzig 1929.

Stevens, W. M., "Fulda scribes at work", in: *Bibliothek und Wissenschaft* 7 (1972), pp. 287-317 mit neun Faksimilia.

----, "A ninth-century manuscript from Fulda", in: *Bodleian Library Record* IX (1973), pp. 6-16 mit drei Faksimilia.

----, "Walahfrid Strabo - a student at Fulda", in: *Historical Papers 1971 of the Canadian Historical Association*, Ottawa 1972, pp. 9-16.

----, vide *Rabani de computo*.

Terminologie de la vie intellectuelle au moyen âge. Actes du Colloque, Leyde/La Haye, 20.-21. septembre 1985, hrsg. Olga Weijers, *CIVICIMA: Etudes sur le vocabulaire intellectuelle du moyen âge* 1, Turnhout 1988.

Traube, L., *Vorlesungen und Untersuchungen*, hrsg. Paul Lehmann, Band III, München 1920.

Vielhaber, K., *Gottschalk der Sachse*, Bonn 1956.

Wattenbach, W., *Das Schriftwesen im Mittelalter*, 3. ed., Leipzig 1896; nachgedruckt: Graz 1958.

Wieland, G., "Medieval MSS: an additional expansion for the abbreviation 'Q'", in: *Monumenta Apuliae ac Japygiae* I (1981), pp. 19-26.

Wolf, R., *Die Geschichte der Astronomie*, München 1877.

CORRECTIONS

The process of writing and printing is complicated and often leaves mistakes in printed texts. Asked why he offered no grand theories or designs for history, Leopold von Ranke is said to have replied that "God is in the details". For the opportunity to put some of those details right, I am grateful and offer the following corrections (cited by essay number and page).

I. Cycles of Time

29, n.4 and 30, n.7 (see also item III 276 and IX 48–9)

Mss Wien Österreichische Nationalbibliothek 387 and München Staatsbibliothek CLM 210 are near twins from within the period 810 to 818. The former manuscript must have been written in Salzburg and the latter copied from it in the area of Kremsmünster and Passau. Both show early changes of numerals for *annus praesens* in diverse places. Bernard Bishoff, *Die südostdeutschen Schreibschulen und Bibliotheken in der Karolingerzeit* II. *Die vorwiegend österreichischen Diözesen* (Wiesbaden: Otto Harrassowitz, 1980), p. 34 (Mondsee Stil, 803–19), pp. 60–61 and 96–7 (Salzburg, die Nachfolge von Stil II, 809–18).

29, n.5: ms Paris BN n.a.lat.1615

This book is a late collection from parts of three codices of separate origins, assembled by the thief, Libri. See Léopold Delisle, *Catalogue des manuscrits des fonds Libri et Barrois* (Paris: Bibliothèque Nationale, 1888).

The codex which is now f.128–193 (Part III) has not been properly distinguished from the other two parts by those who have affirmed dates for it. However, *annus praesens* DCCCXX appears several times (before the numerals were altered by later users). Its origin and provenance have not been determined.

Delisle dated the other codex f.3–127 (Part II) to ca. AD 830 and thought that it was first known at Auxerre and later at Fleury. After 75 years of paleography, it now seems however to have been written in an early script of Fleury or Orléans, and the Fleury *ex libris* appears to be almost contemporary with the text, according to E. Pellegrin, "Notes sur quelques recueils des vies de saints utilisé pour la liturgie à Fleury-sur-Loire au XIe siècle", *Bulletin d'information de l'I.R.H.T.* XII (Paris, 1963), p. 8, n. 6. This book was probably used at Auxerre in the late tenth or early eleventh century.

Unfortunately, many distinguished scholars have cited Delisle's opinion about f.3–127 [Libri 90, olim Orléans 313(266)] without recognising that his remarks could not be applied to the formerly independent codex f.128–193.

2

II. Bede's Scientific Achievement

As published in 1986, the lecture relegated much material to the small type of notes and appendices, and it included an inordinate number of errors; nevertheless, it was reprinted by Variorum in 1994, along with the other lectures from the beginning of the series.

In the present edition, however, some of the texts and corresponding data concerning latitudes and the coastal locations from which Bede may have received information about tides are better presented for the reader. Important recent research has been mentioned, and I have attempted to clarify some points. Perhaps the lecture will be more useful in this revised version.

III. The Figure of Earth

This essay is more trenchant than I preferred. It would benefit from broader and more literate elaboration, especially concerning the *orbis quadratus* and the Macrobius-type model of the globe. One may soon consult "Models of the Earth before 1600" and other articles in the *History of the Geosciences: An Encyclopedia*, ed. Gregory A. Good (New York: Garland, 1995). The texts of many Hellenistic, medieval, and early modern scholars would be much less difficult to understand if the assumption of appropriate models were recognised.

IV. Scientific Instruction

See the review by D.O Cróinín, in *Peritia* (1982) pp. 404–9, especially 408–9.

96, l. 35: Máeldubh

Perhaps this figure was not existent, see M. Lapidge and M. Herren, *Aldhelm, the prose works* (Ipswich: D.S. Brewer, 1979), p. 7 and especially n. 8 (on pp. 181–2).

104, n.5

M. Walsh and D. O Cróinín, Cummian's letter *De controversia paschali* and the *De ratione conputandi* (Toronto: PIMS, 1988).

110, n.60

read: Albi B.mun. 29.

VI. Fulda Scribes

292–3, n. 10

On Fulda catalogues, see now Gangolf Schrimpf, *Mittelalterliche Bücherverzeichnisse des Klosters Fulda*, Fuldaer Studien 4 (1992).

297, l. 12; 305, l. 1; 312, l. 10

The abbreviation marks in p'pri- and q'qui, p'us prius and q'b; quibus are strictly verticle, as also found recently in ms Paris BN Lat.13955. This was not written in a Corbie script, as previously suggested.

303, n. 28
See Herrad Spilling, *Opus Magnentii Hrabani Mauri in honorem sanctae crucis conditum. Hrabans Beziehung zu seinem Werk*, Fuldaer Hochschulschriften 16 (Frankfurt am Main: Verlag Josef Knecht, 1992).

308–9, n. 37
See Bernhard Bischoff, *Lorsch im Spiegel seiner Handschriften* (München: Arbeo-Gesellschaft, 1974).

314, l. 9: ms Leeuwarden, 836 med
read: 836-s.IX med.

VIII. Hrabani *De computo liber*

169, l. 10: Hilarius
read: Hilarus *hic et seq.*

170, n. 18: Milan B.Ambros. H.150
read: Milano B.Ambros. H.150 inf.

172, n. 27
mss ADEPV *hic et seq.*, see Conspectus Siglorum, p. 198.

183, l. 4: 9 July 820
This is one of two ninth-century observations of planetary motions specifically recorded, according to Bruce S. Eastwood, "The astronomy of Macrobius in Carolingian Europe: Dungal's letter of 811 to Charles the Great", *Early Medieval Europe* III/2 (1994), pp. 117–34, especially 131. Nevertheless, Hraban's observations on this date do not conform with modern retro-projections of planetary positions, according to Eastwood (private communication).

184, l. 25: Bede must have assumed
See Cyclus in C.W. Jones, *Corpus Christianorum*, Series Latina CXXIII C (1980).

IX. *Compotistica et astronomica*

28, l. 22: Roger
read: Maurice Roger

36, l. 12 and n. 26
Unfortunately, ms Basel O.II.3 does not contain the expected muster of *capitalis quadratus* which was cited in the 1965 Aachen exhibition catalogue, *Karl der Große. Werk und Wirkung* (Düsseldorf: Schwann, 1965). My thanks to Paul E. Szarmach for verifying this for me in Basel on 15 August 1994.

4

39–43
Search for Latin Euclid has born fruit and a critical edition of the texts will soon be published, but the TKr incipits and manuscripts listed in 1979 proved to be inadequate for locating them. Identification of Definitions, Postulates, Axioms, and Propositions have now been corrected in revision of this essay.

42, l. 4: Complete texts for Definitions ... verification of theorems
There are diagrams with all propositions, but it is disputed whether they are useful for the verification of theorems. It should also be noted that proofs are usually lacking and that therefore some historians doubt whether any of this material has value. These questions are serious and must be discussed elsewhere in full.

47, l. 12: position of planets on 9 July 820
See correction to VIII 183 above.

48, l. 16: those manuscripts, and a reproduction of it from
Delete.

X. Walahfrid Strabo

19, n. 13: have never been
read: have been

This index includes the names of ancient and medieval authors with their writings, and technical terminology. The contents of quotations, tables, footnotes, appendices are not cited in the general index. Dates are expressed in terms of *anni domini*, unless otherwise indicated. Both indices were prepared with the assistance of Mirosław Bielewicz (University of Winnipeg).

4

5

6

Francia, Germania, I.41; II.14, 23; V.125-6,
137; VIII.166, 170
    Pepinides, VIII.173-5; IX.49; X.13;
    see Karl
Fredugisus, abbot of Tours (804-34), VI.292
Fulda, III.276-7; V.137; VI.287-317; VIII.165,
173-7, 183; IX.27-63; XI.363-78
    calendar, IX.34-5; X.16; XI.364
    Cartulary, VI.309-11; VII.16
    Library, VI.292-3; VIII.177-8, 181

Gellius, Aulus (ca.130-80), *Noctes Atticae*,
VI.307; VII.16
Geminos of Rodos (fl. 50?), III.269
Gennadius, *De scriptoribus ecclesiasticis*, IV.94
Geraint, king of Domnonia [Dyfed; Cornall,
Devon] (ca.670-710?), IV.97; V.130
Gerbert of Aurillac (ca.945-1003), *Ratio
numerorum abacique figuras*, IX.36
Gerlandus compotista (ca.1015-1102), V.140-42
Gildas (d.ca.570), IV.100
Godescalc [Gottschalk] (ca.803-867/9),
VIII.165; X.13-14; XI.365, 376-8
Greeks, I.31; II.10-11, 23; III.269; IV.92;
see Alexandria
Gregorius, bishop of Rome (590-604), I.43;
IV.92; V.127; IX.27, 43
Grimalt, abbot of Sankt Gallen (841-72),
XI.364

Hadrian of Africa, abbot of ss. Peter and Paul,
Canterbury (fl.ca.660-72), IV.97;
V.130
Hebrew, I.30, 34; II.11; III.269
Heiric of Auxerre (ca.841-903), *Ars calculatoria*,
V.138-41; VIII.172
Heriger of Lobbes (ca.950-1007), VIII.172
Hermannus contractus [the Lame] of
Reichenau (1013-54), VIII.168
Heron of Alexandria (s.I, A.D.), IX.39
Hertford, Council of (672), IV.97; V.130
Hieronymus [Jerome] (ca. 348-420), IV.92;
VIII.177-8, 187
Hilarianus of Africa, *De ratione Paschae* (397),
IV.88
Hilarus [Hilarius] (fl. 450), I.40; IV.93;
VIII.169-70: see Leo; Victurius
Hild (d. 680), abbess of Streanæshalch [now
Whitby], V.126
Hipparchos of Nicaea (ca.190-120 B.C.), I.29;
II.2, 17-18; III.276; IX.48

Hippocrates of Chios (ca.470-400 B.C.), X.16
Hippolytus of Ostia (170-236), *Cyclus*, I.35-6,
38, 50
historians, history, I.38-42; II.4-5, 10, 27, 29,
41-2; VIII.176, 183, 187-8; IX.33, 49;
XI.375
Honorius, Julius (s.V), III.269-70
Horatius Flaccus [Horace] (65-8 B.C.), I.43;
IX.28; XI.372
Hraban [Rabanus Maurus] (d.856), abbot of
Fulda (822-42), II.4; V.139; VI.287-8,
290, 292, 298, 309; VII.9; VIII.165,
171, 173, 175, 177-8, 180; IX.27, 30, 37,
41, 44; X.13; XI.365, 370, 375-7
    Biblical commentaries, VI.306; VII.16;
VIII.165, 177-9; XI.365
    Cyclus [tabula Paschalis, 10 columns],
VIII.184-6
    De computo liber, IV.99-100; V.136-7;
VI.287; VII.9-16; VIII.165-89;
IX.31-5; X.14-20; XI.364-8, 370-
3, 376, 378
      sources, VIII.168-87
    De natura rerum liber [*De universo*],
XI.365
Huætbercte (Hwaetbert), abbot of Jarrow (716-
42?), II.32; V.127
Hunbert, bishop of Würzburg (832-42), VI.293
Hwit aern/Whithorn, II.26-8, 31
Hyginus (s.II), III.275; IV.101; VII.12;
VIII.177; IX.45-6

Innocentius, bishop of Rome (401/2-17),
I.37, 51
insula Hii/Iona, II.27, 31; IV.84-5, 88, 98;
V.126
Ireland, Irish, I.41; II.23-5, 43; V.83, 85, 89, 91,
93-6, 98-9; V.125-8, 134-6; VIII.170,
175; IX.44
Irenaeus, bishop of Lyons (ca.178-202), IX.28
Isidorus [Isidore], bishop of Sevilla (d.636),
II.4-5, 8-9; III.268, 272-7; IV.92;
V.129-33; VI.295; VIII.171, 179, 181;
IX.27, 37
    De natura rerum [*Liber rotarum*], II.8;
III.268, 272-7; IV.96, 98-101;
V.136, 140; VIII.177; IX.43-5
      versions, III.272-5; IV.100; V.136-7;
IX.44
    Differentiae, IV.98
    Origines sive etymologiae, II.9; III.272-3;
IV.96, 98; VIII.177, 182

CODICES MANUSCRIPTI

Manuscripts mentioned either in the main text or the footnotes of the above writings are cited in this index. However those enumerated in the appendices of item III are not repeated here.

Albi, Bibliothèque publique
29 (So. France, s.VIII ex), III.272; IV.110

Amiens, Bibliothèque municipale
223 (Fulda. second quarter s.IX), VI.303

Avranches, Bibliothèque municipale
114 (Normandy? s. XII^1), V.137; VIII.172

Bamberg, Staatsbibliothek
H.J.IV.22 (No.France, s.IX/X) f.1-16,

Basel, Universitätsbibliothek
F.III.15a (Fulda, s.VIII ex), VI.292-3; IX.44
F.III.15b (A-S minuscule, s.VIII^2) f.1-19, II.28
F.III.15f (A-S minuscule, s.VIII^2), V.149; IX.44
F.III.15k (Benediktbeuern, 820-40), VIII.172, 180
F.III.42 (s.XVI), VI.293
A.N.IV.18, f.1v (Fulda, s.IX^1), III.276; VIII.168; IX.48-9
A.N.IV.18, f.2-10 (Fulda, second quarter s.IX), VIII.168
O.II.3, ff.2-11v (Seligenstadt or Fulda, ca.A.D.836), IX.36

Berlin, Deutsche Staatsbibliothek
Phillipps 1830 [Cat.129] (Laon, s.IX/X), I.41; IV.106
Phillipps 1831 [Cat.128] (Verona, ca.800), VIII.170
Phillipps 1833 [Cat.138] (Fleury, s.X ex), II.23; V.138, 150; VIII.173, 189; IX.56-7

Bern, Burgerbibliothek
250, f.1v-11v (Seligenstadt or Fulda, ca.A.D.836), IX.36, 56

f.1v, 12 (Auxerre? s.IX), VIII.167
f.1r, 12-17v (Fleury, s.X/XI), II.21, 42; VIII.167
347 (Auxerre? s.IX ex), I.30
366 (s.IX^1, corr. Servatus Lupus), VI.297
611, f.94-115 (E.France, pre-Carolingian minuscule, 727), VIII.170
Bongar.87 (Luxeuil, s.IX^2) f.8v-18, IX.40

Cambridge
St. John's College I.15 (s.XII ex), V.151

Trinity College
O.2.45 [cat.1149] (Cerne Abbey, post 1248), V.137
O.7.41 (s. XI^2), V.151

University Library
Kk.1.1 (s.XIII), V.151
Kk.5.16 [Moore Bede] (737-50?), II.31
Addit.4543 (930-1039?), IV.109

Chartres, Bibliothèque de la Ville
19(26) (s.X): destroyed in 1944, V.146

Einsiedeln, Stiftsbibliothek
319 (Einsiedeln, 996), VIII.172
321, p.27-156 (Alsace, s.IX/X), VIII.171

Épinal, Bibliothèque municipale
149(68) (St. Martin Tours, 744/5), XI.371

El Escorial, Real Monasterio de San Lorenzo
R.II.18,
f.9-24v (Spain, 636-90), III.272-3; V.149; IX.43

Erfurt, Wissenschaftliche Bibl. Amploniana
Q.351 (s.XIV), V.151

16

Weimar, Landesbibliothek
    Fragment 414a (Fulda insular minuscule,
      s. VIII ex), V.149

Weissenburg-im-Breisgau, Stadtbibliothek
    Fragment (Fulda, 836-38), V.306, 314

Wien, Österreichische Nationalbibliothek
    387 (Salzburg, 810-18), I.30; VIII.168,
      170; IX.48-9, 62; XI.375
    652 (Fulda, ca.840), VI.303

Windesheim, Stadtbibliothek
    Fragment (Fulda, 836-38), V.306, 314

Wolfenbüttel, Herzog-August-Bibliothek
    105 Gud.lat. [Cat.4418] (Corbie, s.IX/X),
      IX.59

Weißenburg 73 (Fulda s.IX med), VI.297
Weißenburg 86 (St. Martin Tours, 730-50),
    VI.290
Weißenburg 92 and 84 (Fulda, ca.842),
    VI.290, 297, 300, 304, 306, 311,
    314; VII.15

Würzburg, Universitätsbibliothek
    Mp.th.f.46 (Salzburg, ca.800), I.41
    Mp.th.f.61 (Irish majuscule, s.VIII[2]),
      IV.108; VII.15

Zürich, Zentralbibliothek
    Rh.hist.27 (Reichenau, 826-30), XI.376